# Setting the Standard

*Chris Tollefson, Fred Gale, and David Haley*

# Setting the Standard
Certification, Governance, and
the Forest Stewardship Council

**UBC**Press · Vancouver · Toronto

16 15 14 13 12 11 10 09 08 07     5 4 3 2 1

Printed in Canada with vegetable-based inks on FSC-certified ancient-forest-free
paper (100% post-consumer recycled) that is processed chlorine- and acid-free.

**Library and Archives Canada Cataloguing in Publication**

Tollefson, Chris
Setting the standard : certification, governance and the Forest Stewardship Council /
Chris Tollefson, Fred Gale and David Haley.

Includes bibliographical references and index.
ISBN 978-0-7748-1437-9 (bound); ISBN 978-0-7748-1438-1 (pbk)

1. Forest management – British Columbia. 2. Forests and forestry – Certification –
British Columbia. 3. Forest management. 4. Forest Stewardship Council. I. Haley,
David  II. Gale, Fred P. (Fred Peter), 1955- III. Title.

SD131.T65 2008                    634.9'209711                    C2008-902833-3

Canadä

UBC Press gratefully acknowledges the financial support for our publishing
program of the Government of Canada through the Book Publishing Industry
Development Program (BPIDP), and of the Canada Council for the Arts, and
the British Columbia Arts Council.

This book has been published with the help of a grant from the Canadian Federation
for the Humanities and Social Sciences, through the Aid to Scholarly Publications
Programme, using funds provided by the Social Sciences and Humanities Research
Council of Canada, and with the help of the K.D. Srivastava Fund.

UBC Press
The University of British Columbia
2029 West Mall
Vancouver, BC V6T 1Z2
604-822-5959 / Fax: 604-822-6083
www.ubcpress.ca

FSC
Mixed Sources
Cert no. SW-COC-001271
www.fsc.org
© 1996 Forest Stewardship Council

# Contents

# Illustrations

# Preface

Much has written about what are often termed "new" modes of governance and regulation. Within our respective disciplines of law, politics, and economics, and indeed far beyond, scholars have expounded at length upon the meaning and significance of these developments and their implications for government, business, and civil society. To date, however, detailed, case-based research aimed at illuminating, testing, and refining emerging governance and regulatory theory has been relatively rare.

In this book we seek to remedy this by embarking on an in-depth comparative study of an organizational icon of new governance and regulatory theory, the Forest Stewardship Council (FSC). In mounting this analysis, we take as our point of departure one that was defined by our shared interest in the future of forests and forestry in British Columbia. For much of our respective careers, this has been a topic of enduring personal and professional concern. British Columbia is endowed with tremendous forest wealth, and in recent years its conflict-ridden efforts to manage the transition to a more sustainable model of forestry – in particular, one that better balances social, environmental, economic, and First Nations values – have attracted international attention. The lengthy, complex, and difficult negotiations that ultimately culminated in the October 2005 approval of an FSC standard for sustainable forestry in British Columbia in many ways underscore the complexity of this task. Indeed, while the province has undoubtedly been a fertile ground for the ambitious vision of forestry the FSC represents, paradoxically it has also proven to be one of the most daunting venues for FSC standard development.

Drawing on dozens of interviews with many key players within the FSC and beyond, *Setting the Standard* delves into this apparent paradox by means of a detailed examination of FSC-BC standard-development negotiations. It also considers, against a comparative backdrop, the content and implications of the FSC-BC standard for the future of forestry in British Columbia.

In Parts 1 and 2 we examine voluntary forest standard-setting negotiations and outcomes in a range of jurisdictions both within the FSC umbrella and under the auspices of competing certification models. We then seek to integrate these research findings into a broader theoretical frame. To this end, Part 3 develops an extended exposition on regulation and governance within the FSC; we structure this presentation around three governance dimensions – political, regulatory, and institutional – that elucidate the pioneering nature of the FSC as a certification model. In Part 4, we ruminate on the nature and significance of the FSC as an experiment in governance and regulation – an exemplar of what we term "global democratic corporatism" – and on the broader lessons for certification and civil-society-led governance that flow from our research into the FSC-BC case.

Like the standard-setting process at the core of this case study, our project has evolved along a more complicated trajectory and over a longer time period than any of us could have predicted. It originated almost a decade ago as an inquiry into the prospects for reforming the forest tenure arrangements in British Columbia to improve forest practices, protect environmental values, and ensure a more sustainable utilization of the province's abundant forest resources. Efforts to develop an FSC standard for British Columbia were launched during this time. As we tracked these efforts, we grew increasingly intrigued by the potential of forest certification to leverage these and other benefits through the market, even where momentum toward reform in the realm of hard law had stalled. Drawing upon our earlier research into the role of market-based instruments for sustainable forestry, we decided to reconfigure our project to focus on the unfolding FSC-BC standard-setting process as an example of the challenges and opportunities associated with translating the aspirations of a global certification scheme on the ground. The standard-setting process that we set out to chronicle laboured on for the next six years, and at critical points it appeared to be stalemated. Although it was immeasurably more daunting, complex, and time-intensive than anticipated, the FSC-BC process has also proven to be a richer source of empirical insights into civil-society-led governance and regulation than we could have hoped.

In undertaking this project, we have enjoyed support and advice from a remarkable variety of sources. Along the long route to publication, we benefited from academic collaborations with several key colleagues, including Denise Allen, George Hoberg, and Connie McDermott. We especially want to recognize the important research and conceptual contributions of Denise, who refined the discussion in Chapter 9 about the provisions of the FSC final standard with respect to environmental values and laid much of the groundwork for our account in Chapter 10 of indigenous and social networks. We are likewise grateful to friends and colleagues associated with the FSC at all levels, including many who played key roles in the events chronicled in this book. We would especially like to thank Jessica Clogg, Jim McCarthy, and Matthew Wenban-Smith for their comments on an earlier draft. We owe thanks as well to the three anonymous UBC Press reviewers; we benefited enormously from their comments and made extensive changes to the manuscript in light of their observations and suggestions.

This book has been written collaboratively, although we have each taken the lead on particular chapters. Chris Tollefson took overall responsibility for Chapters 1, 3, 8, 11, 13, and 14; Fred Gale for Chapters 2, 4, 5, 9, 10, and 12; and David Haley for Chapters 6 and 7.

Completion of the manuscript would not have been possible without research support provided by students at the University of Victoria, Faculty of Law. Their industry and enthusiasm were an ongoing source of encouragement and helped to push the boundaries of our analysis in numerous ways. Particularly deserving of recognition are the tireless contributions, over several years, of Robert M. Scott. We are also grateful to Ryan Copeland, Devyn Cousineau, Cameron Elder, Keith Ferguson, James Kulla, Alison Luke, Cheyenne Reese, Barry Robinson, Airi Schroff, Hart Shouldice, Matt Synnott, and Tim Thielmann. Thanks are also due to Faculty of Law secretary Rosemary Garton.

We are also grateful to UBC Press (especially Randy Schmidt and Anna Eberhard Friedlander) and to Kris Klaasen and his team at Working Design for their patience and for their editorial and creative contributions to this project as it neared completion.

Research for *Setting the Standard* was supported by a generous grant conferred jointly by the Social Sciences and Humanities Research Council of Canada and the Law Commission under its Economic Relationships in Transition program. It has been published with the help of a grant from the Canadian Federation for the Humanities and Social Sciences through the Aid to Scholarly Publications Program, using funds provided by the Social Sciences and Humanities Research Council of Canada. Fred Gale also acknowledges the support of the Australian Research Council via a Discovery Grant to investigate state responses to voluntary certification systems in Australia, Britain, and Canada.

Finally, we wish to recognize and thank our families. Chris Tollefson is grateful to Hannah and Rory, who have grown up with this project and, at times, shown more patience for it than their father. He is also greatly indebted to his partner, Krista, and to his father, Edwin. Fred Gale expresses his deep appreciation to his wife, Beverly, and his son, Evan, for their love, patience, and understanding over the many years it has taken to bring this book to fruition. The exotic but comic relief provided by a family of pademelons bouncing daily around his Tasmanian garden as the book went to press is also cheerfully acknowledged. David Haley would like to express his gratitude to his wife, Catherine, for her patience and support during the countless hours he devoted to this project over a lengthy period that extended many years into his retirement.

# Abbreviations

| | |
|---|---|
| AAC | allowable annual cut |
| ABU | Accreditation Business Unit |
| AF&PA | American Forest and Paper Association |
| Al-Pac | Alberta-Pacific Forest Industries |
| ANSI | American National Standards Institute |
| ANT | actor-network theory |
| ASI | Accreditation Services International |
| BAT | best available technology |
| BATNA | best alternative to a negotiated settlement |
| BCC | Boreal Coordinating Committee |
| BEC | biogeoclimatic ecosystem classification |
| BMP | best management practice |
| Btk | Bacillus thuringiensis var. Kurstaki |
| CAR | corrective action requests |
| CB | certifying body |
| CCFM | Canadian Council of Forest Ministers |
| CFA | community forest agreement |
| CFPC | Certified Forest Products Council |
| CORE | Commission on Resources and Environment |
| CSA | Canadian Standards Association |
| CSFCC | Canadian Sustainable Forestry Certification Coalition |
| CSS | Canadian Standards System |
| CWG-DRC | FSC-Canada working group dispute resolution committee |
| D#-M | Maritimes standard drafts |
| DFA | defined forest area |
| DRAAC | Dispute Resolution and Accreditation Appeals Committee |
| DRC | Dispute Resolution Committee |
| EBM | ecosystem-based management |
| ECSO | environmental civil society organization |
| ED | executive director |

| | |
|---|---|
| EMS | environmental management system |
| FBC | Falls Brook Centre |
| FERN | Forests and the European Union Resource Network |
| FMU | forest management unit |
| FRA | Forest and Range Agreement |
| FRBC | Forest Renewal BC |
| FRO | Forest and Range Opportunity |
| FRP | Forest Revitalization Plan |
| FSC | Forest Stewardship Council |
| FSC-AC | FSC-Asociación Civil |
| FSC-BC | FSC-British Columbia |
| FSC-IC | FSC-International Center |
| FSC-M | FSC-Maritimes |
| FSC-PC | FSC-Pacific Coast |
| FSC-RM | FSC-Rocky Mountain |
| GDP | gross domestic product |
| HCVF | high conservation value forest |
| IAC | indigenous advisory council |
| ICANN | Internet Corporation of Assigned Names and Numbers |
| IDRP | Interim Dispute Resolution Protocol |
| IFOAM | International Federation of Organic Agriculture Movements |
| IFPA | innovative forestry practices agreements |
| IISD | International Institute for Sustainable Development |
| IP | indigenous peoples |
| IRA | integrated riparian assessment |
| ISC | interim steering committee |
| ISEAL | International Social and Environmental Accreditation and Labelling Alliance |
| ISO | International Organization for Standardization |
| ITTO | International Tropical Timber Organization |
| IUCN | International Union for the Conservation of Nature and Natural Resources (World Conservation Union) |
| IWA | Industrial, Wood, and Allied Workers |
| JMA | joint management agreement |
| LEEC | London Environmental Economics Centre |
| LRSY | long-run sustainable yield |
| MBI | market-based instrument |
| MCEP | Mining Certification Evaluation Project |
| MOF | Ministry of Forests (as of 2005, Ministry of Forests and Range) |
| MPB | mountain pine beetle |
| MRSC | Maritimes Regional Steering Committee |
| MSC | Marine Stewardship Council |

| NAFA | National Aboriginal Forestry Association |
| NDP | New Democratic Party |
| NDTs | natural disturbance types |
| NGO | non-governmental organization |
| NI | national initiative |
| NSAC | National Standards Advisory Committee |
| NWG | national working group |
| ODA | Overseas Development Administration |
| OMNR | Ontario Ministry of Natural Resources |
| P# | Principle # of FSC principles |
| PAS | protected areas strategy |
| PCC | Pacific Certification Council |
| PCWG | Pacific Coast Working Group |
| PEFC | Programme for the Endorsement of Forest Certification |
| PIC | pre-industrial condition |
| PNG | Papua New Guinea |
| PPM | process or production method |
| PPWG | Plantation Policy Working Group |
| RMZ | riparian management zone |
| RONV | range of natural variability |
| RPF | registered professional forester |
| RRZ | riparian reserve zone |
| RTRs | resources and tenure rights |
| SBFEP | Small Business Forest Enterprise Program |
| SC | steering committee |
| SCC | Standards Council of Canada |
| SCS | Scientific Certification Systems |
| SFB | Sustainable Forestry Board |
| SFI | Sustainable Forestry Initiative |
| SFIA | Swedish Forest Industry Association |
| SFM | sustainable forest management |
| SGS | Société Générale de Surveillance |
| SLA | Softwood Lumber Agreement |
| SLIMF | small and low-intensity managed forest |
| SSNC | Swedish Society for Nature Conservation |
| ST | standards team |
| STSC | Sustainable Tourism Stewardship Council |
| SYFM | sustained-yield forest management |
| TAB | technical advisory board |
| TAT | technical advisory team |
| TBT | Technical Barriers to Trade Agreement |
| TFAP | Tropical Forestry Action Plan |
| TFL | tree farm licence |

| THLB | timber-harvesting land base |
| TSA | timber supply area |
| TSL | timber sale licence |
| TSWC | Technical Standards Writing Committee |
| UKWAS | UK Woodland Assurance Scheme |
| UNCED | United Nations Conference on Environment and Development |
| UNCTD | United Nations Conference on Trade and Development |
| WARP | Woodworkers Alliance for Rainforest Protection |
| WCEL | West Coast Environmental Law |
| WFP | Western Forest Products |
| WTO | World Trade Organization |
| WWF | World Wide Fund for Nature (World Wildlife Fund, in North America) |

# Setting the Standard

# 1
# Introduction

The last two decades have witnessed a dramatic shift in the political economy of environmentalism. During this period, environmental civil society organizations (ECSOs) worldwide have invested heavily in a variety of new market-oriented strategies to leverage change both domestically and internationally. This turn to markets reflects at least two abiding realities. One is an emerging appreciation of the barriers to achieving "hard law" solutions posed by the complex, polycentric nature of global politics and the pervasiveness of neo-liberal policy prescriptions in the domestic realm. This disillusionment with hard law domestically and internationally has been accompanied, however, by a nascent appreciation of the potential to achieve change beyond the state and, indeed, beyond established venues for international cooperation by drawing on the respective resources of similarly motivated civil society and business actors and organizations through the vehicle of what is often characterized as "soft law."[1]

Developments in the politics of global forestry provide a compelling case in point. By the early 1990s, a widespread pessimism about the prospect of securing effective international legal protection for the world's forests had set in.[2] Paradoxically, as momentum waned on this front, worldwide concern about the future of global forests was reaching a crescendo orchestrated by highly effective market campaigns aimed at focusing attention on the unsustainable forest practices of some of the world's largest forest products exporters. Buoyed by the success of these campaigns, and by a perception that a formidable coalition of interests could be brought together to promote a new vision of forestry that translated global concerns into market decisions, reformers turned their attention to developing the first international system of forest product certification.[3]

Thus, in 1993, the Forest Stewardship Council (FSC) was born. The FSC's self-described mission is "to promote environmentally appropriate, socially beneficial and economically viable management of the world's forests." To this end, it provides standard setting, trademark assurance, and accreditation services for companies and organizations interested in responsible forestry. To support its work, the FSC has developed a unique and highly participatory governance structure aimed at balancing the interests of its three constituent chambers (environmental, social, and economic), each of which is further divided into two institutionalized sub-chambers (North and South). It has also implemented special rules to ensure that the interests of indigenous peoples are effectively represented.[4] The FSC affiliates are responsible for developing national or regional forest management standards that are consistent with and facilitate implementation of the FSC's ten generic principles and fifty-six associated criteria.

Since its founding, the FSC has been a remarkable success story. Over 100 million hectares of forest in more than seventy-eight countries are now FSC certified,[5] and the distinctive FSC logo appears on thousands of forest products worldwide. A 2004 report that assessed eight competing forest-certification schemes concluded that the FSC is the "most independent, rigorous and credible" system of its kind in the world, a conclusion premised on the FSC's commitment to global democratic governance and the depth and breadth of its standards (FERN 2004). These same attributes, however, have also contributed to various ongoing tensions, challenges, and vulnerabilities. Whether FSC can continue to flourish in the emerging highly competitive certification environment will depend on its ability to continue to be effective and creative in the way it approaches its governance, regulatory (standard-setting), and service delivery responsibilities.

### Theorizing the Evolving Nature of Governance and Regulation

The ascendancy of the FSC as a transnational, civil society vehicle to promote sustainable resource management through the market is highly relevant to, and offers a variety of insights for, emerging scholarly reflections on the evolving nature of governance and regulation. Across a wide range of disciplines, there is a growing sense that we are witnessing fundamental change in the nature of governance and regulation and in the respective roles of government, business, and civil society.

For some time now, there has been skepticism in the realm of political science about the notion of the state as the exclusive "sovereign decision-making unit" endowed with a natural monopoly on the regulation of socio-economic affairs (Mason 1999; Swan 2002; Cashore, Auld, and Newsom 2004). At the domestic level, this skepticism is closely related to the deliberate rolling back of the post-Second World War interventionist state occasioned by the dominance, in more recent decades, of neo-liberal ideology (Considine and Painter 1997; Howlett, Netherton, and Ramesh 1999). The state is also being de-centred by the emergence of growing internal pressures to recognize new interests, including those of local communities and governments, and

Aboriginal organizations. We are thus witnessing a multiplication or pluralization of sites of governance. This, it is said, demands a "reconceptualization
of the political" from its traditional focus on the state to a broader frame of
reference that analyzes the meaning of governance both in relation to the
state and also in terms of the new actors and institutions that operate within
civil society and the market at both the local and national levels (Swan 2002,
13; Cohen and Arato 1994; Rhodes 1996, 1997).

Analogous theories are being offered about developments in the realm of
international relations. In this context, observers contend that conventional
state-centric theories conceal historic transformations that are under way
in the modalities of governance (Rosenau 1992; Gale 1998a; Wendt 1999;
Cutler, Haufler, and Porter 1999; Meidinger, Elliott, and Oesten 2003). Central among these are, once again, a pluralization of the forms of governance
and, closely allied to this, a distancing of governance arrangements from the
state. For many international relations theorists, therefore, a key priority is
to develop new ways of thinking about governance in the global arena that
take into account the emerging role of transnational corporations and civil
society organizations. To this end, "global civil society" has been identified
as an arena of action beyond the state at the international level, populated
by a range of civil society actors actively engaged in global politics (Lipschutz
1992; Shaw 1992). Significant progress has also been made toward elaborating
the means and mechanisms – including transnational advocacy networks
– by which global civil society has increasingly exerted its influence over
domestic law and policy (Keck and Sikkink 1998) and epistemic communities
of engaged scientists and policy entrepreneurs (Haas 1992). Likewise, there
is a growing body of research into how intergovernmental actors like the
World Bank, World Trade Organization, and International Monetary Fund
are responding to challenges to their legitimacy by establishing formal consultative arrangements with international non-governmental organizations
(O'Brien et al. 2000).

The implications of these trends in governance, both domestically and
internationally, for law and regulation are also receiving attention. In this
regard, soft law is often touted as an alternative or precursor to traditional
hard-law solutions that typically rely on the sponsorship or support of
national governments. As Kirton and Trebilcock underscore (2004, 5, 9),
soft-law initiatives consisting of "informal institutions ... that depend on
the voluntarily supplied participation, resources and consensual actions of
their members" offer the potential for timely action when governments are
stalemated. They can also lend additional legitimacy, expertise, and other
resources for making and enforcing new norms and standards and serve as a
vehicle for enhancing civil society participation in governance. At the same
time, soft-law arrangements are vulnerable to the criticisms that they lack the

legitimacy and enforceability of their hard-law counterparts; that they may tend "to promote compromise, or even compromised" standards; and that competition between voluntary standards can create consumer uncertainty and fatigue (6).

In the realm of domestic law, a similar process of critique and re-evaluation of hard law – particularly command-and-control regulation – is under way. Here, often under the auspices of the "smart regulation" approach, scholars have contended that a more creative and context-specific deployment of legal and policy instruments can generate significantly improved environmental performance, particularly on the part of industry leaders as opposed to their laggard colleagues (Gunningham and Grabosky 1998; Gunningham and Sinclair 2002, 189; Fiorino 2006). This literature proceeds from a recognition that economic and political pressures over the last two decades have often precluded or limited deployment of traditional forms of government regulation, giving rise to the correlative need to get more "bang" for the environmental regulatory "buck." A key objective for proponents of this approach is to explore opportunities for "reconfiguring environmental regulation," including measures to enhance the involvement of non-state actors in regulatory affairs through voluntary environmental partnerships and codes of practice and the promotion of more participatory governance arrangements (including certification schemes such as the FSC).

The mantra of "smart regulation" has achieved remarkable currency in business and governmental circles.[6] Indeed, in some cases it has been championed as a justification for abandoning all forms of prescriptive government regulation (frequently, and, we will argue, erroneously, equated with command-and-control regulation) in favour of so-called performance-based regulation (also termed outcome- or results-based regulation). The originators of the smart regulation approach are quick to dissociate themselves from this approach by emphasizing "the residual but nevertheless important role that command and control regulation can and should play in environmental policy" (Gunningham and Sinclair 2002, 203). Moreover, they underscore that the tenets of smart regulation counsel modesty about divining policy prescriptions directly from theoretical ruminations. In their words, "much of our knowledge about policy instruments, and in particular about what works and when, is tentative, contingent and uncertain. This suggests the virtue of adaptive learning, and for treating policies as experiments from which we can learn and which in turn can help shape the next generation of instruments" (203).

Evolving debates about governance and regulatory reconfiguration have also spurred new thinking by economists and business theorists. Particularly robust is the literature focusing on business ethics and social licence. Firms, it is argued, do not and cannot operate within the purely theoretical space of the "market." Moreover, they must meet the rising market demand for ethical behaviour or put at risk their social licence (Gibson 1999). This is particularly true with respect to large transnational corporations that depend on a global division of labour in which goods are produced in one location (often a developing country) for consumption in another (often a developed country). Braithwaite and Drahos's magisterial investigation (2000) into global business regulation underscores the institutional and strategic considerations that confront businesses as they seek to meet a broader array of investor and consumer requirements that go beyond conventional considerations of price, quality, and service to embrace such matters as employee health, environmental sustainability, and fair business practices.

As the research agendas of scholars from various disciplines converge on the task of understanding the implications of the evolving nature of governance and regulation in a globalizing world, the need for rigorous cross-disciplinary work grows in importance.[7] The continuing work of Gunningham et al. on smart regulation provides an encouraging illustration of the potential for synergy between law and political science. Two other volumes also tackle these issues from a cross-section of disciplinary perspectives. One of these, edited by Kernaghan Webb, explores the interrelationship of voluntary codes, private governance, the public interest, and innovation, drawing primarily on Canadian case studies (Webb 2004). Another important contribution is a volume co-edited by Kirton and Trebilcock (2004) that reflects on the complex and evolving relationship between hard and soft law, employing illustrations from various international case studies. The FSC experiment figures prominently throughout this body of work, with full chapters devoted to it in these latter two volumes.[8]

## Concepts and Definitions
We embark on this project mindful of the richness and complexity of the story of the FSC-BC final standard that we have set out to chronicle and of the diverse implications of that story (and the outcomes it has yielded) in terms of the converging literatures just addressed. Thus, at one level, this is a book about forest certification and, in particular, the challenges confronted by participants in the FSC-BC regional process to develop a standard that would define what the laudable but often abstract aspirations articulated in FSC's ten principles and their accompanying criteria mean in British Columbia. In every sense, and at every step of the way, developing this standard was an unavoidably political process, from its rather modest origins in 1996, to the machinations surrounding its approval by the FSC as a preliminary standard in 2003, to its accreditation as a final standard in late 2005.[9]

As important and instructive as the process itself is the emerging final standard, which now guides FSC certifications across the province. The most comprehensive and arguably one of the most "rigorous" forest-certification standards ever drafted, its meaning and implications have provoked considerable controversy both within and beyond the FSC system. One of our key goals in *Setting the Standard* is to offer a comparative assessment of the FSC-BC final standard. To this end we propose to evaluate the BC standard against those in place in Sweden, the United States, and elsewhere in Canada. We will address the rationales for these comparators and our comparative methodology shortly.

At another level, this is a book that aspires to contribute to the evolving debate over the future of governance and regulation. It is no coincidence that, as we have elaborated above, academic work in a range of disciplines is converging on these two closely related topics. In both the international and domestic realms, a confluence of developments create the need to rethink our understanding of "governance" and "regulation." We argue that, as an experiment in global democracy and as an organization heavily and intimately engaged in the production of soft law, the Forest Stewardship Council provides a rich source of material for researchers interested in testing and refining these evolving concepts.

**Governance Theory**

Until recently, in both academic and popular discourse the terms "government" and "governance" have tended to be used synonymously. Across a broad range of disciplines, however, there is an emerging consensus that these concepts should be treated as analytically distinct. Indeed, it has been argued that government and governance should be regarded as representing two poles on a continuum of governing types (Pierre and Peters 2000; Rhodes 2000). At the "government" end of this continuum, it is contended, are strong, highly centralized states of the type ascendant in the generation following the Second World War – states heavily reliant on command-and-control-style regulation. Positioned at the other pole are "pure" governance arrangements in which the "business of government" has effectively been assumed by self-organizing and coordinating networks of social actors. Such arrangements are characterized by more flexible and fluid forms of governance and regulation (Schout and Jordan 2005; Stoker 1998; Jordan, Wurzel, and Zito 2005). In this book, we propose to define "governance" as an umbrella term to denote arrangements for "steering and coordinating the affairs of interdependent social actors based on institutionalised rule systems" that depart from the traditional strong government paradigm posited above (Benz, quoted in Treib, Bähr, and Falkner 2005, 5).

Frequently, scholars seek to portray the magnitude of this departure by invoking the public-private distinction. Accordingly, "public governance" is envisioned as a much closer relative to "government" than "private governance." Where new governance arrangements belong within this public-private governance typology is often said to depend on the nature and extent of state involvement in such arrangements. Thus, private governance arrangements entail modest or non-existent state involvement, while public governance arrangements typically feature a much more active and engaged state presence (Abbott and Snidal 2001).

Of late, however, there is an emerging recognition of the need for more multi-dimensional and nuanced tools to analyze new forms of governance. An approach that has spawned a particularly robust literature is concerned with understanding the political dynamics inherent in such arrangements through an analysis of the social and organizational networks that are at play. Although there is a plethora of modes of network analysis, they share a common interest in the *political dimension* of governance: who wields the power to decide and why (Keck and Sikkink 1998; Dicken et al. 2001; Raustiala 2002; Slaughter 2004)? Another flourishing area of research into the new governance phenomenon addresses the *regulatory dimension* of such arrangements (Gunningham and Grabosky 1998; Gunningham and Sinclair 2002; Coglianese and Lazar 2003; Fiorino 2006; Meidinger 2006b). Key areas of research in this approach include questions of how and to what extent these arrangements can be regarded as or are analogous to law-making institutions: what is the nature of the regulation being generated (hard versus soft law; the forms of applicable standards, etc.); and what are the nature and range of values that are being regulated (Meidinger, 2000)? A third and as yet relatively undeveloped mode of analyzing emerging forms of governance is through attention to what might be termed the *institutional dimension*. Those using this approach seek to understand the institutional architecture that houses internal political networks and facilitates the generation of regulatory outputs. A particularly promising vehicle for pursuing such an analysis is comparative constitutionalism a methodology that assesses the extent to which a prevailing institutional architecture emulates or departs from norms of liberal democratic practice (Dorsen et al. 2003).[10]

Employing the broad working definition of "governance" offered by Treib and set out above, in Part 3 of the book we undertake an in-depth analysis of FSC governance with a view to exploring the insights that derive from each of three dimensions: the political (Chapter 10); the regulatory (Chapter 11); and the institutional (Chapter 12). Drawing on these insights, in Part 4 (Chapters 13 and 14) we tackle the task of providing an affirmative description of the meaning and theoretical significance of the FSC in governance terms. We contend that the FSC represents, in many ways, a unique governance form that defies conceptual categorization based on orthodox state-market or private-public dichotomies. In this, we part company with other scholars of the FSC who have tended to portray this iconic illustration of new governance as a form of "private governance" or "non-state market-driven" regulation (Lipschutz 2000; Meidinger 2000; Cashore, Auld, and

Newsom 2004; Pattberg 2005; Rhone, Clarke, and Webb 2005). In contrast, we characterize the FSC model as an ambitious experiment in what we term *global democratic corporatism:* a governance regime that is at once global in scope, democratic in practice, and corporatist in design.[11]

### Regulatory Theory

Just as traditional theories of government must explain the process by which laws and regulations are negotiated, formulated, promulgated, and legitimated, understanding the nature of emerging forms of governance arrangements requires the development of an analogous analysis that de-centres prevailing notions of "regulation" as the exclusive function of the state (Swan 2002). Accordingly, flowing from our definition of "governance," we define "regulation" in this broader context as referring to an *institutionalized rules system that sets and enforces standards aimed at influencing the behaviour of interdependent social actors in order to promote mutually agreed-upon values, principles, or objectives.* In this view, regulation serves as the vehicle through which the overarching shared values are translated into directives, procedures, and requirements that become the operating "code of practice" for parties to the governance arrangement.

As we shall see in the FSC-BC case, both the means by which this process of translation occurred and the content of the resulting standard were a source of considerable contention and debate. One of our key objectives in *Setting the Standard* is to understand why the province of British Columbia, seemingly such a receptive and fertile ground for FSC's vision, proved to be such a challenging terrain for realizing its regulatory mission. To presage what follows, we will argue that in large measure this intractability was due to competing visions of the ultimate role and purpose of the FSC. This process was complicated further by the FSC's relative youthfulness as an organization and by the absence of established rules to guide standard setting at the regional, national, and international levels. At the same time, we argue that the difficulties encountered in "setting the standard" in British Columbia reflect an ongoing organizational struggle to grapple with the meaning, implications, and respective merits of competing *forms of standard.*

As is the case in regulatory reform exercises elsewhere, the FSC standard-setting process proceeded from an explicit desire to eschew *prescriptive* regulation, commonly associated with traditional state-centric command-and-control models, in favour of a performance-based approach that specified outcomes as opposed to prescribing the means (via approved management systems or technologies) for their achievement. The presumed superiority of performance-based over prescriptive regulation has become a staple of popular political wisdom, a phenomenon we explore in detail in Chapter 11 as part of a broader discussion that addresses the distinctions between these forms of regulation and the contextual factors that determine their relative effectiveness and suitability.

In our view, the commonly espoused dichotomy between prescriptive and performance-based regulation is both misleading and false. *Prescriptive regulation*, as we propose to use this term, connotes regulation that constrains the actions of, or discretion exercised by, a party in relation to a regulated activity. In this sense, *all forms of regulation are prescriptive*. Indeed, if they did not impose such a constraint, they would not constitute "regulation." Although regulations can be distinguished in terms of their relative prescriptiveness, we would argue that the key variable that elucidates their nature and impact is the stage of the production process at which the constraint is imposed. Thus, what we would term *management-based standards* impose constraints at the pre-production stage; *technology-based standards* constrain production methods and processes; and *performance-based standards* constrain production outputs (Coglianese and Lazar 2003; Chapter 11 in this volume).

**Methodology and Structure**
This book takes as its point of departure the FSC-BC standard-development process and the draft, preliminary, and final standards that emerged from this process. Our decision to embark on a comprehensive exploration of the BC case is driven by several considerations. One of these is the province's importance to the world economy as a forest products exporter. Canada is the world's largest softwood exporter, accounting for a quarter of total global production. Over half of this production comes from British Columbia. Exports of BC forest products to major world markets – in the United States, Europe, and Japan – constitute a critical element in global timber flows. If FSC were to take root among major producers in British Columbia, it would have significant ramifications for the global timber market, much as the certification of forest companies in Sweden has had for the European market.

The province has also attracted world attention for the intensity of the debate surrounding, and the nature and implications of governmental responses to demands for fundamental changes to, the province's long-standing forest management strategy of high-volume annual cuts and the liquidation of old-growth forests. Over the last decade, the province has been embroiled in a fractious debate on how to maintain and enhance its competitive position as a world producer while simultaneously making the transition to a more sustainable and equitable model of forest management. In grappling with this latter task, successive governments have struggled to respond to and appease growing demands for reform by a diverse range of interests including First Nations, environmentalists, communities, non-timber economic interests (i.e., tourism operators, outfitters), and small producers. The future of the industry is arguably now more uncertain than it has ever been, due in large measure to questions surrounding the meaning and implications of seemingly never-ending legislative and policy experiments (see Chapter 3).

British Columbia is also broadly regarded as a critical testing ground for the viability of the FSC as both a "brand" and an institution. Since its establishment, the FSC has been closely connected to the unfolding debate within the province with respect to forest policy reform. Many of the FSC's founding and most active members hail from the province and were energetically engaged in the development of the BC standard. As the FSC has grown and evolved, new and sometimes incongruous visions of the organization have emerged. A key institutional challenge for the FSC has been, and continues to be, how to reconcile these competing visions. The crucible within which this challenge has presented itself is the development of the BC standard (see Chapter 4). The controversy surrounding the BC standard-development process, and the apparent elusiveness of a resolution to these differences, raises serious institutional/governance issues for the FSC as a whole. Moreover, the lack of a final accredited BC standard until November 2005 has created a serious bottleneck in the province's production of FSC-certified products.

Our book is doubly comparative. To provide context for our study, we examined the FSC standards currently in place in five comparable jurisdictions (see Chapter 5). We have selected one national standard (Sweden) and four regional standards (two each from the United States and Canada). The first Swedish standard is particularly instructive due to its status as a first-generation standard developed in 1996-97 during the early years of the FSC, prior to the finalization of FSC's generic principles and criteria.[12] We also examine regional standards for two American jurisdictions contiguous to British Columbia. The standards for the Pacific Coast and the Rocky Mountain regions stand in sharp contrast to the BC standard in terms of both the process by which they were developed and their substantive content. The final comparators are the Canadian standards developed for the Maritimes and Boreal regions. Again, these reveal some fascinating parallels and differences in both the standard-development process and substantive outcomes. The Maritimes standard-development process concluded in 2000 and culminated, as the BC process did, in controversy, prompting a formal inquiry by FSC-AC. In some senses, the Boreal process, which began more recently, represents an attempt by the FSC to avoid some of the pitfalls associated with earlier standard-development initiatives, including those in British Columbia and the Maritimes.[13]

We have also undertaken a comparative analysis of the FSC relative to selected competitors and analogues in the broader certification world (see Chapters 2 and 13). In terms of competitor regimes, the global comparators we consider are the International Organization for Standardization's (ISO's)

14000 Environmental Management System (EMS) and the Programme for the Endorsement of Forest Certification (PEFC). Because the PEFC is an umbrella body for nationally based schemes, we also examine and discuss two PEFC member organizations: the Canadian Standards Association (CSA) and the American Forest and Paper Association's Sustainable Forestry Initiative (SFI). To round out our analysis, we also include a non-forestry comparator: the Marine Stewardship Council (MSC), an organization that is commonly considered the FSC's sister organization in the fisheries context.

*Setting the Standard* is organized in four parts. Part 1 provides a sense of the broad context within which the FSC-BC final standard was developed, exploring the dynamics and politics of the standard-development process. To this end, Chapter 2 explores the emergence of the FSC as a transnational organization and considers, both in the BC context and beyond, the challenges and issues facing the FSC as it seeks to secure its long-term viability as a brand and as an institution. This chapter also introduces our institutional comparators, the ISO and PEFC, as well as FSC's sister organization, the MSC. Chapter 3 provides an overview of the complex politics of forest policy in British Columbia since 1945 and identifies some of the key factors that have made the province both a fertile ground for the vision that the FSC represents and a challenging terrain for FSC standard development. Based on in-depth interviews with many of the key participants,[14] Chapter 4 chronicles the political dynamics surrounding the FSC-BC standard-development and approval process, providing a detailed history of the evolution of the final standard. And, in Chapter 5, we introduce readers to the five comparator jurisdictions we have chosen to employ in Part 3, offering a comparative analysis of the processes by which each of these jurisdictions negotiated and developed their respective FSC standards.

Part 2 considers, compares, and critiques the substantive content of the FSC-BC final standard. It comprises four chapters, each of which focuses on a key cluster of values that all FSC standards are mandated to address: tenure, harvesting, and customary rights (Chapter 6); community and workers' rights (Chapter 7); indigenous peoples' rights (Chapter 8); and environmental values (Chapter 9). Each chapter follows a common template that assesses the context of the debate with respect to the relevant issues, provides a substantive analysis of the FSC-BC final standard (on its own merits and in comparative context), and reflects on the implementation challenges that lie ahead.

Part 3 is an extended exposition and analysis of regulation and governance within the FSC system, drawing extensively on our research into the FSC-BC case. It is structured to highlight the three dimensions of governance (politics, regulation, and institutions) introduced above, which we contend reveal the unique nature of the FSC governance model. Chapter 10 explores the *political* dimension of governance within the FSC, employing a political "network analysis." Using this methodology, we consider how and to what extent six identifiable informal governance networks within the FSC (certifiers, donors, large producers, indigenous peoples, environmentalists, and social activists) are represented within and exercise influence over FSC decision-making processes. Chapter 11 investigates the *regulatory* dimension of governance within the FSC system, giving particular attention to the practical and conceptual challenges associated with crafting effective standards. To this end, it advocates a context-specific approach to standard setting that is closely attuned to the relative merits of, and conceptual distinctions between, management-, technology-, and performance-based standards. It also contains an analysis of the manner and extent to which the FSC-BC standard relies on these distinct forms of standards. Finally, Chapter 12 considers the *institutional* dimension of governance within the FSC regime, through an analysis of the FSC's constitutional arrangements, including its membership and electoral arrangements; its federated political structure; and the manner in which its "legislative," "executive," and "judicial" arms operate and interact, we employ a comparative constitutional methodology that considers the extent to which FSC governance institutions and processes reflect liberal democratic norms and practices.

Part 4 integrates and elaborates on theoretical insights developed in Part 3 and reprises the analysis and conclusions offered elsewhere in the book. Chapter 13 is a theoretical rumination on the nature and significance of the FSC as a governance regime on its own merits and relative to analogous certification systems. The comparative analysis employed in this chapter is structured around the three key dimensions of governance (politics, regulation, and institutions) addressed in Chapters 10, 11, and 12. Such an approach, we contend, offers a more nuanced understanding of the FSC system than is found in the extant governance literature, which tends to be framed around more familiar private-public or state-market dichotomies. Based on this analysis, we argue that, in governance terms, the emerging FSC regime is *sui generis:* a first-instance illustration of what we term "global democratic corporatism." Finally, in Chapter 14, we reprise the conclusions that flow from our research into the FSC-BC case (in terms of the standard-development process and of the final standard emerging from that process) and offer some observations on the broader lessons that can be drawn from our work with respect to the evolving art of governance and regulation in the burgeoning certification context.

# Part 1
# Developing the FSC-BC Standard

# 2
# The Rise and Rise
# of Forest Certification

The forest-certification movement was born in the late 1980s out of a mounting frustration with the failure of national and intergovernmental processes to halt tropical deforestation and forest degradation. In its early days, the movement attracted a diverse range of allies, including academics supportive of an incentive-based approach to environmental regulation; an assortment of leading environmental civil society organizations (ECSOs) (including World Wide Fund for Nature, Greenpeace, and Friends of the Earth) that perceived the movement's potential as a means of protecting vulnerable forests, particularly those in the South; and an impressive array of donors, including prominent US-based charitable foundations (notably the Ford and MacArthur foundations) and even several European governments (UK, Dutch, Austrian, and the European Union).

Through the 1990s, as the certification movement grew and diversified, some curiosities emerged. Although certification was initially developed to improve tropical forest management in the developing world (where standards were, and are, woefully low), industry take-up has proven to be much more robust in the temperate and boreal forests of North America and northern Europe. Also, while many governments and industry associations vociferously opposed certification in the early days – claiming that it was unnecessary, costly, and impractical – by the mid- to late 1990s they were taking a different tack, in many cases developing their own national or industry-led certification schemes.

These curiosities arise in part from the practical implementation of forest certification in highly charged global, national, and local contexts. Indeed, the emerging experience with forest certification over the past decade has highlighted not only how politicized the process of standards development can become but also the immense technical and administrative challenges associated with this new form of governance and regulation. This chapter commences with a discussion of the economic rationales for certification and an overview of varying forms of certification and labelling. We then chronicle the history of forest certification leading up to the establishment of the Forest Stewardship Council (FSC), providing a detailed account of the FSC's formation, organizational structure, and governing principles and criteria. Next, we provide an overview of the FSC's main competitors and analogues in the certification world. We conclude with a synopsis of the key issues that arise in the forest-certification context, many of which are addressed in greater detail later in the book.

## The Economics of Certification and Labelling

Forest management is typically pursued through a combination of regulatory, institutional, and market-based instruments (Tollefson 1998). Regulatory instruments are often regarded as being synonymous with the "command and control" mechanisms employed by governments to regulate forest industry practices (Stanbury and Vertinsky 1998). A good example of this regulatory approach to forest sector governance is British Columbia's former *Forest Practices Code*. Under this code, the province identified and defined appropriate forest practices, assigning a watchdog role to an independent Forest Practices Board.

Institutional instruments to achieve good forest management differ in that they seek to alter the inputs to, and/or structure of, decision-making processes. Inputs are altered, for example, when governments restructure forest ownership arrangements by increasing the number of smallholders and reducing the number of large companies or vice versa. Oligopolistic ownership and management operations generate a markedly different forest political economy than where ownership is dispersed among numerous smaller actors. The precise impact of such changes on forest management is much debated. Some argue that large, integrated companies have the resources and expertise to do a better job, while others contend that smaller tenures are inherently more sustainable because they operate locally and are more sensitive to community desires and public pressure. Another institutional mechanism involves the creation of forest-planning advisory bodies. In the 1990s, for example, the BC government launched the Commission on Resources and Environment (CORE), which was designed to be a more inclusive and transparent forum for the negotiation of forest and resource rights than the relatively closed "iron triangle" government-industry-union model that preceded it (see Chapter 3).

Market-based instruments (MBIs) also play a key role in the regulation of forest management. Market-based instruments alter producers' incentives, generating rewards for producers who engage in environmentally sustainable production. Such rewards can take the form of higher prices, lower taxes, preferential treatment in procurement contracts, increased market share, or some combination of these that directly benefit a company's bottom line. Examples of MBIs include green taxes, ecological subsidies, and consumption

charges for previously "free" goods (see Stanbury and Vertinsky 1998, 50-53).[1]
Economists argue that MBIs should be preferred to regulatory instruments
because desired objectives can be achieved more effectively and efficiently
from the perspective of the firm, and at a lower cost to the public purse. In
British Columbia, potential MBIs for environmentally sustainable forestry
exist in stumpage rates, differential tax rates on managed and unmanaged
forestland, and subsidies for silviculture activities, including replanting,
thinning, and pruning.

When considering MBIs, economists focus much of their attention on
"green" taxes and subsidies. Institutional economists, however, challenge
a central assumption of mainstream economic theory – the concept of perfect information – arguing that imperfect information leads to suboptimal
outcomes because consumers are misled in their purchases by a lack of (or
distorted or incorrect) market signals (see Rametsteiner 2000, 18-19; see also
Stiglitz 1997). Within this literature, certification is thus seen as a powerful
vehicle to correct market failure.

Increasingly, civil society organizations have sought to shape consumer
preferences through public demonstrations, corporate boycotts, media ex-
posés, and scientific reports. According to institutional economists, consumer
preferences are not exogenous to the market (as is assumed in much conventional economics). Instead, demand is seen as an endogenous function that is
manipulated by actors through various non-price mechanisms, which depend
in part on the amount and quality of the information consumers receive.

Information, therefore, is central to empowering consumers to act on their
environmental and social preferences. Information is vital not only because
it enables consumers to distinguish between otherwise "like" products, but
also because it constructs the environmentally and socially aware consumer.
Certification schemes enable consumers to act on their environmental and
social preferences, which, in the absence of such schemes, remain latent.
However, because information helps constitute consumer preferences, it has
become highly politicized and contested. In large part, the "certification
wars" of the past decade (Humphreys 2006) were contests over who should
communicate what information and in what form; they often pitted gov-
ernment- and industry-supported schemes aimed at reassuring the public
about the environmental and social sustainability of forest practices against
competing information sources offered by environmentalists, indigenous
peoples, and other activists.

## Typologies of Certification and Labelling

In the forest sector, environmental certification has been defined as "a process
which results in a written certificate being issued by an independent third-
party, attesting to the location and management status of a forest which is
producing timber" (Baharuddin and Simula 1994, 9-10). This deceptively

simple definition raises a host of questions: Who should be accredited to assess whether a certification should be granted? What standards should govern the granting of certification? How should these standards be set? What kind of certification label should be used to communicate with consumers?

### First-, Second-, and Third-Party Certification

A threshold design question in the development of any certification scheme is: Who should be empowered to assess compliance with the relevant standard? The conventional answer distinguishes between first-, second-, and third-party certification schemes, defined in terms of the relationship between the developers and users of the scheme. In first-party schemes, forest companies are responsible for evaluating the consistency of their forest management practices against a given standard. Because companies have an economic stake in the result, however, first-party schemes are vulnerable to abuse. In particular, companies are likely to overstate their compliance with a given standard (Read 1991).

Credibility concerns associated with first-party schemes have triggered the development of second- and third-party certification schemes. Compliance oversight under second-party certification schemes rests with industry associations or forest product customers rather than with the individual firm. Although these schemes create stronger incentives for compliance, they suffer from many of the same deficiencies as first-party schemes. Industry associations are beholden to their members and tend to be reluctant to impose costs on them by developing onerous standards or curbing managerial discretion. These systems can be effective in weeding out unscrupulous operators, but it is difficult for them to promote substantive social and environmental practices. Also, compliance is often poorly monitored, resulting once again in exaggerated claims. These problematic features of second-party certification were evident in the early days of the American Forest and Paper Association's (AF&PA) Sustainable Forestry Initiative (SFI) (Gale 2002).

Increasingly, therefore, third-party certification schemes have grown in popularity. Here, the body empowered to conduct the certification audit (the certifying body) is independent from the company seeking certification. While this does not guarantee objectivity and independence – the Enron and WorldCom cases dramatically revealed the symbiotic relationship that can develop between large companies and their auditors – it does minimize the possibility of collusion. By the end of the 1990s, therefore, most actors involved in forest policy came to regard third-party certification as a key prerequisite of sustainable forest management.

### Management-, Technology-, and Performance-Based Standards

The certification literature distinguishes between management-, technology-, and performance-based standards (see Rametsteiner 2000, 84; see also

Chapter 11 in this book). Management-based standards are used extensively in conjunction with environmental management systems (EMS). In their purest form, management-based standards are not concerned with the outcome of the production process; instead, they seek to regulate a firm's activities during the pre-production or planning stage. Thus, for example, a company's EMS might specify that it conduct an environmental impact assessment before proceeding with a proposed project or activity. In this context, the certifier's task is to determine whether the assessment was done in conformity with company-established procedures. For the purposes of this audit, actual environmental impacts are, strictly speaking, irrelevant. In contrast, technology-based standards represent a regulatory intervention into the production process, typically by prescribing what technology should be employed. In relation to dioxin emissions, for example, a technology-based standard might prohibit the use of incinerators unless they have been fitted with scrubbers of a certain specification to remove toxic elements. Performance-based standards are concerned almost exclusively with outcomes and deliberately refrain from regulating how firms can or should seek to achieve this end – for example, a standard that required the company to ensure that its incinerators did not emit dioxins into the designated air shed. The certifier's task here would be to assess whether the company has complied with this performance-based prohibition.

Management-based approaches assume that by improving planning at the firm level, improved outcomes in terms of environmental sustainability will ensue. Technology-based standards specify the product or system to be used and are often pejoratively associated with inflexible "command and control" government regulation. Although this need not be the case, the general consensus in the certification literature is that technology-based standards should be avoided because they curb the firm's capacity to innovate by determining for itself how best to achieve desired outcomes. Performance-based approaches directly audit a company's practices against pre-established outcomes. All other things being equal, performance standards are generally preferred because they specify the outcome but leave the manner of its achievement to the firm. In practice, however, applicable regulatory standards often combine performance-based and management-based requirements and may even include – at least implicitly – some technology-based elements.

## Negotiating Standards

Whether management-, technology-, or performance-based standards are adopted, much turns on how they are developed. Standards that are developed by forest companies or industry associations often lack credibility because of a perception of bias. More credible certification schemes typically involve participation by a range of stakeholders, though there is considerable diversity in how this is implemented. AF&PA consulted widely within the industry before adopting its SFI scheme; but at the outset there was little participation from other stakeholders, a problem which was recognized by its expert group in 2000 when it called for wider stakeholder consultation and representation (AF&PA 2000).

Likewise, when the Canadian Standards Association (CSA) developed its Sustainable Forest Management System, it recruited representatives from a range of interests, including the forest industry, indigenous peoples' organizations, and the environmental movement. Some observers contend, however, that despite these recruitment efforts, ECSO participation in the process was too limited (Elliott 1999, 305-6).

The Forest Stewardship Council seeks to ensure balanced decision making by involving relevant interests under a tripartite structure that represents economic, environmental, and social interests. However, it too has been criticized, particularly by government and industry, for under-representing certain political and economic interests.

## Labelling

A key question is whether and how a company that has secured certification should be able to communicate that achievement in the marketplace. One option is for companies to treat certification strictly as an internal benchmarking exercise. This is a logical approach when companies perceive that the costs of publicizing the results of their certification audits will exceed the benefits. Most companies, however, view certification as an important achievement that they should publicize. Such publicity allows them to obtain the intangible benefits of a green corporate image, even if more tangible benefits of a price premium and increased market share do not occur. Once achieved, therefore, certification leads logically to an interest in product labelling. Labelling involves placing a mark or logo on the product that signals to clients and consumers that the product meets one, several, or all of the criteria of a particular eco-label.

The FSC combines certification and labelling under a single scheme. FSC-certified companies are entitled to use the FSC logo (Figure 2.1) on their products to signal their achievement to clients and consumers. However, because a product can change hands several times as it is transformed from a raw log in the forest to a manufactured product in the marketplace, the introduction of a label requires that the wood from a certified forest be tracked through the product chain from forest to retail outlet. This tracking process is known as "chain of custody" auditing. A separate process exists to certify companies in the timber chain, ensuring that the mix of certified and uncertified timber either does not occur or only occurs in proportion to the percentage-based claims made on the label (i.e., 75 percent of the product is made from certified timber).

*Figure 2.1*

**The Forest Stewardship Council's logo**

*Source:* Forest Stewardship Council.

Logos, such as the one used by the FSC, are known as Type I eco-labels (IISD/UNEP 2000, 47). Their purpose is to communicate to consumers that a product is environmentally appropriate according to one or more criteria. A large number of Type I eco-labels exist today, many sponsored by national governments. One of the earliest, launched in 1977, is Germany's Blue Angel eco-label. This label is sponsored by the German government; companies that meet designated product criteria are authorized to place the Blue Angel logo on their products. Canada's Environmental Choice program operates in much the same manner. Type I labels signal that a product is "environmentally friendly," but in many schemes this simply means that the product is superior to others across a single dimension (such as energy efficiency, biodegradability, or toxicity). Only a small number of national schemes employ life-cycle assessments that take into account the net impact of all the materials and processes involved in the production process.

In contrast, Type II eco-labels communicate claims about the product that have been made by manufacturers, importers, and distributors but have not been independently verified. As such, Type II labels raise a variety of credibility-related concerns. Moreover, such labels usually relate to a single characteristic of the product; information about the "eco-footprint" associated with the production process and end use is almost always excluded.

Type III eco-labels typically provide a simple list of a product's ingredients or component elements. The value of Type III labels tends to be limited since consumers rarely possess the means necessary to assess one product against another. This is exacerbated by the fact that such labels provide little or no information on the production and processing methods used. Perhaps the most valuable function of such labels is consumer protection; for example, such labels enable purchasers with allergies to avoid consuming products that contain identifiable allergens.

## The History of Certification and Labelling

There is nothing particularly novel about labelling products so that consumers can discriminate between them. This is, after all, what marketing and branding are all about. Companies develop brand names and logos to foster consumer loyalty for otherwise "like" products. Such marketing schemes link the brand or logo with particular features of the product and/or the company (such as quality, value-for-money, service) in an effort to secure and protect the company's market share.

Although motivated by different considerations, small-scale agricultural producers adopted a similar approach in the 1970s to market organic produce. Before long, a host of local and regional organic labels had emerged, generating disputes over what constituted proper organic production and who was entitled to use the organic logo (Guthman 1998). These disputes proved difficult to resolve due to the fragmented and dispersed nature of the organic agriculture movement and the absence of an oversight body to coordinate and harmonize standards.

As the organic movement grew in size and sophistication, so too did momentum to develop common standards and certification protocols. By the mid-1980s, standard-setting initiatives were underway in several different countries that were linked internationally through the International Federation of Organic Agriculture Movements (IFOAM). The federation has since developed a set of international standards for its accredited certifiers to use when auditing agricultural operations that are seeking organic certification. The organic agriculture movement grew relatively slowly until the 1990s, when it experienced a significant expansion as a result of several highly publicized food scares in the United States and Europe. While organic producers still, for the most part, do not compete head-to-head with industrial producers, there is growing evidence to suggest that industrial agriculture interests are taking competition from their organic counterparts more seriously by, among other things, launching their own certified organic product lines (Guthman 1998).

In the forestry context, the primary focus of activists during much of the 1980s and early 1990s was tropical deforestation. Among the activists' concerns was the replacement of tropical forest with plantation agriculture (as a consequence of government-sponsored development programs), as well as reckless industrial logging leading to biodiversity loss, soil erosion, and riparian destruction. In 1983, a number of international ECSOs, including the World Wide Fund for Nature (WWF), Friends of the Earth, and Survival International, brought these issues to the attention of the newly established International Tropical Timber Organization (ITTO) (Gale 1998b). At around the same time, the UN's Food and Agriculture Organization in conjunction with the World Bank, the World Resources Institute, and the United Nations

Development Programme launched a major new initiative, known as the Tropical Forestry Action Plan (TFAP), to promote sustainable forestry and curb tropical deforestation. At the outset, there was a great deal of hope that these two initiatives would be able to halt the destruction and degradation of tropical rainforests. By the late 1980s, this optimism had all but evaporated as it became increasingly clear that TFAP was faltering and that the ITTO, hamstrung by political compromise, was unable to take decisive action.

In response to these developments, a small group of British tropical timber activists began to explore the idea of using certification and labelling to improve tropical forest management. Leading members of this pioneering cohort of forest certification proponents included Koy Thompson, a forest campaigner with Friends of the Earth (UK), and Tim Synnott of the Oxford Forestry Institute. In 1988, Thompson and Synnott developed a feasibility proposal that was endorsed by the UK's Overseas Development Administration (ODA). ODA forwarded their proposal for funding to a 1989 meeting of the ITTO, where it ran into a storm of criticism from tropical-timber-producing countries and members of the timber industry who claimed that eco-labels would be a barrier to trade, encouraging consumers to substitute temperate for tropical timber (see Gale 1998b, 158-77). Thompson and Synnott's proposal was substantially "reformulated" at the 1989 ITTO meeting in a way that managed to address the concerns of producing-country members.

At this point, the evolution of forest certification begins to bifurcate. Along one path, the ITTO developed a strategy to fund a series of consultant reports on the potential of forest certification. One early report on "incentives" for sustainable forest management was prepared by the Oxford Forestry Institute, but it largely ignored the option of certification and labelling, focusing instead on the feasibility of imposing a national levy (Oxford Forestry Institute 1991). A second study, commissioned by the British government through the London Environmental Economics Centre (LEEC), was more holistic. Completed in 1993, the LEEC study concluded that certification and labelling could be a modest yet positive incentive for sustainable tropical forest management. LEEC proposed that governments consider sponsoring national certification schemes, but this suggestion once again met with strong opposition from developing countries and the forest industry assembled at the ITTO. As a group, the producing countries were so opposed to certification and labelling that they forced interested ITTO members to debate the matter in a special meeting organized outside of the ITTO's regular session. At this juncture, the most the ITTO could do was monitor the issue of certification and labelling, an activity it pursued by commissioning regular consultancy reports on the topic (Ghazali and Simula 1994, 1996, and 1998; Eba'a Atyi and Simula 2002; Pinto de Abreu and Simula 2004).

While these initiatives were unfolding at the ITTO, certification was also beginning to make headway in North America. In 1989, the Rainforest Alliance launched its SmartWood certification program and, two years later, a second US-based certifier came on the scene: Scientific Certification Systems (SCS). Not long after, both SmartWood and SCS were certifying forest operations using "in-house" standards. As the number of certifiers grew (the Soil Association and Société Générale de Surveillance (SGS) also began to certify operations in the early 1990s), so too did concerns over the proliferation of "sustainable forest management" standards.

Meanwhile, Herman Kwisthout, a bagpipe maker in Britain and founder of the Ecological Trading Company, was raising questions about the sustainability of the tropical timber he was importing from developing countries for his business. After pursuing his concerns with the Rainforest Foundation and the WWF-UK, Kwisthout began to promote the idea of an international forest monitoring agency in 1990. His concept, the earliest precursor of what was later to become the FSC, proved both prescient and catalytic (Synnott 2005). By this juncture, the ITTO consultancies were beginning to provide a theoretical rationale for forest certification, and the SmartWood Program, SCS, and other certifiers were demonstrating its practical feasibility. What remained to be achieved was integration of the separate certifier-based programs into a more coherent, global forest-certification system.

The spark for this global initiative came from a little-known California-based organization called the Woodworkers Alliance for Rainforest Protection (WARP): a group concerned about tropical deforestation and determined to purchase forest products from sustainable sources. WARP endorsed Kwisthout's idea for an international monitoring agency and established a working group to develop an implementation strategy.[2] During 1991 and 1992, this working group convened a series of meetings during which it was proposed that this new global organization be named the Forest Stewardship Council (Synnott 2005, 13). By late 1992, the FSC was provisionally established with an interim board, funding from the WWF, and logistical support donated by the Rainforest Alliance's SmartWood Program.[3]

The challenge confronting the FSC interim board was daunting: to create a global certification network linking interests in North and South America, Europe, Asia, and Africa. To this end, in 1992 and 1993, it embarked on negotiations with a range of stakeholders and certifying bodies that culminated in the founding of the Forest Stewardship Council at a stormy meeting in Toronto, Ontario, in October 1993. Some environmental and indigenous groups were angry that industry representatives had been invited to attend and were eligible to vote (Hammond 1993). They worried that industry would take control of the FSC and dilute its standards and perpetuate business-as-usual forestry. This concern extended to the draft FSC Principles and

Criteria document, which was viewed by some environmental and indigenous leaders as being too industry-friendly. While at times it appeared the meeting would end in deadlock, eventually a compromise was reached: an organizational structure designed to preclude industry dominance was approved, and a draft set of governing principles and criteria was accepted on the basis that they would be subject to ongoing revision and review.

These developments were closely monitored by the mainstream forest industry. By the late 1980s, the industry had become concerned about the competitive implications of forest certification, and in the early 1990s, many industry leaders were convinced of the strategic necessity to spearhead development of alternative, industry-friendly certification schemes. In North America, the Canadian Pulp and Paper Association donated a million dollars to the CSA to develop a Canadian scheme based on an environmental management standards approach adapted from the International Organization for Standardization (ISO). At about the same time, the American Forest and Paper Association launched what is now the Sustainable Forestry Initiative (SFI). In Indonesia, efforts were underway to reinvigorate the development of a national eco-label, culminating in the establishment of Lembaga Ekolabel Indonesia. And, in the United Kingdom, the Forestry Commission (the governmental agency responsible for UK forests) was under pressure to develop a British standard. After FSC-UK developed a draft standard in 1998, and as a consequence of the policy entrepreneurship of a small number of key individuals acting within the peculiar structural features of Britain's forest sector, stakeholders translated the FSC-UK standard into a British national standard – the UK Woodland Assurance Scheme (UKWAS) in 1999, making the UKWAS both the prevailing national FSC *and* a British government forest management standard.

Today there is a plethora of forest-certification schemes worldwide. Before discussing the FSC's major competitors, we will turn our attention to the FSC's structure and operation.

**Forest Stewardship Council**
The FSC's founding assembly created an institution with several unique attributes. One was its organizational structure, which allocated members to different chambers depending on their predominant interests. On its establishment in 1993, two chambers were created, one for environmental and social interests and the other for economic interests. Initially, environmental and social interests were allocated 75 percent of the voting rights, with the remainder assigned to economic interests. This organizational structure was revised at the first General Assembly meeting in 1996, when the environmental and social chambers were separated. Today, the governing body of the FSC (FSC-AC) is composed of three chambers representing environmental,

economic, and social interests. Each chamber holds one-third of the total vote, with a super-majority of 66 percent required to pass General Assembly resolutions. The official name of the FSC network is FSC-Asociación Civil (FSC-AC). In the balance of the book, we will use FSC-AC to refer to the network, its board of directors, and its international secretariat in Bonn, Germany (FSC-International Center), which is responsible for policy and standards, and network and stakeholder relations. FSC-AC also encompasses FSC's certifier accreditation arm, Accreditation Services International (ASI), and its branding and trademark arm, FSC-Global Development Company.

## Membership

A defining characteristic of the FSC is that it is a membership organization. Any non-governmental group or individual may join, providing they agree with its mission and its principles and criteria. As of December 2006, the FSC had 643 members. Table 2.1 provides a general overview of FSC membership; however, several general points deserve to be highlighted.

First, while in the past northern members have tended numerically to dominate the FSC, of late that gap has closed. As of 2006, 330 (51 percent) of the FSC's total membership (643) were from the northern, industrialized countries, while the remainder were from the South. Another feature highlighted in Table 2.1 is the under-representation of social interests relative to their economic or environmental counterparts. As of 2006, only 17 percent of FSC members belonged to the social chamber, compared to 40 percent for the environmental chamber and 43 percent for the economic chamber. Social chamber members include indigenous peoples' groups, unions, welfare organizations, development agencies, and so forth. To date, it has proven difficult for the FSC to recruit participants to the social chamber. Notwithstanding these difficulties, there are some powerful organizations included in the current membership, including, from Canada, the Shuswap Nation Tribal Council; the Pulp, Paper and Woodworkers of Canada; the Industrial, Wood, and Allied Workers of Canada; and the United Fishermen and Allied

*Table 2.1*

**Breakdown of FSC membership (as of December 2006)**

|  | Northern members | | Southern members | | Total | |
|---|---|---|---|---|---|---|
|  |  | (% total) |  | (% total) |  | (%) |
| Economic chamber | 161 | (25%) | 117 | (18%) | 278 | (43%) |
| Environmental chamber | 113 | (18%) | 141 | (22%) | 254 | (40%) |
| Social chamber | 56 | (9%) | 55 | (9%) | 111 | (17%) |
| Totals | 330 | (51%) | 313 | (49%) | 643 | (100%) |

*Source:* FSC 2006.

Workers Union. Third, while the FSC is often perceived as being dominated by environmental interests, it is interesting to note that economic interests are strongly represented, constituting 43 percent of the total membership, with a dominant position in the North. Finally, the relative power of the North and the economic chamber may be somewhat understated by these raw membership numbers. Given that the FSC is now located in Bonn, Germany, as well as the relative disparity in wealth between northern and southern members, the former tend to be over-represented at FSC General Assembly meetings and the latter are under-represented. Although the FSC does endeavour to subsidize the attendance of southern members, its ability to do so is constrained by somewhat limited finances.

## Governance

The governing body of the FSC is its General Assembly, which has met at approximately two-year intervals since 1995. These meetings provide members with the opportunity to influence the organization's policy direction through the development of resolutions for debate and adoption. In addition to influencing the FSC's overall policy direction, members also elect its nine-member board of directors. The board serves as the FSC's executive committee, convening regularly between General Assembly meetings to interpret resolutions, decide policy, approve national and regional standards, and ensure effective organizational implementation. At the outset, in order to prevent industry domination, only two positions on the board were reserved for representatives from the economic chamber, with the balance to be drawn from the environmental and social chambers. A few years after the FSC was founded, however, a motion to permit all three chambers equal participation on the board was passed overwhelmingly at the 2002 General Assembly. In addition to chamber representation, the board must also ensure equal North/South representation over time. As of 2005, five board members were from the South (Argentina, Bolivia, Colombia, Ecuador, and South Africa) and four from the North (Germany, New Zealand, Sweden, and the United Kingdom).

From 1995 to 2003, the FSC was headquartered in Oaxaca, Mexico, where it established a small secretariat to manage the organization's business. In February 2003, the FSC moved to Bonn, Germany, as a consequence of recommendations made in a 2001 Change Management Team report that argued in favour of greater organizational devolution and a more substantial presence in the North. Although the original idea of locating in a country in the South was admirable, the operational consequences were that the organization became marginalized by virtue of its location. In pursuit of devolution, the FSC has continued to establish "national working groups" (NWGs) in different countries, including Canada, Chile, Germany, Ghana, the United States, and Vietnam. Guidelines for the establishment of NWGs were developed in 1996, the same year the Canadian working group (FSC-Canada) was formed. FSC-Canada consists of a small secretariat in Ottawa, managed by an executive director who reports to an eight-member FSC-Canada Board. The board is made up of two members from each of FSC-Canada's four chambers, which include a First Nations indigenous peoples' chamber in addition to the regular environmental, economic, and social chambers. FSC-Canada is formally responsible to FSC-AC for the promotion and development of the FSC in Canada, including the development of regional standards, such as those now in place in the Maritimes, Great Lakes-St. Lawrence, British Columbia, and Boreal regions.

## Finance

Prior to 1990, donors to sustainable development and environmental causes focused on national efforts by NGOs to protect and conserve wilderness areas. In the 1990s, donors shifted their focus from national to global approaches and from community- and government- to market-based strategies. To a considerable extent, FSC-IC was "founded" by donors to test the viability of this new approach, and without such financial and strategic support third-party forest certification would never have evolved beyond a fringe concept (see Chapter 10).

While the FSC is working to reduce its reliance on support from charitable foundations, the majority of which are based in the United States, foundation funding remains a key and, at this point, indispensable revenue source. Foundation funding has likewise played a crucial role in establishing and running FSC-Canada. Table 2.2 outlines FSC-Canada's revenues from all sources since 2001. On average, foundation grants constituted more than 90 percent of FSC-Canada's total annual revenues from 2001 to 2004, when a substantial drop in revenues necessitated significant restructuring and downsizing.

*Table 2.2*

**Breakdown of FSC-Canada revenues 2001-6**

| Year | Total revenue (C$) | Foundation grants | Donations | Other |
|---|---|---|---|---|
| 2001-2 | 707,258 | 93% | 6% | 1% |
| 2002-3 | 909,893 | 92% | 5% | 3% |
| 2003-4 | 1,191,855 | 94% | 5% | 1% |
| 2004-5 | 341,496 | 84% | 15% | 1% |
| 2005-6* | 468,976 | 67% | 23% | 3% |

*Source:* FSC-CAN, Annual reports, various years.
* 2005-6 year as reported, although does not add up to 100 percent.

*Figure 2.2*

---

**The FSC's principles, with criteria for Principle 1**

Principle 1:   Compliance with laws and FSC principles
    *1.1*   Forest management shall respect all national and local laws and administrative requirements.
    *1.2*   All applicable and legally prescribed fees, royalties, taxes and other charges shall be paid.
    *1.3*   In signatory countries, the provisions of all binding international agreements such as CITES, ILO Conventions, ITTA, and Convention on Biological Diversity, shall be respected.
    *1.4*   Conflicts between laws, regulations and the FSC Principles and Criteria shall be evaluated for the purposes of certification, on a case-by-case basis, by the certifiers and the involved or affected parties.
    *1.5*   Forest management areas should be protected from illegal harvesting, settlement and other unauthorized activities.
    *1.6*   Forest managers shall demonstrate a long-term commitment to adhere to the FSC Principles and Criteria.
Principle 2:   Tenure and use rights and responsibilities
Principle 3:   Indigenous people's rights
Principle 4:   Community relations and workers' rights
Principle 5:   Benefits from the forest
Principle 6:   Environmental impact
Principle 7:   Management plan
Principle 8:   Monitoring and assessment
Principle 9:   Maintenance of high conservation value forests
Principle 10:   Plantations

---

*Source:* FSC Principles and Criteria for Forest Stewardship, FSC-STD-01-001 (Version 4-0) EN, approved 1993, amended 1996, 1999, 2002. http://www.fsc.org/fileadmin/web-data/public/document_center/international_FSC_policies/standards/FSC_STD_01_001_V4_0_EN_FSC_Principles_and_Criteria.pdf.

## Principles and Criteria

In drafting its guiding principles and criteria, FSC representatives drew on the experience of the organic movement and state-sponsored eco-labelling schemes, as well as on intergovernmental efforts to develop guidelines for sustainable forest management, such as those produced in 1990 by the ITTO (ITTO 1992) and the Helsinki and Montreal processes that commenced in 1990 and 1993, respectively. The result was agreement on ten principles and fifty-six criteria that were designed to ensure "environmentally appropriate, socially beneficial and economically viable" forest management. The FSC's ten principles are set out in Figure 2.2, together with the criteria for Principle 1 to illustrate how the system works. (The Appendix presents FSC's complete Principles and Criteria document.)

Like other certification systems, FSC's is hierarchical (see the next section), so its principles and criteria apply to all FSC-certified forest management operations. However, because they are often phrased in general language, their meaning in particular contexts must be elaborated through the drafting of indicators and, sometimes, verifiers. These will normally be spelled out in the relevant FSC national or regional standard; where such a standard has yet to be approved, certifying bodies are free to develop and employ their own.

## Organizational Structure

The structure adopted by the FSC to manage its certification system is outlined schematically in Figure 2.3, which highlights several important features about the organization. First, the FSC is an accreditation body and, as such, does not certify forest operations itself. Private companies with expertise in forest certification and labelling become accredited to FSC's accreditation arm, Accreditation Services International (ASI); in turn, these companies develop their own proprietary certification procedures and processes, which can vary widely. Some certifiers use pre-assessments and check lists; others prefer a more qualitative approach. As of 2007, fifteen certification organizations are accredited to ASI, with all but two offering both forest management and chain-of-custody certification. Moreover, all but two have their headquarters in the North, with only VIBO and SGS QUALIFOR operating from the South, from Mexico and South Africa respectively (see Table 2.3).

Second, as noted in the previous section, forests are certified to an FSC standard in two different ways. If a forest operation wishes to become FSC certified and there is no national- or regional-level standard, the operation is certified under FSC's generic principles and criteria as interpreted by the certifying body the company manager selects. In Australia, for example, Hancock Forest Plantations was certified by SmartWood, which used FSC's generic principles and criteria, but interpreted them within the Australian context. This form of FSC certification is not uncommon. There are many jurisdictions where there are no national or regional FSC standards, either because the FSC has a marginal presence in the country or because there have been difficulties and delays in negotiating the final national standard. In Papua New Guinea, for example, even though a national working group was established in 1996 and a draft standard was agreed upon in 1999, no final national standard has been approved (Bun and Bewang 2006).

Where a national or regional standard has been developed, there is a second route to certification. National and regional standards elaborate FSC's generic principles and criteria in the form of indicators and, sometimes, verifiers. Indicators are designed to spell out performance outcomes as precisely as possible. Verifiers let the certifying body know what information to collect to determine whether the indicator has been achieved. There are usually several indicators for each criterion and, as we have seen, several criteria for each FSC principle. This hierarchical approach is used in most approaches to standards development – as one descends the hierarchy, the actions that

*Figure 2.3*

The organizational structure of the FSC

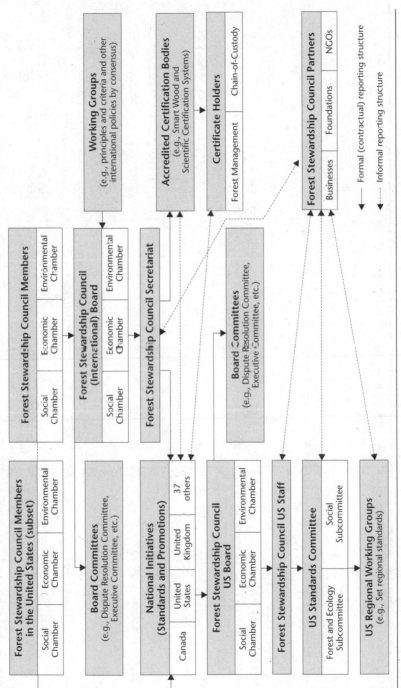

*Source:* Meridian Institute 2001.

*Table 2.3*

**FSC-accredited companies offering certification and chain-of-custody auditing**

| Name of organization | Country of headquarters | Scope of accreditation |
|---|---|---|
| BM Trada Certification | United Kingdom | Chain of custody |
| Centre technique du bois de l'ameublement (CTBA)* | France | Chain of custody |
| Certiquality | Italy | Chain of custody |
| Control Union Certification BV (formerly SKAL) | The Netherlands | Management and chain of custody |
| Det Norske Veritas Certification AB | Sweden | Chain of custody |
| Eurocertifor – Bureau Veritas Certification (BV) (formerly BVQI) | France | Management and chain of custody |
| EuroPartner | Russia | Management and chain of custody |
| Fundación Vida para el Bosque AC (VIBO) | Mexico | Management and chain of custody |
| GFA Consulting Group GmbH | Germany | Management and chain of custody |
| Instituto per la Certificazione ed i Servizi per Imprese dell'Arrendemento e del legno (ICILA) | Italy | Management and chain of custody |
| Institut für Marktökologie (IMO) | Switzerland | Management and chain of custody |
| KPMG Forest Certification Services Inc. | Canada | Management and chain of custody |
| Scientific Certification Systems (SCS) | United States | Management and chain of custody |
| QUALIFOR, SGS South Africa | South Africa | Management and chain of custody |
| SmartWood, Rainforest Alliance (SW) | United States | Management and chain of custody |
| Woodmark, Soil Association | United Kingdom | Management and chain of custody |
| Swiss Association for Quality and Management Systems (SQS) | Switzerland | Management and chain of custody |

*Source:* FSC 2007.

\* CBTA merged with Association Forêt Cellulose (AFOCEL) in June 2007. The new company is now named Institut Technologique Forêt, Cellulose, Bois-construction, Ameublement (FCBA).

are to be performed are described more and more precisely. The assumption is that if a company is meeting all the specified indicators, it is probably also meeting all the criteria and principles and, therefore, will qualify for certification. However, in many instances certifiers discover that a company's performance varies widely across indicators, so the certifier must judge whether, on balance, the operation is certifiable or not. Where a company is doing many things well, but some things poorly, the certifier can issue corrective action requests (CARs) to the applicant, indicating what changes it expects the company to make. Certifying bodies can certify operations in the expectation that such changes will be made over the coming years, or, if the certifier believes the changes are critical, it can make certification conditional on the completion of the CARs.

It is important to bear in mind that the FSC does not directly certify forest operations; rather, certifying bodies accredited to ASI are licensed to interpret applicable FSC certification requirements. This devolution of responsibility has been a source of controversy among some environmental groups, who claim that the system creates a conflict of interest (Counsell and Lorass 2002). These groups argue that certifying bodies have a vested interest in promoting lowest-common-denominator standards because the less demanding a standard, the easier it will be for a forest operation to become certified, enhancing market demand for certifier services. They also argue that, because certifying bodies are paid by the company, they have a vested interest in helping their clients become certified and, hence, in overlooking or downplaying non-compliance. These criticisms are familiar, because they also apply to most auditing arrangements in which companies pay accountants to audit their financial performance during the year and provide a signed statement that the accounts are a true and accurate representation of the company's finances. Although high-profile cases such as Enron and WorldCom can undermine popular confidence in the audit function, the system operates adequately much of the time. The consequences of poor performance, for example, can have significant consequences on auditors, as the demise of the multinational accounting firm Arthur Andersen attests.[4]

Under the FSC model, the performance of certifying bodies is monitored in two ways. First, any certified operation is under the constant scrutiny of local stakeholders; dissatisfaction at the local level can be relayed to the national and/or international bodies for redress. They are also reported on the FSC-Watch website. Certifying bodies are aware of this fact and, in difficult cases, will usually require operations to meet a significant number of CARs before being certified. Second, certifying bodies are accredited to the FSC, which is a valuable part of their company's good will. They are unlikely to recklessly certify a single operation in order to profit in the short term if this puts at risk their FSC accreditation and business over the long term. Thus, even though there have been a number of questionable certifications over the past ten years, for the most part the system appears to have operated effectively and efficiently.[5]

## Other Certification Schemes

To provide a general context for our consideration of the FSC case and, more specifically, to preface our contention, elaborated in Chapter 13, that both within the forest-certification context and beyond the FSC represents a *sui generis* governance form, we now turn our attention to the origins and institutional arrangements of the FSC's main competitors in the forest-certification sector, as well as those of its sister "stewardship model" organization, the Marine Stewardship Council (MSC). In the section that follows we discuss forest certification under the ISO, PEFC, SFI, and CSA regimes before reviewing arrangements under the MSC.

### Competing Forest-Certification Schemes

During the 1990s, in large measure as a strategic response by forest industry interests to the emergence of the FSC, a variety of competing forest-certification regimes were born. In the North American context, these included AF&PA's Sustainable Forestry Initiative, a CSA scheme sponsored by the Canadian Pulp and Paper Association, and the ISO 14001 system. In recent years, as we discuss below, SFI and CSA have associated themselves with what has become FSC's principal global competitor, the Programme for the Endorsement of Forest Certification (PEFC).

### *International Organization for Standardization (ISO)*

ISO is a global organization composed of the national standards-setting bodies of member countries. Thus, for example, the Standards Council of Canada (SCC), the body responsible for accrediting Canada's national standards-setting bodies including the CSA, is a member of ISO. So too are its Australian (Standards Australia), American (American National Standards Institute), British (British Standards Institute), and Japanese (Japanese Industrial Standards Committee) equivalents. The structure of these national bodies varies enormously, from governmental to statutory and associational arrangements. One important distinction is between organizations that accredit standards (such as the SCC) and those that develop standards (such as the CSA). The SCC is a member of ISO; the CSA is not. The latter provides standards development and related services to its members and at the request of the SCC.

Historically, business has tended to dominate ISO standard-setting processes due to the incentives associated with having proprietary product-related features or specifications embedded in national or global standards.[6] For most of the twentieth century, the principal focus of the SCC and other standards-setting bodies was the development of technical standards to

*Figure 2.4*

**Environmental management system model**

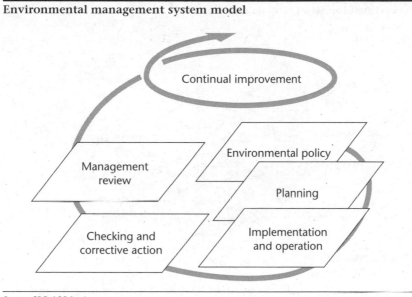

*Source:* ISO 1996, vi.

facilitate trade and commerce. In the last thirty years, however, issues of safety came to the fore, requiring broader stakeholder consultations. Thus, starting in the 1980s, ISO moved into the field of "management standards," establishing its popular ISO 9000 quality management standards. To secure ISO 9000 certification, companies were required to demonstrate that they had instituted an ISO-approved management system.

Subsequently, building on its 9000 series, ISO pioneered the development of an environmental management system (EMS) approach to standard setting in its ISO 14001 series, placing it squarely at the management end of the management-performance continuum. The cyclical "feedback" nature of the EMS model is set out in Figure 2.4. Under this approach, a company must formulate a comprehensive environmental policy that serves as a benchmark and driver for implementing and enhancing its EMS (ISO 1996, 6). A company must also develop an EMS implementation plan, which includes a review of current operations that covers legislative and regulatory requirements, identifies significant environmental aspects of the company's operation, examines existing environmental management practices and procedures, and incorporates feedback from investigations of previous incidents (ISO 1996, 7). The output of the planning process is a program to implement an EMS that designates line responsibility for implementation, trains staff, documents program operations and consequences, and ensures emergency preparedness.

*Figure 2.5*

**ISO-certified forests in Canada, 2006**

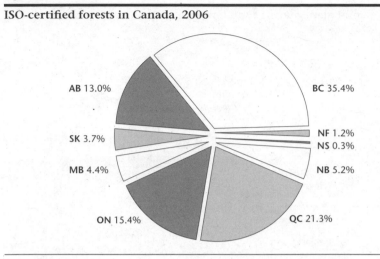

*Source:* CSA International.

In implementing an EMS program, companies are obliged to monitor their performance to identify areas of non-compliance and take appropriate corrective and preventive action. To this end, they must undertake regular EMS audits. The EMS cycle also entails regular management reviews. Unlike EMS audits, these are designed to assess environmental outcomes against EMS objectives through a rolling process of review that considers audit results, the suitability of objectives and targets in light of changing circumstances, and the concerns of relevant interested parties (ISO 1996, 10).

The ISO 14001 system differs substantially from the other approaches discussed below in that it is purely a management system approach that deliberately eschews use of performance standards. This allows companies to establish their own performance measures and targets. In the forestry context, for example, some companies will identify clear cutting as a major environmental challenge to be addressed, while others will identify it as an acceptable forest management tool. The only "absolute requirement for environmental performance" under the ISO 14001 approach is a commitment "to compliance with applicable legislation and regulations and to continual improvement." Thus, firms carrying out similar if not identical activities but achieving markedly different environmental performance can both be in full compliance with ISO 14001 requirements (ISO 1996, vi). Because of its lack of performance requirements, ISO certification is generally regarded as one of the most straightforward and least costly certification models for most operators. In Canada, as Figure 2.5 depicts, over 132 million hectares of forest were slated to be ISO certified by the end of 2006, considerably more than is certified by any other competitor system.

*Programme for the Endorsement of Forest Certification (PEFC)*
The Pan-European Forest Certification Council was formed in 1999 as a consequence of Scandinavian – notably Finnish – resistance to the development of FSC national standards. Following a meeting of representatives of eleven mainly northern countries in Paris, it became an umbrella organization for European certification schemes.[7] Since then the organization has expanded rapidly, renaming itself the Programme for the Endorsement of Forest Certification in 2003 and attracting national certification members from around the world. During this period, the magnitude of PEFC-certified forests has grown exponentially: currently it certifies over 187 million hectares of forest worldwide (PEFC 2006b). A majority of the national schemes currently participating in the PEFC are from northern countries, including the Finnish Forest Certification Council, PEFC-Norway, PEFC-Germany e.V., and the PEFC Council of Ireland. North American representatives include the CSA and the SFI. However, the PEFC is working hard to increase its representation in the developing world, where associated schemes include Chile's CertForChile and Cerflor in Brazil (PEFC 2006c).

Currently, the PEFC operates primarily as an accreditation body, endorsing schemes that are developed nationally through various stakeholder processes and deploying principles drawn from various "inter-governmental processes for the promotion of sustainable forest management," including the Helsinki Process in Europe and the Montreal Process in North America (PEFC 2006d).

*Sustainable Forestry Initiative (SFI)*
One of the earliest competing schemes to emerge following the founding of the FSC was the Sustainable Forestry Initiative. Spearheaded by a variety of AF&PA member organizations, the SFI was launched in 1995. The SFI model is structured around thirteen objectives that are akin to the FSC's ten principles. These objectives range from broadening "the implementation of sustainable forestry by ensuring long-term harvest levels based on the use of the best scientific information available" (Objective 1), through protecting "the water quality in streams, lakes, and other water bodies" (Objective 3) and managing "the visual impact of harvesting and other forest operations" (Objective 5), to promoting "continual improvement in the practice of sustainable forestry" and monitoring, measuring, and reporting "performance in achieving the commitment to sustainable forestry" (Objective 11) (AF&PA 2002).

These objectives are elaborated through performance measures and indicators. The former are "a means of judging whether an objective has been fulfilled"; the latter are "those indicators integral to conformance with the SFIS [Sustainable Forestry Initiative standard]." A typical form of performance measure is compliance with applicable laws (SFB 2004, 5). Associated

with each performance measure are a number of indicators. For Performance Measure 3.1, which requires members to "meet or exceed all applicable
federal, provincial, state, and local water quality laws," indicators include
requirements that the company have in place "(i) [a] program to implement
state or provincial BMPs [best management practices] during all phases of
management activities; (ii) contract provisions that specify BMP compliance;
(iii) plans that address wet-weather events ... ; (iv) monitoring of overall BMP
implementation" (6).

Although the terminology is different, the hierarchical approach adopted
in the SFI standard is similar to the approach adopted by the FSC. And though
it does not actually accredit audit companies itself, the SFI does require its
certifying bodies to be "environmental management system (EMS) registrars
and accredited by the American National Standards Institute or the Standards
Council of Canada" (SFB 2004, 24), leading to some overlap between SFI-
and FSC-accredited companies (KPMG can audit a firm to either standard,
for example).

The SFI's scheme also departs from the FSC model in several respects. First,
the SFI permits its members to obtain first- and second-party certification
(which it terms "verification"), in addition to third-party certification. Second,
SFI's standards have been developed largely by the US forest industry, with
environmental and social interests being under-represented (Gale 2002). Third,
although the Sustainable Forestry Board (SFB), the entity that manages the SFI,
claims that it now represents a "balanced array of interests," the organization
has, in the past, been dominated by industrial forestry interests. In recognition
of this, the SFI has embarked on various initiatives to create a more independent governance structure, including establishment of the SFB.

In its early days, the SFI's institutional structure was relatively simple: in
effect, the organization was run as a committee of the AF&PA. In the face of
continued criticism that the organization was compromised by its close association with the forest industry, an independent board (the SFB) was established in July 2000 with a mandate to "oversee development and continuous
improvement of the Sustainable Forestry Initiative (SFI) Program Standard,
associated certification processes and procedures, and program quality control
mechanisms" (AF&PA 2006). The SFB is constituted as a fifteen-member board,
ostensibly equally balanced between "5 environmental non-profit CEO's,
5 forest industry CEO's and, and 5 members of the broader forestry community" (SFB 2006a). Of those listed as board members in 2006, however,
only two appear to be representatives of the environmental community.[8]
Notably, none of the broader-based environmental membership organizations, such as the WWF, Friends of the Earth, Sierra Club, or Greenpeace,
are represented on the SFB board. Moreover, the SFB has been deliberately
structured as a foundation rather than a membership-based association,
which raises issues of accountability for users of the SFI scheme.

*Figure 2.6*

**The growth of forest certification in Canada, by scheme, 1999-2006**

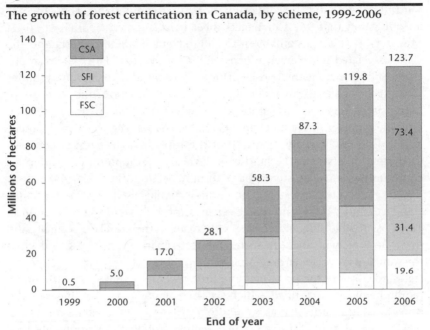

*Source:* CSFCC (Canadian Sustainable Forestry Certification Coalition) 2007.

At the outset, only AF&PA members were eligible to join the SFI. In 2000, however, eligibility was extended to Canadian companies as well. Following this change, the SFI has seen significant growth in Canadian certifications, which now cover more than 31 million hectares (see Figure 2.6).

*Canadian Standards Association (CSA)*
Another forest-certification scheme that has seen significant growth in recent years is the CSA's Sustainable Forest Management System. The origins of this scheme date back to 1993, when the Canadian Pulp and Paper Association, in conjunction with forest groups across Canada, formed the Canadian Sustainable Forestry Certification Coalition (CSFCC) and hired the Canadian Standards Association to consult with stakeholders to develop a certification scheme in the forest sector (Cashore, Auld, and Newsom 2004).[9] The CSFCC hired the CSA because it is one of the few civil society organizations accredited by the Standards Council of Canada (SCC) to develop standards for the Canadian Standards System (CSS). Established in 1919, the CSA develops standards that are intended to "reflect a national consensus of producers and users – including manufacturers, consumers, retailers, unions and professional organisations, and governmental agencies" (CSA 2003).

The CSA's forest standard builds on the EMS approach pioneered by the ISO (outlined above) and also contains performance components. These include an obligation that forest managers meet the Canadian Council of Forest Ministers' (CCFM) sustainable forest management (SFM) criteria and the CSA's associated SFM elements for a defined forest area (DFA).[10] For example, Criterion 1 concerns the conservation of biological diversity and requires that this be achieved "by maintaining integrity, function, and diversity of living organisms and the complexes of which they are part" (CSA 2003, 44). The CSA identifies four elements related to this criterion, with Element 1.1, Ecosystem Diversity, stating that the forest manager must "conserve ecosystem diversity at the landscape level by maintaining the variety of communities and ecosystems that naturally occur in the DFA" (44).

Both FSC and CSA systems incorporate management and performance standards and require third-party certification by an independent auditor. The most notable differences between them are in their level of specificity and in the arrangements made to ensure interested parties are represented when the standard is developed and implemented. In terms of specificity, the FSC's standard contains ten principles and fifty-six criteria that set out the broad requirements to be certified, which are then elaborated in specific national or regional indicators that allow auditors to assess field-based performance. In contrast, the principal forest management obligations under the CSA standard are more structurally fragmented and diffusely expressed.[11] Steps have been taken to make the CSA standard clearer and more user-friendly, but efforts to integrate management, performance, and participatory components into the standard have created a rather unwieldy document that lacks the overriding hierarchical architecture evident in both the SFI and the FSC. Moreover, although performance obligations are clearly identified and expressed, the terminology employed is often vague and discretionary.

The first version of the CSA standard, produced in 1996, was criticized by observers who felt the process was dominated by industry and lacked the participation of major environmental groups (Elliott 1999).[12] When it revised the standard between 2000 and 2002, the CSA made efforts to overcome this deficiency by using a "matrix" approach to representation. This provided for four categories of participation within the CSA's SFM Technical Committee, with members coming from industry; the professions, academia, and practitioners; environmental and general interest groups; and government/regulatory categories. Several environmental groups participated, including the Canadian Wildlife Federation, but most major environmental organizations did not.

Some groups remain skeptical about the organization's inclusivity. In 2002, Pollution Probe reported that, despite the CSA's efforts to engage ECSOs throughout the 1990s, their participation has lagged because of a number of factors. These included a lack of funding support for such groups, the ECSOs' lack of faith that voluntary regulatory initiatives were valid or effective, and the persisting perception that the CSA emphasized the international trade dimensions of standardization at the expense of environmental and social factors.[13] Many ECSOs alleged that their input on CSA committees was ignored, despite the matrix system (Pollution Probe 2002, 32).

Despite these criticisms, the CSA scheme has expanded across Canada quickly, growing from about 3.5 million hectares certified in 2000 to more than 73 million hectares in 2006 (Figure 2.6, p. 41). A recent report places the total amount of CSA-certified forest in British Columbia at almost 33 million hectares, with several major corporations listed as certified, including Abitibi-Consolidated, Ainsworth Lumber, Canfor, Tolko Industries, and Western Forest Products (Abusow International 2007). These certifications have occurred on both tree farm licences and forest licences, the latter requiring the collaboration of a large number of partners, including the BC Ministry of Forests.[14]

A core feature of the CSA standard, especially the revised version, is an extensive public consultation requirement. In contrast to the FSC process, in which national and regional indicators are negotiated via an interest-balancing process, the CSA standard effectively devolves the development of indicators to the local, DFA, level. In Clause 5, "Public Participation Requirements," it specifies that the company seeking certification will "openly seek representation from a broad range of interested parties," provide them with "relevant background information," demonstrate that "efforts were made to contact Aboriginal forest users and communities," and ensure that the consultation process follows a set of "Basic Operating Rules" (CSA 2003, 12-13). The overall goal of this consultation is to give the public "an opportunity to be involved proactively in the management of the DFA" (12).

Because negotiations within the DFA play such a crucial role in the CSA approach, the potential exists for managerial fragmentation across the landscape as differently structured negotiation processes result in different visions and values becoming embedded in forest management plans. Such fragmentation could occur within a single forest company operating in several different DFAs, as well as across forest companies. It is not possible to say whether such variation is, in fact, occurring, as no published studies are yet available comparing the visions, plans, and indicators for CSA-certified DFAs. If there is considerable local variation in forest management objectives and practices between adjacent DFAs, this would raise questions about whether the CSA standard is achieving its "standardization" objectives. Conversely, a lack of variation could undermine the claim that the CSA is responsive to community differences. The tension is similar to the one the FSC faces with respect to harmonizing regional and national standards.

## Marine Stewardship Council (MSC)

The final institutional comparator to be considered originated in 1996 as an attempt to translate the FSC "stewardship model" into the marine fisheries context. This initiative was championed by way of a strategic partnership between the WWF and Unilever. Although the FSC and the MSC share somewhat analogous visions, the MSC has adopted a distinct approach to governance.

The FSC is committed to a democratic, membership-based model, an approach the MSC has deliberately eschewed in favour of a more "insulated," foundation-style structure. This decision was significantly influenced by advice from its consultants, Coopers and Lybrand, who were reputedly "horrified" by the "chaotic" nature of the 1996 FSC General Assembly they attended as observers (Synnott 2005, 25).[15] Since 1999, the MSC has operated as a not-for-profit, international organization fully independent of the WWF and Unilever, with the majority of its funding coming from a range of cháritable foundations and private organizations

Much as the FSC does in the forestry sector, the MSC certifies the sustainable performance of fisheries on a global scale. To this end, it has developed principles and criteria for the promotion of sustainable fisheries and has applied these to certify several operations, including the western Australian rock lobster fishery (Australia's most valuable fishery), the New Zealand hoki fishery, and the Alaskan salmon fishery.[16]

When it was first created, the MSC encountered widespread skepticism from many quarters, including fisheries managers, industry representatives, and several environmental organizations (Potts and Haward 2001). In particular, critics raised concerns about the degree and nature of consultation over the MSC's founding principles and criteria. In response to these concerns, the MSC reviewed its governance arrangements in June 2001 and made a series of organizational changes. The current structure of the MSC is set out in Figure 2.7.

The board of trustees is the executive decision-making body within the MSC. It consists of up to fifteen members and is the final decision-making authority. It works in close association with the technical advisory board (TAB) and the stakeholder council. The TAB has fifteen members and advises the board on such matters as setting and review of the MSC standard, logo licences, and chain-of-custody certification. The board appoints members to the TAB who, in turn, appoint their own chair. The council consists of between thirty and fifty members representing diverse interests within the organization, including conservation, industry, academia, and developing nations. It fulfils the role of a participatory forum and representative authority and submits its views directly to the board (Gale and Haward 2004).

From this description, it is clear that the MSC's structure has a very different organizational logic than that of the FSC. Indeed, its organizational arrangements more closely resemble those of the SFI than the FSC. Both the

*Figure 2.7*

**Marine Stewardship Council governance**

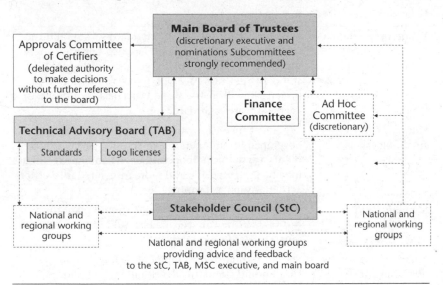

*Source:* Potts and Howard 2001.

MSC and the SFI are governed by boards that have been subject to criticism on the grounds that they are not representative (Constance and Bonanno 2000; Ponte 2006, 15-17). The SFB appears to have made some headway toward addressing these concerns, but the MSC remains vulnerable to criticism. Accordingly, even as the MSC ostensibly embraces the values of fairness, inclusiveness, impartiality, independence, reliability, professionalism, openness, and accountability, the composition of its board suggests it remains dominated by business. Of the ten board members in 2007, seven hail from the corporate sector and only one has strong environmental credentials (MSC 2006). Moreover, because the MSC is a foundation and not a membership-based association, its board is insulated from the wider concerns of the global fisheries "polity," whose ability to influence board decisions is restricted to making submissions to the stakeholder council, which, in turn, is responsible for representing these views to the board. The MSC is more responsive and accountable than it was before its recent restructuring, but it remains to be seen whether the new arrangements will be adequate to counter ongoing criticisms of industry dominance.

### Issues in Forest Certification

In this section, we offer some introductory observations on a range of key issues in forest certification – summarized in Table 2.4 – that we revisit at various points in subsequent chapters.

*Table 2.4*

**Issues in forest certification**

| Issue | Key questions raised |
| --- | --- |
| Vision of the standard | How high a bar should the standard set? Should the standard align itself with other existing certification models or should it seek to recognize only "gold standard" performance? Or, to use the terminology preferred by industry, should this be a "boutique" or a "general store" standard? |
| Standards and smallholders | Whatever the vision, should a single standard be developed for all operators or should special arrangements apply to small operations, where management may be less intensive and more integrated into other activities such as farming? |
| Standards and plantations | What exactly is a "plantation"? Should plantations be certified? If they are certified, should it occur under a separate plantation standard? |
| Financing standards development | Should standards development be largely voluntary? How should standards development be financed? How can standards development be insulated from the influences of funders? |
| Negotiating standards development | Who should negotiate the standards? How should standards be negotiated (i.e., via stakeholder bargaining or through facilitated mediation)? |
| Form of the standard | Should standards be management-, technology-, or performance-based? What are the differences between and relative merits of these various forms of standards? What is the relationship between the form of a standard and its prescriptiveness? |
| Harmonization of standards | What does "harmonization" mean? Should it be procedural or substantive? What process should be used to ensure that standards in one jurisdiction equate to standards in a neighbouring jurisdiction? |
| Mutual recognition of standards | Under what circumstances should mutual recognition occur? What are the dangers inherent in adopting this principle? |
| Auditing of standards | Who should be entitled to audit a standard for compliance? Should certification schemes directly audit companies for compliance or can this function be contracted out under licence? Is there a conflict of interest between private auditors and their clients in relation to forest certification? What can be done to minimize any such conflict of interest? |

▶

◄　*Table 2.4*

| Issue | Key questions raised |
|---|---|
| Single versus step-wise standards | Should the standard specify only a single threshold to be reached by a forest operation? Or should a "stepwise" approach be adopted – bronze, silver, gold – to encourage poor performers to engage in the process? |
| Standards and international trade law | Are voluntary standards subject to emerging trade-law disciplines under the World Trade Organization's Agreement on Technical Barriers to Trade? If so, what should voluntary standards organizations such as the FSC do to ensure compliance? |
| Standards, legitimacy, and democracy | If standards are developed without government participation, can they be deemed "legitimate" and "democratic"? Does the fact that standards are "voluntary" render these concerns moot or should even "voluntary" standards be subject to a democratic test? |

## Vision

A key issue in FSC standard setting is how and where to set the performance bar. Large-industry interests frequently argue that it is important to avoid developing a "niche" or "boutique" standard: in other words, one that imposes performance requirements that are feasible for small-scale or community forestry operations but too costly for larger operators to adopt. Such interests, typically, advocate for a "mainstream" or "general store" standard that will maximize the potential for take-up across the forest industry. Many environmental and social justice representatives reject the boutique-general store analogy as a false and self-serving dichotomy. They contend that only a rigorous, credible standard backed up by market pressure will succeed in securing the transformative benefits that the FSC's vision of "environmentally appropriate, socially beneficial and economically viable management of the world's forests" exists to promote (see Counsell and Lorass [2004, 25-26] for a discussion of these issues in the context of the debate over the FSC's fast-growth strategy).

## Smallholders

Another key issue is how the standard should treat different types of forest operators. Small, part-time, poorly capitalized operations that are integrated into other activities such as farming are clearly different from large, full-time, highly capitalized, expert-driven forestry. How should a standard treat smallholders? One approach is to develop a single standard that treats everyone equally on the basis that all operators, regardless of size, should meet the same requirements. Another approach is to adopt a single standard and include clauses that modify the application of the standard "subject to scale and intensity" or other similar wording. A third approach is to develop separate standards for smallholders and large-scale forestry. The approach adopted can have significant implications not only for standards development but also for the relative take-up of certification by different operators.

### Plantations

Many environmentalists and eco-foresters believe that plantations are not forests and should not be certified. Plantations are, however, "sources of wood," and they obviously compete directly with other sources of wood in the form of semi-natural and natural forests. Defining the term "plantation" is difficult because some plantations display almost all the characteristics of a regrown natural forest. In other cases, however, the distinction is clear. Even when plantations are clearly identified, questions arise as to whether they should be certified at all, especially if they have been created recently by clearing naturally forested land. If they are to be certified, should they be subject to special conditions and, if so, which conditions? The issue of plantations threatened to derail the formation of the FSC before it was established. It has recently resurfaced for further debate within the FSC via the establishment of a Plantations Advisory Group. In the BC context, as we shall see in Chapter 9, the issue of plantations also became contested, despite the fact that there are, in fact, not many straightforward examples of plantation forestry in the province.

### Financing Standards Development

Developing regional or national standards can be a resource-intensive proposition requiring substantial inputs, including technical reports and advice, high-level stakeholder consultations, and significant public feedback. Frequently, however, standard-development processes are under-resourced, especially within the FSC system, which relies heavily on volunteer labour. When industry and government have chosen to invest in standards development, this funding has typically been channelled to industry-dominated national standards, leaving the FSC to fall back on grants from foundations and charities. These organizations have supported the FSC significantly during its first decade, but such funding cannot continue forever, which raises difficult questions about financing standards development, renewal, and harmonization over the long term.

### Standards Negotiation

The negotiating arrangements adopted to develop a standard clearly influence its final outcome. There are numerous negotiation models to choose from, ranging from direct, relatively unmediated, stakeholder bargaining to indirect, centralized, mediated and facilitated standards development, where stakeholders participate at arm's length from each other. Should one model automatically be preferred over another or are different models appropriate in different contexts? Once a model is chosen, who should be invited to participate at the negotiating table? Of the huge number of potential stakeholders, should all, some, or only a select few be allowed to participate? If all stakeholders are not involved, what criteria should be used to select the participants, and who should do the selecting? And when a group has been formed, how should its members relate back to their constituencies? Should there be a formal reporting relationship, or should representatives take responsibility for their own reporting arrangements? In Chapter 4 we discuss in detail how these issues were addressed during the development of the FSC standard for British Columbia.

## Forms of Standard

It is often contended that the optimal standard is one that minimizes "prescriptiveness" and, therefore, that performance-based standards should be preferred. We argue that when assessed against the definition of "prescriptive" we employ in this book, all standards are inherently prescriptive in that they constrain decision making at the firm level. In this sense, management-, technology-, and performance-based standards can all be seen as prescriptive regulatory constraints, albeit arising at different stages of the production process.

As we shall see, forest-certification schemes adopt differing approaches to standard setting. For example, most commentators would agree that the ISO tends toward the management side of the continuum, whereas FSC standards tend to be more performance-based. Moreover, even within the FSC, regional and national standards can and do differ significantly in terms of the form of standard they adopt. Another emerging trend is hybrid standards that seek to integrate the respective benefits of management- and performance-based approaches. The complex art of crafting standards in the forest context and beyond is covered in detail in Chapter 11.

## Harmonization

Standards typically evolve out of national and regional negotiations, with the result that the applicable standard in one jurisdiction will vary, sometimes significantly, from its counterpart in a neighbouring jurisdiction. In terms of relative production costs, this can create competitive advantages and disadvantages for firms that are competing in the same markets. Standards, therefore, need to be harmonized across regions, yet it is not clear how harmonization should be defined and how it should be achieved. Harmonization can be viewed either as a process or as a substantive achievement. Viewing it as a process involves establishing a set of consultative arrangements whereby input is obtained from stakeholders in neighbouring jurisdictions, with their comments taken into consideration as the standard is developed. However, unless such stakeholders are sitting at the negotiating table, their views are liable to be sidelined by those actually engaged in the negotiations. Another approach is to conduct an independent evaluation of a draft standard with respect to neighbouring jurisdictions to ensure it does not impose unnecessary requirements on firms falling within its jurisdiction. The problem here is that any standard reflects a delicate balance among local stakeholders, and altering any single component after the fact risks undoing the entire package. Harmonization, then, is fraught with difficulties, creating significant dilemmas for those engaged in standards negotiation.

## Mutual Recognition
The idea of mutual recognition is that national standards developed through national processes should be deemed as equivalent to each other, enabling purchasers in one country to have confidence that certified timber from another country has been produced according to reasonable standards of "sustainable forest management." Industry and governments have been the main drivers of mutual recognition of forestry standards. There are significant dangers inherent in mutual recognition given that many national standard-development processes have been flawed in terms of stakeholder consultation and in terms of the comprehensiveness and rigour of the emerging standard. Unless there is an independent body that can assess standards and ensure that only those that meet minimal standards are, in fact, mutually recognized, the process of mutual recognition will quickly lead to a race to the bottom, where the lowest of the available standards will establish the context for all others. The Programme for the Endorsement of Forest Certification (PEFC) has become the de facto organization for carrying out such assessments. Should the FSC entertain the notion of mutual recognition – implicit, for example, in the relationship between the FSC-UK standard and the UK Woodland Assurance Scheme (UKWAS) – or should it maintain its commitment to a uniform FSC standard in an effort to prevent a race to the bottom?

## Auditing Standards
A substantial industry has developed over the past several decades to audit compliance with mandatory requirements and voluntary standards. Large companies, such as Societé Générale de Surveillance (SGS), have specialized in providing auditing services to large multinational corporations in forestry and other sectors. The emergence of this new service industry has raised questions about whether determining compliance to a standard should be contracted out to independent auditing companies, or whether it should remain an in-house responsibility of the certifying body. Each

approach has its own problems. As we have seen, the FSC system has been criticized for contracting out the auditing function to accredited certifiers – like SmartWood, Scientific Certification Systems, SGS, the Soil Association, and others – because there is the potential for a conflict of interest between the certifier and the client, with the former indulging the latter because the latter is paying for the auditing service. This argument leads logically to the conclusion that the FSC should retain an in-house auditing function. This could raise badly needed revenue for the FSC, but it does not overcome the conflict-of-interest problem. It simply moves its locus from private companies to the FSC, which would then face the temptation to dilute standard requirements so it could grow the business to pay for its own development. Another issue is who should be accredited to audit standards. Each certification scheme adopts slightly different arrangements in this regard, and there is the potential, perhaps, for a breach of trade practices law if any standard explicitly excludes an auditor accredited to one standard from qualifying to be an auditor under another standard.

## Single versus Stepwise Standards

Any standard imposes costs on a company when it obliges the company to do things it would not otherwise do. Ironically, the further the company is from the desired practices set out in a standard, the less financial incentive there is for the company to adopt that standard because of the costs of re-forming existing practices. The logical response to this problem is to adopt a stepwise approach to standard implementation, whereby a company that cannot currently meet the required "gold" standard is nonetheless encouraged to engage in certification by taking the first "bronze" step. The problem with the stepwise approach to certification is ensuring that the bronze step is the first in a series toward achieving certification at the gold level and does not become a new, de facto, minimal benchmark for certified wood in the marketplace. Industry and governments have been enthusiastic about stepwise certification, but others have resisted it for precisely this reason. On the other hand, research in developing countries (Cashore et al. 2006) suggests that ecoforestry-oriented and community-based operations are having a difficult time meeting the FSC standard, and this is leading to renewed interest within the organization in the development of a stepwise approach, albeit one with strict procedures in place to ensure that the steps are completed in a timely and incremental fashion.

## Voluntary Standards and Global Trade Law

Initially it was argued that because certification was voluntary, it would not be subject to emerging international trade rules under the World Trade Organization (WTO) and particularly the Agreement on Technical Barriers to Trade (TBT). In part as a response to continuing uncertainty about this complex issue, the FSC and other similar eco-certification and labelling schemes co-founded the International Social and Environmental Accreditation and Labelling (ISEAL) Alliance in 1999. A key function of ISEAL is to provide strategic advice to its member organizations with respect to trade-related issues (see Chapter 11).

### Standards, Legitimacy, and Democracy

At an even more abstract level, observers question whether voluntary standards are "legitimate" and "democratic," and, if they are, what makes them so? Critics from a broad spectrum of political opinion have occasionally argued that voluntary standards are "undemocratic." For some, this critique is based on the premise that the only institution entitled to regulate in a national space is the state, and any competing institution that seeks to undertake this role is somehow usurping democracy. Other critics argue that the real threat to democracy is that these emerging non-state actors lack appropriate accountability to the general public and tend to be captured by vested interests. Debates such as this, of course, are highly dependant on one's definition of "democracy," a quintessentially "contested concept." Moreover, it can be argued that the state's monopoly on the right to govern – as well as its ability to do so – has been dwindling in recent years. As a result, the stage may now be set for various new players to take on what were formerly state governance and regulatory roles. Further reflections on this topic are offered in Chapters 12 and 13.

### Conclusion

Forest managers can choose from a growing range of certification options and models. Not surprisingly, decisions about certification are driven by strategic considerations related to a firm's perceptions of the strengths, weaknesses, opportunities, and threats posed by certification in general, as well as the costs and benefits of specific schemes. The result of these analyses varies substantially among companies, judging by the relative popularity of different certification and labelling schemes. In the early 1990s, many companies adopted a "wait and see" approach because they believed that there was little to be gained from becoming certified early and that the benefits derived might not be worth the costs incurred if forest certification turned out to be a fad. Over the past decade, however, it became apparent that most forest companies regard certification as a basic operating requirement and are currently seeking or have secured certification under one or another of the schemes outlined in this chapter.

For many companies, becoming certified means adopting ISO's 14000 EMS approach. The ISO approach is internationally recognized in the marketplace, is cost-effective to implement, retains managerial autonomy, and does not require major organizational adjustments. These features of ISO stand in sharp contrast to the FSC's performance-based scheme. The FSC not only requires forest companies to meet an internationally established benchmark for good forest management, it also incorporates into that benchmark substantial ecological and social content that simultaneously reduces managerial discretion and creates new compliance obligations. Industry therefore tends to view the FSC as more onerous and costly than other certification systems.

Regardless of these differences, a key measure of the success of any certification regime is market acceptability. And a key to market acceptability is a scheme's credibility in the eyes of clients and consumers. This is where FSC competitor schemes are vulnerable, because none currently rival the level of environmental and social support enjoyed by the FSC. Although some PEFC schemes claim to have such support, a detailed comparative study finds them weak in many relevant criteria (Gale 2002). Of the two PEFC schemes examined here, the SFI was developed in-house by AF&PA with minimal input from environmental, social, and indigenous groups, and, although it has sought to restructure itself significantly to address this criticism, it remains heavily dominated by the US forest industry. Likewise, the CSA has sought to recruit mainstream environmental groups into its technical committee. However, the architecture of its standard (based on the ISO EMS approach) and its bias in favour of industry stakeholders has constrained its ability to broaden its base. ISO also has a reputation for being industry dominated; this is partly a legacy of its organizational history and partly due to its dependence on national organizations for the development of international standards. Statutory bodies such as the SCC perceive their mission in terms of facilitating trade in Canada's interests and inherit a consultative model that is poorly adapted to the development of standards that go beyond technical concerns to issues of safety, hazardousness, and environmental and social benefit. Even the FSC, despite its good reputation with environmental, social, and indigenous groups, has faced significant challenges establishing credibility with two segments of the forest industry – large, integrated multinational corporations on the one hand, and small forest operators on the other. To understand how these and other challenges were managed by proponents of the FSC in British Columbia, it is important to understand the province's evolving forest policy context, which constitutes the topic of our next chapter.

# 3
# The BC Forest Policy Context

For nearly two decades, the BC forest sector has been mired in crisis, its afflictions many and varied. As a heavily export-dependent producer, the forest industry has seen its traditional markets eroded and its competitive edge dulled. In the late 1990s, BC coastal producers saw their hitherto robust and growing markets in Japan and other Pacific Rim countries virtually disappear as Asian economies entered a deep and prolonged recession. Efforts to secure new markets or develop existing ones were hampered by a costly and prolonged lumber export dispute with the United States, the province's largest customer, and concerns, particularly in European markets, about the sustainability of provincial forest practices fanned by international market campaigns mounted by domestic environmental groups. To compound this gloomy picture, regulatory costs have been rising, and historically low sectoral capital investment has been insufficient to replace obsolete coastal processing facilities (Pearse 2001). The depth and duration of the crisis has forced many forest sector businesses into bankruptcy protection and prompted several front-page corporate mergers. The result, for much of this period, has been mill closures and production shutdowns becoming regular staples of the nightly news.

As if this were not enough, the beleaguered forest sector has also been buffeted by natural forces, including a pine beetle epidemic that has killed 530 million cubic metres of pine as of 2006 and may eventually destroy almost 80 percent of the pine in the timber-harvesting land base, as much as 1 billion cubic metres (BC Ministry of Forests and Range 2007; Eng et al. 2005, 26, 29, 44; McGarrity and Hoberg 2005). This unprecedented epidemic is forcing widespread emergency salvage logging but will ultimately result in serious wood shortages that will spell ruin for the economies of many interior communities. Furthermore, devastating wildfires in the summer of 2003 significantly reduced the available cut in many forest districts and tapped Forest Service coffers that had already been heavily cut.

The new millennium has also seen significant realignments in global timber supply/consumption relationships. Wood supplies from traditional suppliers (such as western and eastern Europe) and emergent producing regions (including Siberia, the Russian far east, New Zealand, Chile, Brazil, and other Latin American countries) are growing significantly, and China is emerging as not only an important market for logs but also as a formidable global competitor in manufactured forest products markets (White et al. 2006). These phenomena are accompanied by changing patterns of consumption as consumers switch to more readily available and less costly lower-grade wood products and non-wood substitutes. For the foreseeable future it appears that real prices for pulp, lumber, and manufactured forest products will continue to drop in an increasingly competitive global marketplace.

In response to mounting market competition, the global industry is undergoing a period of rationalization as firms jostle to establish and maintain a competitive advantage. Industry consolidation is increasing, new technology is being developed and introduced, and high-cost plants are being closed down as more competitive facilities are modernized and expanded – a trend that has significantly improved the efficiency and competitive position of the BC Interior industry in recent years.

The crisis in the forest sector was a major factor in the landslide victory of the BC Liberal Party that took office in May 2001, winning an extraordinary seventy-seven out of seventy-nine seats in the provincial legislature and vanquishing the New Democratic Party (NDP), which had governed the province for most of the preceding decade.[1] This election victory was, in part, the result of the Liberals' radical platform of forest policy reform aimed at restoring the international competitiveness of the sector. Among other things, the Liberals promised to reduce regulatory costs, promote industry investment by reducing uncertainty, and appease American trade complaints by moving toward a more market-driven forest economy. The BC government implemented a broad suite of reforms to attain these goals, but, due to weakness in the US housing market and a strong Canadian dollar, it is too early to tell if the reforms will achieve their desired effect. Moreover, the reform package has come under sustained criticism for compromising the interests of rural communities hit hard by declining forest industry fortunes, ignoring the constitutionally protected rights of First Nations, and heightening the risk to environmental amenities.

But the roots of the troubles facing the BC forest sector run deeper and are more complex than the foregoing sketch might suggest. By the early 1990s, there was a consensus that the prevailing model of forest policy, dominant for the latter half of the twentieth century, was no longer viable. The overarching imperative within this model has been termed the "liquidation-conversion project" (Wilson 1998, 181). This is a management strategy premised on the goal of liquidating the province's supply of old-growth timber and converting the logged areas to second-growth forests with a "normal" structure (Dellert 1998).² As support for this model dwindled in the face of mounting criticism from environmentalists, labour, First Nations, and other interests dependent on conserving values associated with old-growth forests (both "timber" and "non-timber" values, to use the lexicon of forest resource economists), the NDP government led by Premier Mike Harcourt, which came to power in 1991, embarked on a range of forest policy initiatives.

To appreciate the challenges and opportunities that confront the development of an FSC standard for British Columbia, an understanding of the complex and ever-changing landscape of BC forest policy over the last decade is critical. In the first section, we inventory the province's timber and non-timber assets, consider the significance of the forest sector as an economic driver, and briefly discuss the manner in which rights to timber are allocated. We also underscore two important characteristics of forestry in British Columbia: its export dependence and the distinction between the coastal and interior forest industries. In the second section, we chronicle the rise of the "liquidation-conversion" project and of the "development coalition" – players, institutions, and processes – that supported it in the years following the Second World War. In the third section, we reflect on what we term the "first wave" of contemporary forest policy reform, a broad suite of reforms introduced by NDP governments during the 1990s. In the final section, we conclude with an overview of a "second wave" of even more sweeping forest policy reforms implemented by the Liberal government of Premier Gordon Campbell in the wake of its election to power in 2001 and subsequent re-election in 2005.

### The Health and Wealth of BC Forests

Historically, the intensity of the debates surrounding forest policy in British Columbia reflects an enduring political verity that provincial governments ignore at their peril: that, as a public-owned asset, the province's forests are broadly regarded as a heritage that belongs to all British Columbians. Only 4 percent of British Columbia's 60 million hectares of forestland is in private hands; title to 95 percent of this area is held by the provincial government. The remaining 1 percent – mainly in national parks and Indian reserves – falls under federal jurisdiction (see Figure 3.1).

However, the constitutionality of the Crown's claim to forestland ownership is by no means unchallenged. This legal uncertainty arises from the fact that much of the province's forestland overlaps with the traditional territories of BC First Nations and, as such, is subject to ongoing litigation or treaty negotiations aimed at defining the scope and nature of existing Aboriginal rights and title. As the "land question" inches toward final resolution, it now appears inevitable that title to significant tracts of provincial forestland is

*Figure 3.1*

**Forestland ownership in British Columbia**

Provincial
95%

Federal
1%

Private
4%

*Source:* Canadian Council of Forest Ministers, National Forestry Database, 2005.

likely ultimately to vest in First Nations' hands. It is also inevitable that the provincial Crown (and potentially also the federal Crown) will be liable to pay compensation to reflect the value of timber harvested on First Nations' traditional territories without their consent.[3]

Not only are BC forests predominantly in public hands, they are also the province's predominant geographical feature. Over 63 percent (about 60 million hectares) of British Columbia is forested. Of this amount, 88 percent (53 million hectares) is capable of producing commercially valuable wood. Of the land capable of sustaining a commercial timber harvest, about 47 percent (or 25 million hectares) is available for logging over time. The remainder of the potentially commercial forest is either inaccessible, uneconomical to harvest, or not available due to environmental constraints. Thus, of the 60 million hectares of forested land in the province, about 42 percent (25 million hectares) is managed for commercial timber production, about 48 percent (29 million hectares) is potentially commercially valuable but not available, and 10 percent (6 million hectares) is in parks and protected areas (British Columbia Ministry of Forests 2003b, 2, 11).

Over the last quarter century, debate over the future of BC forest policy has gained international attention due to the efforts of an increasingly robust domestic environmental movement. This attention is attributable, in large measure, to the fact that British Columbia is home to globally significant reserves of mature temperate rainforest. Currently, approximately half of the 7.6 million hectares of coastal rainforest is old growth (British Columbia Ministry of Forests 2003b, 5-6). However, this figure tends to understate the impact of more than a century of industrial timber extraction. In landscape terms, less than 20 percent of coastal watersheds in excess of

5,000 hectares remain pristine. Most of these untouched watersheds are on British Columbia's remote central and north coasts.

Most international environmental campaigns to date have been directed at protecting coastal old-growth forests, but concerns are now being voiced about the future of the province's interior forests, where road building and intensive logging are having serious impacts on forest regeneration and species habitat. Moreover, as the BC economy has diversified and developed greater reliance on tourism and other non-timber forest uses, the challenge of managing competing claims to provincial forests has become exponentially more complex.

The BC forest industry has traditionally responded to these competing claims by invoking its status as the province's leading economic driver. In 2005, the industry (comprising logging operators; saw, board, and planing mills; and pulp and paper producers) contributed 41 percent of the value of shipments from the BC manufacturing sector (British Columbia Ministry of Forests and Range 2006a) and accounted for 38 percent of the province's exports (Canada, Industry Canada 2006). It is also a significant source of provincial revenue, contributing well over a billion dollars to government coffers each year in direct revenues (stumpage and other direct charges) and makes up an important component of the province's tax base (CCFM 2008). Its contribution to provincial gross domestic product (GDP), however, is more modest than these statistics suggest. Its direct contribution to GDP is 8 percent (down from over 10 percent forty years ago), although government sources claim that, when spinoff effects are included, the industry's actual GDP contribution is closer to 15 percent (British Columbia Ministry of Forests 2003b, 4). The industry's significance to the province is also declining in terms of employment. Direct employment in the sector has declined steadily since the 1960s and currently hovers around 5 percent of the provincial total. In fact, the number of workers employed in the industry has declined from 92,000 in 1990 to 80,000 in 2004, even as harvest levels have remained about the same (CCFM 2005).

The wealth of British Columbia's industrial forest sector is based on the production and export of commodities with a relatively low value – softwood lumber, pulp, and newsprint. Products that undergo further manufacturing – often referred to as "value-added" products – although not insignificant in absolute terms, represent a relatively small proportion of total sector output. In 2005, wood products, mainly lumber, accounted for about two thirds of British Columbia's shipments of forest products; the balance comprised pulp and paper, mainly raw market pulp, newsprint, industrial paper, and paperboard products. In that same year, British Columbia's forest products sector contributed 39 percent ($12.4 billion) of Canada's balance of trade in forest products. The province's forest exports included softwood lumber (45 percent), wood pulp (19 percent), and newsprint (5 percent) (CCFM 2008).

Another key feature of the BC forest sector is the dominant role played by a handful of large companies, often referred to as the "majors." Membership in this elite group has historically fluctuated, but its control of a lion's share of the provincial allowable annual cut (AAC) has remained more or less constant. As of 2003, five companies control over 40 percent of the provincial AAC (Walter 2003, n21). As a result of a recent merger with one of its largest competitors, the province's biggest forest company – Canadian Forest Products – controls almost half of the AAC that currently rests in the hands of the majors, or 18 percent of the total provincial AAC (Walter 2005).

Since the earliest days of European settlement, one of the leading forest policy questions facing governments throughout Canada has been how to transfer timber growing on publicly owned lands to the private wood-products manufacturing sector. Pre-Confederation governments, faced with small populations, meagre budgets, and vast, valuable public timber resources, devised and implemented licensing arrangements under which rights to harvest public timber, usually within a defined geographical area, were transferred to the private sector in return for payments – in the form of royalties, land rents, and licence fees – and the assumption of some management responsibilities; the land itself remained publicly owned. Today, licensing arrangements that transfer certain rights in Crown forests to the private sector, and require varying degrees of responsibility for forest management, encompass a majority of Crown forestlands in Canada and account for most of the timber harvested from these lands (Haley and Luckert 1990). In British Columbia, Crown forest tenures account for close to 90 percent of the annual provincial timber harvest.

The BC *Forest Act* contains provisions for eleven different forms of tenure arrangement under which harvesting rights can be granted.[4] These include tree farm licences (TFLs), forest licences, timber sale licences (TSLs), timber licences, woodlot licences, community forest agreements (CFAs), community salvage licences, free use permits, licences to cut, road permits, and Christmas tree permits. The last five tenure types are usually grouped together as "miscellaneous" tenures and are of little consequence to the certification debate.

Currently, two forms of large-scale industrial Crown forest tenure – TFLs and forest licences – dominate the industrial forest landscape in British Columbia. In 2008, area-based TFLs and volume-based forest licences[5] accounted for 20 percent and 58 percent of the provincial AAC, respectively.[6] The next largest share of the AAC (17 percent) was allocated to TSLs to be sold by BC Timber Sales, which replaced the Small Business Forest Enterprise Program.[7]

The only Crown forest tenures capable of supporting small-scale forestry are woodlot licences and CFAs. Woodlot licences, which are designed to encourage small-scale, sustainable forest management by individual citizens and small firms, accounted for about 1.3 percent of the AAC in 2008, and CFAs accounted for 0.6 percent (British Columbia Ministry of Forests and Range 2008a). The average size of a TFL, measured in terms of AAC, is 526,000 cubic metres, and the average size of a forest licence is 129,000 cubic metres. To put this in perspective, the average AAC allocated to a woodlot is less than 1,500 cubic metres (British Columbia Ministry of Forests and Range 2008a).

No discussion of the BC forest sector would be complete without reflecting on the fact that, in many ways, the province has two forest industries. The structure and prospects of the coastal and interior forest industries are distinct; indeed, in some ways they are polar opposites. The regions differ considerably in terms of their climate, forest types, ecosystems, and topographies. Much of the Coast consists of high, rugged mountains bisected by deep, steep-sided river valleys; most of the Interior's commercial forests occupy gently rolling plateaus. These physiographic differences have important consequences for logging costs, which are considerably higher on the Coast, and for the environmental impacts of industrial harvesting activities. On the Coast, a complex system of tidal waterways make it possible to transport logs over great distances at a relatively low cost, allowing processing plants to be located many hundreds of kilometres from the forests that supply their raw material. In contrast, log transportation in the Interior, which is mainly by road, is more costly, resulting in manufacturing centres that rely on logs from more localized "timbersheds." Both the Coast and the Interior have large areas of primary, or old-growth, forest, but it is the Coast, with its towering forests of Douglas fir, Sitka spruce, western red cedar, grand fir, and western hemlock, that has attracted the attention of the global environmental movement.

In the province's early years, the coastal industry was the lifeblood of its economy. Much of today's capitalized wealth within the province was created by exploiting this coastal timber abundance. In comparison, the interior industry is much younger, only beginning to seriously rival its coastal counterpart within the last thirty years. Currently, 75 percent of all lumber produced in the province comes from the Interior. The region is also beginning to challenge the historical dominance of the Coast in pulp and now accounts for approximately a third of provincial production. The Coast and Interior have different market orientations as well. The Interior mainly serves the continental market, depending heavily on rail transportation, and the Coast relies on marine transportation to serve a variety of offshore markets in Europe and Asia, as well as the US eastern seaboard.

Today, as the coastal industry is beset by environmental campaigns, high logging costs, a shortage of accessible timber, increasing competition in traditional markets, and aging processing plants, the mood of the Interior industry is more optimistic. Interior producers have largely managed to avoid the wrath of environmentalists – although campaigns involving the boreal and sub-boreal forests are gathering momentum – they enjoy significantly lower per-unit logging and manufacturing costs than their coastal counter- parts, and in recent years they have significantly improved the efficiency of their processing facilities through rationalization and new investment. Currently, the interior industry enjoys extraordinarily low wood costs as a result of the enormous volumes of timber being salvaged from lodgepole pine forests ravaged by the mountain pine beetle. However, within ten to fifteen years, little merchantable lodgepole pine will remain throughout much of the region, resulting in high wood costs, mill closures, and economic hard- ship for many forest-dependent communities.

## The Development Coalition and the Liquidation-Conversion Project, 1945-90

In broad relief, the history of industrial forestry in British Columbia has followed a trajectory not unlike that commonly observed in other forest- dependent jurisdictions worldwide. Initially, forest development is largely unregulated. The overriding concern of governments during this stage was to capture economic and social benefits through rapid development of the resource. In the face of resource abundance and, to some degree, as a result of public policies, forest rents are kept low. In order to attract and harness private investment, governments typically provided companies with long- term rights to a fibre supply in return for the economic benefits they recycled into the economy by establishing and maintaining processing facilities.

As the inherent limits to this liquidation paradigm emerge, public confi- dence in the industry is undermined and an alternative model typically takes hold. Under this model, which draws on the "potent symbolism of sustained yield" (Wilson 1998, 79), the state takes on the responsibility of maximizing long-term production of timber by seeking to achieve an equilibrium that balances timber growth and harvest rates. To this end, managers undertake timber supply inventories and develop complex resource models to manage the transition from old growth to second growth by identifying an AAC that is theoretically sustainable in perpetuity. Under this model, the liquidation of old-growth forests and their conversion to even-aged regrown forests is planned.

British Columbia embarked on the liquidation-conversion process in 1947, when the government of the day amended the *Forest Act* to implement the recommendations of the Sloan Royal Commission (Sloan 1945), which had concluded that the "cut and run" practices of the past should be rem- edied by a new system of secure forest tenures. Today's tree farm licences owe their origin to reforms proposed by the head of the commission, Chief

Justice Gordon Sloan. Forest licences – the other dominant Crown forest tenure in British Columbia – evolved, between 1947 and 1979, from small-scale competitive timber sale licences as an increasingly capital-intensive and concentrated forest industry demanded greater tenure security. These forms of tenure met industry concerns about security insofar as they were of relatively long duration (twenty-five years for TFLs; twenty years for forest licences) and, since 1979, were "evergreen" in nature; a term that connotes the ability of the holder to have government "replace" the tenure with a new one on a regular basis, indefinitely. The tenures were also designed to satisfy government and worker interests by allowing the Ministry of Forests (MOF) to impose "appurtenancy" conditions on tenure holders. These conditions obliged companies to establish mills and meet local processing requirements.[8]

For over two decades – from the early 1950s to the mid 1970s – support for the liquidation-conversion model was widespread and largely unquestioned. For much of this period, particularly during the 1960s, the BC economy boomed. Capital investment reached all-time highs with the construction of numerous resource-processing facilities, hydroelectric projects, and major highways. The boom was, in large measure, driven by the prosperity of the forest sector, and annual timber production more than doubled between 1950 and 1970 (Wilson 1998). The boom was particularly pronounced in the BC Interior, which saw its share of provincial timber production jump from one-quarter of the provincial total to almost half. With a succession of industry-friendly governments in Victoria, much of the province's forest resources came to be controlled by a handful of large companies under the new tenures implemented in the wake of the Sloan Commission.

Varying concepts have been used to describe the political détente between government, industry, and labour that underpinned the growth of the forest industry between 1950 and the late 1970s. Economic geographer Roger Hayter contends that Sloan's recommendations "provided BC's recipe for Fordism ... a forest to be dominated by Big Business that in close alliance with the provincial government, emphasized the forest as timber supply for mass production. Moreover, labour unions quickly accepted this recipe to create an implicit (and classic Fordist) deal among (big) business, (big) government and (big) labour" (2000, 49).

In his acclaimed analysis of post-Second World War BC forest policy, Wilson (1998) characterizes this political constellation in somewhat different terms – as the "development coalition," the existence of which, by the 1960s, had become fully apparent:

> The provincial government offered secure rights to the resource along with an implicit assurance that, in administering its responsibilities, it would remain sympathetic to industry's needs and interests. In return, companies

supplied capital, committed themselves to management obligations derived
from the government's goals of community stability and sustained yield,
and agreed to pay taxes and stumpage charges. It was taken for granted that
· the provincial government would plough revenues derived from the forests
back into forest protection and some reforestation; into the development
of the transportation, energy and community infrastructure required by
industry; and into programs that would help ensure an adequately skilled
and reasonably docile workforce. (81)

Both Hayter and Wilson trace the beginning of the end for this political
bloc, and the vision of forestry it championed, to the early 1970s. In 1972,
for the first time in the province's history, a new left-leaning party took over
the reins of power in Victoria. The NDP government of Dave Barrett was
short-lived – after thirty-nine months the free-enterprise Social Credit Party
regained power.
   Despite its short term in office, the NDP interregnum during the mid-
seventies is notable in at least two respects. Perhaps its most enduring accomplishment was to double the size of the provincial park system, a testament
to the emerging power of a fledging BC environmental movement. It was
also a period during which, for the first time, leading industry supporters
of the development coalition were put on the defensive. They were forced
into this posture by the government's stated desire to revisit forest tenure
and related stumpage arrangements to secure for the province an enhanced
return on forest development. Public debate on the issue culminated, in
1975, in the appointment of a Royal Commission headed by Professor Peter
Pearse.
   Like the Sloan Commission before it, the Pearse Commission made recommendations that subsequently formed the basis for legislative reforms
(Pearse 1976). Central to Pearse's recommendations were greater security
for Crown forest tenure holders, the reintroduction of competition for a
substantial proportion of Crown timber, increased incentives for private-
sector management of public timber, and management of public forests for
a greater diversity of products.
   In 1978, a new *Forest Act* and a *Ministry of Forests Act* were proclaimed.[9]
The *Ministry of Forests Act* established an improved system for reporting on
the state of the province's Crown forests. It included periodic timber supply
reviews and, for the first time, explicitly required the Ministry of Forests to
manage Crown forestlands for multiple products – specifically timber, range,
fisheries, wildlife, water, and outdoor recreation – having regard for the long-
term economic and social benefits to the province. The *Forest Act* introduced
a new and more transparent process for determining AACs, created new
requirements for public participation in certain management decisions, and
set aside a proportion of timber sale licences for small businesses under the

SBFEP. For the most part, however, the tenure system remained essentially intact.

During the ensuing years, in spite of the commitment to multiple use and the planning procedures that accompanied it, timber remained the dominant use of Crown forests. The continuing sway of the liquidation-conversion project became particularly apparent during the 1983-84 economic recession, when relaxed harvesting and management standards – euphemistically known as "sympathetic administration" – were allowed in an attempt to maintain economic momentum.

In 1987, the provincial government, buffeted by criticism from small producers and environmentalists that the new legislation had failed in any meaningful way to address the problems that had led to the establishment of the Pearse Commission, responded by announcing further incremental forest policy reforms. It implemented a new stumpage system that was designed to exempt British Columbia from the terms of the US-Canadian Memorandum of Understanding on the softwood lumber trade. As well, 5 percent of tenure holders' timber-harvesting rights was reallocated to an expanded SBFEP. To discourage merger activity and to make more timber available for reallocation, an acquired company could be required to surrender 5 percent of its authorized AAC. Finally, a schedule was drawn up to dramatically expand the TFL system using 1982 amendments to the *Forest Act* that allowed forest licence holders to surrender their licences for area-based tenures.

Widespread and vigorous public opposition to the expansion of the TFL system took the government by surprise. Following province-wide public hearings, the initiative was abandoned. In its place, a "permanent" Forest Resources Commission was appointed, chaired by a former senior civil servant, Sandy Peel, to examine the effectiveness of TFLs as a form of tenure and, more generally, to advise the government on the state of the province's forestland base and recommend how its management could be improved. The commission's report fanned the flames of controversy by proposing dramatic forest tenure reforms that would have reduced the AAC of major tenure holders from 85 to 50 percent of the provincial total and reallocated these rights to communities, First Nations, and woodlot operators (Haley and Leitch 1992). The commission also reiterated support for unimplemented recommendations contained in the Pearse report that advocated the province move from an administered stumpage system toward a model based on prices established through a competitive bidding process. Although the commission's "radical thinking" had no direct influence on government forest policy, and indeed guaranteed the commission would eventually be disbanded, observers claim that its approach "opened space for moderate reformers within the MOF" who argued that the days of the liquidation-conversion project were numbered (Wilson 1998, 79).

Meanwhile, pressure to jettison the project was mounting on a variety of fronts outside government. By the end of the 1980s, it had become apparent that the government's efforts to contain the environmental movement were failing dismally. Increasingly sophisticated and organized, the movement was now engaged in what amounted to an open public relations war with government. In valley-by-valley skirmishes over the fate of pristine wilderness watersheds slated for logging, British Columbia's war in the woods began to attract international attention. This growing awareness reflected rising global concern about environmental causes generally. A high-water mark in international environmental consciousness was the much-celebrated report of the World Commission on Environment and Development (Brundtland 1987) that popularized the term "sustainable development" and stimulated governments into action worldwide. The BC environmental movement was quick to embrace and champion the Brundtland Commission's suggestion that nations secure 12 percent of their land base as protected areas, pointing out that less than 6 percent of the province was currently protected.

Although wilderness protection remained the environmental movement's main issue, many of its key thinkers were at work developing alternative models of forestry and land-use planning. Core elements of the emerging critique were the need to move from corporate to community-based forestry and the correlative need to move from high-volume commodity production, such as basic construction lumber and newsprint, to the production of higher value-added forest products, such as finger-jointed lumber, floor tiles, furniture, and fine paper, that required further processing within British Columbia. For this vision to be realized, critics argued, fundamental changes were required in two areas of provincial forest policy: the AAC and tenure.

BC First Nations played a critical role in the crystallization of this oppositional bloc. Contrary to the pattern prevalent elsewhere in North America, colonial settlement in what is now British Columbia occurred without negotiated resolution – in the form of land- and rights-based treaties – of the continuing rights of the indigenous population. For much of British Columbia's history, provincial governments found it expedient to view these rights as extinguished. Support for this historical interpretation was dealt a severe blow by the Supreme Court of Canada in its landmark 1973 decision in the *Calder* case, which affirmed the prospect of enduring Aboriginal title to lands unceded through treaties.[10] The *Calder* case did not alter provincial policy – the government still refused to recognize Aboriginal rights and title – but First Nations were emboldened by the decision to press for protection of what they considered to be their traditional territories. During the 1980s, there were several well-publicized attempts by First Nations to advance this agenda in response to government-authorized logging operations. Most significant, from a legal perspective, was a precedent-setting effort by the Nuu-chah-nulth First Nation to secure an injunction to stop logging on Meares Island. The Nuu-chah-nulth argued that logging would do irreparable harm to their pending claims for Aboriginal title to what is commonly known, due to subsequent logging protests, as Clayoquot Sound. In response to the growing protests and First Nations blockades of forestry operations, the provincial government established the Ministry of Native Affairs in 1988 and set up the BC Claims Task Force in 1990. The task force recommended the establishment of a treaty commission to coordinate the land claims and treaty negotiation process: the BC Treaty Commission was subsequently created in 1992 (Howlett 2001).

Corporate strategies also played a role in undermining the broad-based and unquestioned support formerly enjoyed by the liquidation-conversion project. As wages of unionized forest industry workers climbed during the post-Second World War period, major forest companies invested in capital-intensive technologies that would reduce the labour intensity of production. This trend accelerated following the 1982-83 recession. As workers accustomed to secure jobs and handsome pay packages saw their ranks thinning and their bargaining power being eroded, anxieties and tensions in forest-dependent communities escalated. Slowly, but surely, Big Labour began to have doubts about the future of the Fordist social contract that had been a fixture in British Columbia since the mid-1940s.

With the election of an NDP government in 1991, there was hope for a ceasefire, if not a permanent resolution of the war in the woods; for significant progress toward settlement of First Nations' long-standing claims to rights and title; and for a new deal for forest workers and communities increasingly uneasy about the future of the postwar social contract.

### Greening Forest Policy: The First-Wave Reform Agenda of the NDP Years, 1991-2001

Expectations for the NDP government of Premier Michael Harcourt, elected to power in the fall of 1991, were exceedingly high. Although faced with a daunting list of priorities, a significant factor working in the Harcourt government's favour was that the BC forest industry was in the boom stage of its business cycle; a juncture during which government's bargaining power relative to industry's has historically been at its strongest (Cashore et al. 2001a).

The Harcourt government pursued an agenda containing a variety of components. One of the government's most successful initiatives was an ambitious protected areas strategy (PAS) that by decade's end had succeeded in more than doubling the amount of protected area within the province, exceeding Brundtland's famous 12 percent target.[11] However, Harcourt took pains to reassure supporters in the labour movement that his government

remained strongly committed to protecting the social contract, boldly pledg-
ing "not one forest worker will be left without the option to work as a result
of a land-use decision" (Hoberg 2001, 212).

In support of this pledge, and capitalizing on the profitability of the sector,
the government doubled stumpage rates, placing a substantial part of the
increased income under the trusteeship of a newly created Crown corporation
– Forest Renewal BC (FRBC). Chaired by a prominent former forest union
leader, with a broadly based board of directors that included representatives
from environmental groups and First Nations, FRBC was required to reinvest
the stumpage funds it received in the forest sector through forest restoration,
intensive silviculture, worker retraining, value-added manufacturing, and
community economic development. Hoberg (2001) describes the move to
create FRBC as "brilliant in concept: the government got industry to agree
to a significant cost increase, creating the financial means to cover at least
some of costs [sic] created by the government's environmental initiatives ...
keep[ing] its precarious coalition of environmentalists and labour together for
the time being, without provoking industry opposition" (214, 215). However,
FRBC quickly ran into implementation problems, stemming largely from
a lack of capacity to prudently manage such a large investment portfolio.
FRBC's image problems were exacerbated later in the decade when the next
NDP government "raided" its funds in an effort to address general revenue
shortfalls caused by the onset of the forest crisis referred to in the introduction to this chapter.

The Harcourt government also deserves high marks for innovation in
managing disputes over land-use decisions, particularly for its creation of the
Commission on Resources and Environment (CORE) in 1992. The commission was given the task of developing land-use plans for a number of regions
around the province, using consensus-based decision-making processes that
involved a broad range of interests.

A major stumbling block confronted by CORE, and one that played a role
in its eventual demise, was the understandable reluctance of First Nations to
participate in a negotiation process that they felt treated them as an interest
group like any other. Without full First Nations buy-in (in retrospect, an un-
realizable prospect), discussions about how to zone regions at the landscape
level by preferred use, took on an aura of unreality.[12] This was particularly
true where the lands in question overlapped with First Nations' traditional
territories that were subject to litigation or treaty talks.

CORE was also plagued by another structural limitation inherent in the
Harcourt government's approach to forest policy. This was the government's
decision, made early in its term of office, not to tackle what many forest
advocates considered to be the top priority: tenure reform. This decision
was highly pragmatic and driven by a concern about maintaining industry

support for (or at least limiting active opposition to) other elements of the government's reform agenda, notably the stumpage increases that allowed FRBC to proceed. As a result, the government's only tenure reform initiative was a pilot program for community forest tenure, initiated in 1998, that was limited in scope and made no demands on the AAC allocated to existing licensees (Haley 2002).

The Harcourt government was also determined to make headway toward resolving the long-standing First Nations land question. It invested significant resources in the BC Treaty Commission process, through which it sought to funnel unresolved title claims. A key component of this strategy was to discourage Native organizations from resorting to title litigation, an avenue that was not only enormously costly but also put government and First Nations into an adversarial relationship. A principle the treaty commission adopted to achieve this goal stipulated that First Nations could only participate in the treaty process if they refrained from litigation. This principle proved to be a source of considerable frustration for First Nations, particularly because, under the treaty process, "interim measures" protections (which offered the prospect of exempting from forestry development the traditional territories that were the subject of treaty negotiations) were only available once treaty negotiations had reached the agreement-in-principle stage. In more general terms, the government sought to modestly enhance the level and nature of MOF consultation with First Nations around logging decisions. Such consultations had been virtually non-existent in the past. However, here too government efforts did not go nearly far enough to satisfy First Nations. As a result, in many instances, First Nations went to court arguing that governmental consultations did not satisfy the constitutional requirements that were being elaborated and, in many cases, broadened by the courts. In several of these cases, the MOF's consultation processes were judicially criticized as being inadequate.[13]

Despite this broad range of initiatives, the war in the woods did not abate. Indeed, in the spring of 1993, it escalated dramatically following a controversial land-use decision by the NDP cabinet that would have allowed logging in one of the province's best-known environmental hot spots: Clayoquot Sound. By summer's end, this decision had become an international cause célèbre. Over the course of the summer, more than eight hundred protesters challenging the decision had been arrested for blocking MacMillan Bloedel logging trucks. To increase international pressure on the government, environmental organizations took their cause on the road, visiting forest-products buyer groups in Europe in an effort to persuade them to boycott BC forest products until the government committed to implementing more sustainable land-use decision-making processes and logging practices. Ultimately, with the local First Nations having joined in the fight, the government was forced to back down, but not before its commitment to "green" BC forestry had lost considerable credibility.

In an effort to rehabilitate its reputation on this front and forestall the impact of environmentalist market campaigns, the government enacted the province's first legislated code of forest practices in 1995. Under the Forest Practices Code (the "Code"), companies were required to undertake a variety of new planning requirements with a view to identifying landscape or "higher level" objectives that would then be embedded in operational-level logging plans. These objectives became legally binding when they were included in operational-level logging plans, prescriptions, and contracts.

After introducing the new Code with considerable fanfare, the government promised that it would be strictly enforced and that foreign buyers could now be assured that British Columbia had the best forest practices in the world. Environmentalists did not agree. They contended that the legislation gave the MOF broad and unfettered discretion and, as a result, represented little more than a repackaged version of the status quo. The Code also proved to be enormously unpopular with business and labour. They criticized the Code's prescriptive and inflexible nature, its cost implications – estimated, on average, to add 20 percent to the cost of logging operations (Haley 1996) – and its downward impact on the AAC. Under heavy pressure, and based on the chief forester's estimate that compliance with the minimum requirements set out in Code guidebooks would lead to a reduction in provincial AAC of 6 percent, the government committed to capping the impact of the code at the 6 percent level – a commitment that the MOF began to apply as it oversaw the Code planning requirements. Over time, as MOF funding cutbacks undermined the government's promise to rigorously oversee Code compliance, and as the drafting and implementation of wildlife habitat and biodiversity guidelines (which potentially would have had significant AAC implications) were delayed, any support for the Code within the environmental movement evaporated. Consequently, what the government had initially conceived as a centrepiece of its forest policy reforms became marginalized and politically vulnerable, lacking champions either within or outside government circles.

In short, although the Harcourt agenda for forest policy reform was ambitious and, on balance, reasonably well conceived and well executed, its net effect in terms of producing an enduring transition from the liquidation-conversion paradigm was, at best, rather modest. Some progress was made in the areas we have discussed above – employment transition strategies, land-use planning, consultation with First Nations, and forest practices oversight – but two fundamental areas of forest policy, tenure and AAC, remained essentially unchanged. Had the forest sector boom that greeted Harcourt when he came to office continued, or had he remained in the premier's office for a second term, the story might have been different. Indeed, when the decision not to proceed on this front was made public, government insiders reputedly told proponents of a fundamental tenure overhaul to be patient – tenure reform would be on the second-term agenda.

However, when Harcourt retired from politics in 1996, and the forest industry descended into a down cycle, this first wave of forest policy came to a crashing halt. Harcourt was replaced as NDP leader and premier by Glen Clark, who was much more closely aligned with big labour. To the surprise of many observers, Clark managed to lead the NDP to another term in office, but forest policy reform took a back seat to a spate of other issues, including ever-worsening economic conditions and a continuing round of scandal and recriminations. The one new forest policy under Clark was the Jobs and Timber Accord of 1997, which, the government promised, would create 22,400 new forest jobs by 2002 by overtly linking company access to AAC to ambitious job-creation targets. The accord was spectacularly unsuccessful, as a market downturn resulted in a loss of 7,500 forest jobs by 2001 (Hoberg 2001). When the next provincial election took place that year, with yet another leader at the helm of the NDP, it was clear that a change of government was imminent. It was equally clear that a new wave of forest policy reform was about to crest.

### Restoring International Competitiveness? The Second-Wave Reform Agenda of the Campbell Liberals, 2001-8

In the provincial election of May 2001, the BC Liberals were swept into power with an unprecedented mandate. A key plank of their campaign platform was a wide-reaching plan to restore the competitiveness of the BC forest sector by enhancing business certainty and reinvigorating flagging investment. In the Liberals' view, this meant implementing ambitious reforms to re-establish access to US markets for BC forest products and forging a "new relationship" with First Nations (British Columbia Ministry of Aboriginal Relations and Reconciliation 2008).[14]

As an initial step, the government commissioned Peter Pearse, who had led the Royal Commission on Forest Resources in the mid-1970s, to examine the state of the coastal forest sector. In his report, Pearse painted a picture of an industry in crisis (Pearse 2001).

Market access had become a pressing priority as the Canada-US Softwood Lumber Agreement (SLA) was due to expire a month before the election was called. When the SLA lapsed, the US government instituted countervailing tariffs on the export of Canadian softwood lumber, exacerbating the serious market challenges already facing BC producers. The magnitude of the tariffs sent shockwaves through the BC industry. In May 2002, tariffs were exacted at the US border at a rate of 27.22 percent of the shipment value, declining to 10.8 percent by December 2005, taking approximately $5.3 billion in total from Canadian companies. The American action was based on allegations that the existing tenure and timber-pricing arrangements unfairly subsidized Canadian lumber producers relative to their American counterparts. These complaints were recycled versions of arguments the United States had relied on, during the previous twenty years, in several skirmishes over the rules governing the Canadian-US softwood lumber trade, which, at $10 billion per annum, was the single largest commodity trade across any border in the world (Clogg 2003). Nor were the complaints without substance (Sizer 2000; Gale and Gale 2006).

The Canadian government responded to the imposition of these tariffs by litigating the matter before numerous trade-dispute bodies, including the World Trade Organization, a North American Free Trade Agreement panel, and the US Court of International Trade. Despite Canada's repeated legal victories, the United States continued to levy import duties on Canadian soft-wood lumber. It was not until April 2006 that the protracted dispute reached a tentative resolution, when the parties settled on a framework agreement that obligated the United States to repay about 80 percent of the more than $5 billion in duties it had collected since 2002. Despite initial opposition at home from the forest industry and affected provincial governments, the Canadian government garnered enough support to push the deal ahead, signing the new Softwood Lumber Agreement on 12 September 2006. The agreement, which came into force on 12 October 2006 and extends over the next seven years, removes US tariffs on lumber, but requires Canadian companies to pay export taxes of between 5 and 15 percent when the price of lumber drops below specified levels.[15]

If reopening US markets was the Liberal government's number one priority, making peace with First Nations at home was a close second. Ironically, the need to address First Nations issues arose in part due to the Liberals' own folly: namely, their ill-advised campaign promise to stage a province-wide referendum on the principles and future of treaty negotiations (see Chapter 8). First Nations responded bitterly to this announcement and, after the Liberals were elected, sought unsuccessfully to derail the referendum in court. Although the Liberals interpreted the referendum results as supporting their harder, bottom-line-oriented approach to negotiating treaties, proceeding with this initiative came at a high price in terms of their relationship with First Nations. Many predicted that, with the Liberals in power, the treaty process, which had been moving at a glacial pace since it was established in 1992, would grind to a halt.

Frustrated with the process, and with the inadequacy of interim protection for traditional lands and resources, First Nations turned to the courts to seek redress for their grievances. Many of the ensuing cases alleged that the provincial government had breached its constitutional duty to consult First Nations in the allocation or renewal of resource development tenures (often forest licences) on their traditional territory. Several of these cases led to high-profile victories, the most notable being a BC Court of Appeal decision that criticized the failure of the BC government and Weyerhaeuser Ltd. to consult with the Haida before renewing a TFL on Haida territory in the Queen Charlotte Islands (or Haida Gwaii, the Haida name for the islands).[16] As judicial decisions progressively clarified the definition of and requirements for constitutional consultation, and as First Nations steeled themselves for an ongoing series of showdowns with what they had come to regard as an unsympathetic, if not hostile, administration in Victoria, reconciliation with First Nations slowly but surely moved upward on the Liberal government's agenda.

In the meantime, the perceived need to break the softwood lumber trade stalemate between Canada and the United States gave the Liberal government the political capital necessary to embark on what has been characterized as the most significant set of forest policy reforms since the implementation of the Sloan report's recommendations over fifty years ago. This package of reforms included far-reaching and closely linked changes to both the provincial tenure and timber-pricing systems.

The centrepiece of the reform initiative was the Forest Revitalization Plan (FRP), announced in March 2002 (British Columbia Ministry of Forests 2002a). The FRP left the existing structure of major tenures essentially intact, but it brought about a significant redistribution of tenure rights by means of a one-time, 20 percent take-back of AAC from all tenures with an AAC of more than 200,000 cubic metres. Tenure holders required to surrender AAC were eligible for compensation from a reserve fund of $200 million: affected workers and communities received compensation from a fund of $75 million. Half of the take-back was allocated to First Nations, woodlot licences, and community forests, with the remainder auctioned off under the auspices of a new government program known as BC Timber Sales. Prices established through these auctions of standing timber – which, when combined with the volume previously sold competitively under the SBFEP, make up more than 20 percent of the provincial AAC – would, in turn, be used to establish stumpage rates for timber not sold on competitive markets on a province-wide basis. New market-based stumpage determinations, which came into force on the Coast in 2004 and which were implemented in the Interior in 2006, are designed to respond to a perennial argument advanced by American producers that government-administered stumpage determinations constitute an unfair subsidy to the BC forest companies.

Reforms introduced under the FRP also aimed to render BC forest policy more market based and, as such, less vulnerable to US trade complaints by eliminating a variety of restrictions that formerly encumbered the rights of major tenure holders. Formerly, to advance goals associated with British

Columbia's forestry social contract, tenure holders could only subdivide or transfer their tenure rights after they had received government permission. Although permission to transfer was usually granted, earlier legislation on such transfers allowed the government to impose a 5 percent AAC take-back that was used to redistribute fibre supply to support emerging programs such as the SBFEP. Government also placed timber processing, appurtenancy, and minimum-cut requirements on tenure holders to ensure stable employment in forest-dependent communities. The FRP abolished all of these restrictions.

Several of the Liberal government's new programs and policy reforms were meant to shore up First Nations support for its agenda of forest policy reform. In the run-up to the Forest Revitalization Plan, the government invested considerable energy negotiating "direct award" tenure agreements with several First Nations. These agreements were widely publicized as breakthrough "interim measures" that demonstrated the sincerity of the government's commitment to move forward at the treaty table. As of January 2006, the province had signed a hundred interim forest and range agreements, providing First Nations with access to 15.7 million cubic metres of timber and $115 million in shared revenues (British Columbia Ministry of Forests and Range 2006b). The FRP set aside 40 percent of the AAC take-back (8 percent of the total provincial AAC) for First Nations. The FRP is vague on how this 8 percent is to be allocated, but the amount was apparently based on the percentage of First Nations people in the general rural population.

The Liberal reform package also contained a variety of other initiatives aimed at fulfilling the promises contained in the party's election platform. Many of these initiatives have figured prominently on the forest industry's wish list for much of the last decade. These included the replacement of the NDP's much-maligned Forest Practices Code with a less-prescriptive, results-based regime administered under the *Forest and Range Practices Act*.[17] Under the new act, tenure holders are required to prepare Forest Stewardship Plans that address objectives set by the government in such areas as visual impact, water quality, and biodiversity. The FRP also introduced the Working Forest Policy, with the goal of protecting the land base for the forest industry. The policy defined the "working forest" as all Crown forestlands outside parks and protected areas, a total of 45 million hectares (British Columbia Ministry of Forests 2003b), and proposed the introduction of resource-specific targets to delineate the proportions of working-forest areas devoted to forestry and other resource uses. In the face of widespread public criticism that the Working Forest Policy biased future land-use planning initiatives in favour of timber production, the government shelved the initiative, announcing in July 2004 that it would allow the existing land and resource management planning process to "establish the amount of forest land that will be available for timber harvest over the long term" (BC Ministry of Sustainable Resource

*Table 3.1*

**Highlights of forest policy in British Columbia, 1945-2006**

| Initiative categories | The liquidation-conversion project, 1945-90 | The NDP years, 1991-2001 | The Liberal years, 2001-8 |
|---|---|---|---|
| First Nations | • Held that Aboriginal rights had been extinguished<br>• In response to growing protest and blockades, created Ministry of Native Affairs in 1988 and BC Claims Task Force in 1990 | • Increased resources devoted to the BC Treaty Commission<br>• Adopted position that First Nations could only participate in the treaty process if they refrained from litigation<br>• Modest increase in the level and nature of consultation requirements | • Held province-wide referendum on principles and future of the treaty process<br>• Approached negotiations with an increasingly hardline stance, causing many First Nations to seek judicial solutions, which, in turn, increased the constitutionally mandated requirements of consultation<br>• Negotiated a series of "direct award" tenure agreements with First Nations, with $115 million in shared revenues allocated over first 4 years and 8% of AAC reallocated to First Nations |
| Legislative | • *Forest Act* amendments in 1947 created secure, long-term tenures and initiated sustained yield policy<br>• New *Forest Act* and *Ministry of Forests Act* in 1978 required management for multiple use and periodic timber supply review and AAC calculation | • Enacted the Forest Practices Code | • Replaced the Forest Practices Code<br>• Enacted the *Forest and Range Practices Act*<br>• Enacted the *Forest Revitalization Act* as part of the new policy direction under the Forest Revitalization Plan |
| Institutional | • Sloan Royal Commission on Forest Resources (1945) recommends long-term secure tenures | • Created Forest Renewal BC (FRBC), whose purpose was to reinvest stumpage revenue into the forest sector | • Abolition of FRBC<br>• Creation of the Forest Investment Account to take the place of FRBC |

| | | | |
|---|---|---|---|
| | • Sloan Royal Commission on Forest Resources (1956)<br>• Pearse Royal Commission on Forestry (1976) recommends greater security for tenure holders and reintroduction of competition for rights to Crown timber<br>• Introduction of Small Business Forest Enterprise Program (1979) | • Formed Commission on Resources and Environment (CORE), whose mandate was to make land-use planning decisions through broad consensus-based processes | • Creation of the BC Forest Revitalization Trust (BCFRT), which provides compensation to workers and contractors negatively affected by the 20% tenure take-back |
| Tenure, AAC, and timber pricing | • Introduction of long-term volume- and area-based tenures<br>• Move away from competitive bidding in awarding tenures | • Essentially unchanged<br>• Community forest pilot program that allocated less than 0.1% of the provincial AAC to municipalities and First Nations | • One-time 20% AAC take-back from major tenure holders<br>• Half of this amount is allocated to promote fibre access for First Nations, woodlots, and community forests<br>• Half is auctioned under the BC Timber Sales program<br>• Prices established by the auction are then used to establish stumpage rates province-wide<br>• Abolition, through the Forest Revitalization Plan, of many of the restrictions on tenure rights, such as the 5% AAC take-back required on tenure transfers<br>• Increase in the AAC |
| Environmental protection | • *Forest Act* (1978) incorporates environmental values in forest legislation for the first time | • Protected areas strategy doubled the area of protected forest in the province<br>• Cabinet decision allowing logging in Clayoquot Sound precipitated political controversy | • Results-based Forest Stewardship Plans (criticized by environmentalists and Forest Practices Board) |

Management 2004). The reforms also included the abolition of FRBC, which was replaced by the Forest Investment Account – a fund designed to be more accountable to industry interests.

Notwithstanding the tenure take-back, industry support for the reform package has been strong, no doubt in large measure due to the government's characterization of the take-back as a one-time measure and the industry's assessment of the compensatory value of other aspects of the package. Perhaps equally predictable, however, has been the reaction of other stakeholders.

Various First Nations organizations, including the Union of BC Indian Chiefs and the First Nations Summit, passed resolutions calling on the government not to implement the reforms. They contended that the package was developed without consultation and would infringe on constitutionally guaranteed Aboriginal rights and title. They also argued that the one-time allocation of 8 percent of the AAC to First Nations would not meaningfully accommodate Aboriginal rights and title, pointing out that if it were allocated equally by band, this would amount to less than 20,000 cubic metres of wood supply for each community (Clogg 2003). Moreover, the First Nations contended that the absence of guidelines setting out how these new tenures would be allocated appeared "to be designed to generate competition between First Nations and pressure to accept [tenure arrangements] that could fundamentally limit a First Nation's ability to exercise and defend their aboriginal title and rights" (3). First Nations also claimed that changes to requirements for government approval of tenure transfers, subdivisions, and replacement were aimed at shielding the government from lawsuits brought by First Nations to defend the integrity of their traditional territories.

Environmental organizations also actively campaigned against the reforms. They argued that the new results-based forest practices legislation, in combination with debilitating cuts to the Ministry of Forests, would lead to a massive delegation of control over forest practices to the industry. In an initial review of Forest Stewardship Plans prepared by industry, the Forest Practices Board, an independent body responsible for auditing forest practices of both government and licensees and for investigating complaints lodged by the public, found that these plans, for the most part, contained commitments that were vague, unquantifiable, and exceedingly difficult for government to enforce (British Columbia Forest Practices Board 2006).

Arguably the most adversely impacted stakeholders – yet, at the same time, the most quiescent in the face of these reforms – were forest workers and forest-dependent communities. In the interest of laying the groundwork for a new deal with the Americans, the most obvious casualty has been the social contract. By legislatively forgoing its right to embed social-contract goals in forest decision making – by removing tenure transfer and subdivision fetters, and abolishing minimum annual harvests and appurtenancy and processing requirements – the BC government has abandoned a fundamental tenet of post-Second World War provincial forest policy. Some will argue – with some justification – that this tenet did little to stabilize local forest-sector employment and was, in practice, abandoned long ago. Nonetheless, in this regard the reforms truly live up to the "New Era" billing the Liberals gave their election platform, representing a significant retreat by government from hands-on engagement in decisions about where, when, and for whose benefit BC forest resources are developed.

## Conclusion

Forest policy in British Columbia has evolved in response to changing economic, social, and political conditions. Table 3.1 (pp. 74-75) summarizes some of the significant policy changes discussed in this chapter. Following a period of largely unregulated liquidation of the province's forests under one-time timber sales licences in the period before the Second World War, the 1945 Sloan Royal Commission led to the granting of the first long-term area- and volume-based tenures in British Columbia. The requirement of sustained-yield management introduced the first significant regulation of AACs in the province. Over the next fifty years, the forest industry came under increasing pressure as environmental values evolved (particularly with respect to the protection of old-growth forests) and legal recognition of Aboriginal rights and title was consolidated. This led to a first wave of forest policy reform in the 1990s, including the ill-fated Forest Practices Code. Before long, as we have chronicled, a constellation of factors – including growing global competition, escalating trade disputes with the United States, and the emerging pine beetle epidemic – gave rise to renewed pressure for forest policy reform. Even more sweeping than the first, the longer-term implications of this second wave of reform remain unclear.

Meanwhile, as successive provincial governments embarked on these much-publicized policy adventures in the "hard law" domain, many of the province's leading forest policy players were busy in the realm of "soft law," tackling the dauntingly complex task of translating the Forest Stewardship Council's vision of sustainable forest management into practice in the British Columbia context. It is to the story of the development of the FSC-BC final standard that we now turn.

# 4
# Hard Bargaining: Negotiating an FSC Standard for British Columbia

As the discussion in Chapter 3 underscores, the last decade has been a tumultuous one for the BC forestry sector and for BC forest policy. Both the forest industry and the policy environment within which it operates have been forced to change in fundamental ways. For much of this period, therefore, the focus of the ensuing struggle between various interested parties and stakeholders was over the nature and the pace of these inevitable changes.

As these negotiations unfolded at the provincial level, a parallel debate was occurring within the BC chapter of the Forest Stewardship Council (FSC-BC) over the development of a BC regional standard. In this chapter, we provide a detailed account of the negotiations that led to the development of the FSC-BC final standard. We chronicle the standoff that developed between large industry and other participants over the provisions of the third draft of the standard (D3) that resulted, ultimately, in the formulation and implementation of a new approach to standards accreditation by FSC-AC. In the concluding section of this chapter, we offer some reflections on the motivations and interests of some of the key players involved in the "hard" bargaining over the form of this standard – a bargaining process that, as we shall see, amply merits this adjectival description in terms of both its intensity and its intractability.

## Setting the Standard in British Columbia

As described in Chapter 2, there are two routes to certification under an FSC standard. Where there is no locally developed national standard, the certifying body (CB) will interpret FSC-AC's generic principles and criteria and apply them to the forest management unit to be certified. The alternative approach attempts to minimize the degree of variation among CBs by developing a national or regional standard that elaborates, in the form of indicators (and sometimes verifiers), how FSC-AC's generic principles and criteria are to be interpreted within a specific location. Although most countries have developed a single national standard, a persuasive argument was made in Canada and the United States in the mid-1990s that regional standards would be more appropriate given differences in size, ownership, ecology, and administration. As a consequence, in addition to British Columbia, a large number of regional standards exist in draft or final form across the continent. In Canada, regional standards are in now in place for the Maritime provinces (FSC-Maritimes covers the Acadian forest region that includes Nova Scotia, New Brunswick, Prince Edward Island, Newfoundland, and Labrador); for Canada's boreal region (FSC-Boreal incorporates a continuous belt of forests from British Columbia in the west through Alberta, Saskatchewan, Manitoba, Ontario, and Quebec to Newfoundland and Labrador in the east); and for the Great Lakes-St. Lawrence region (FSC-Great Lakes incorporates the southern parts of Ontario, Quebec, Manitoba, and New Brunswick). In the United States, nine regional standards have been approved by FSC-US, covering the forested areas of the entire country.[1]

Any exercise in regional standard setting is an exercise in leadership. One individual, one organization, or one like-minded group has to get the process under way. In British Columbia, this initiating role was played by an ad hoc group formed in 1996 that comprised a handful of environmental and "ecoforestry" activists with connections to the Victoria-based Ecoforestry Institute. The group was coordinated by Lara Beckett (née Lamport) and included Greg Utzig, Mark Haddock, Tamara Stark, John Brink, and Herb and Susie Hammond.[2] The group worked largely through email to develop the first draft of the BC regional standard (D1), and although it was not deliberately exclusive, some core interests – First Nations, large industry, labour – were absent or under-represented.

Recognizing the limitations of this volunteer arrangement for the task at hand, in late 1998 the group decided to establish a more formal organizational structure (Rhone, Clarke, and Webb 2004, 252-53). As a result, it formed an interim steering committee (ISC), and, with funding from Greenpeace and the World Wildlife Fund, the ISC hired four consultants in late 1998 to complete the first draft. In February 1999, it appointed Marty Horswill as the FSC-BC coordinator of standards development. The first minuted meeting of the ISC was held in the same month (FSC-BC ISC February 1999).[3]

Two major tasks confronted the ISC at this juncture. The first was to make a decision regarding the public release of D1. Members of the ISC were concerned that D1 would be negatively received by important interests, especially large industry, workers, and First Nations, and it agreed to pre-release the draft to these groups to head off possible objections. The ISC's second task was to replace itself with a duly constituted steering committee (SC) as soon as possible. In May 1999, the FSC-BC membership, representing business, environmentalists, social interests, and First Nations, nominated candidates and elected two representatives from each of the four chambers. This newly constituted SC took over from the ISC in June 1999 and held its first meeting (a teleconference) a month later (FSC-BC ISC July 1999).

### Getting Down to Business

FSC-BC's first SC consisted of eight members, two from each of the four chambers, as well as alternates (summarized in Table 4.1). Several of the members were new to FSC, so the early work of the SC involved orienting them to the organization and explaining its role in the FSC network. The SC also made several substantive decisions soon after its formation. On organizational matters, the SC agreed to a two-track approach, making preparations, on one hand, to incorporate as a legally registered society within British Columbia while clarifying, on the other hand, its relationship with FSC-Canada. The coordinator of standards developments was directed "to prepare a set of draft incorporation documents for presentation to the next SC meeting with a view to FSC-BC becoming a registered society" (FSC-BC ISC August 1999). A second important organizational decision was the establishment of the SC's Alternates Policy, which enabled individual SC members to nominate someone to represent them in their absence. Alternates played a vital role in the development of the FSC-BC final standard, as they were regularly called upon to make decisions in the absence of SC members over the six-year standards negotiation period.

One of the early items of business at the SC's first face-to-face meeting was to determine the process for negotiating the second draft of the standards (D2). To assist in this discussion, the SC hired a facilitator, Ross McMillan of Boreray Praxis Consulting. After considerable debate over the range of options, it was agreed to establish a standards development team (later abbreviated to standards team or ST), "whose members have an appropriate range of technical skill and experience as well as an aptitude for teamwork," and whose tasks included "consulting with stakeholders, obtaining technical advice where necessary, actually re-drafting the standards document and making recommendations to the Steering Committee regarding standards development issues as they arise" (FSC-BC ISC August 1999) (see Table 4.1 for members).

The SC also agreed that the ST would have equal representation from each of the chambers. FSC members and stakeholders were informed that they could nominate as many potential candidates to the ST as they wished but that the SC would vet those names and make the final decision on its composition. The process of establishing the ST took some time. A call went out in late November 1999 setting 15 December as the deadline for nominations. The BC coordinator of standards development allocated individual nominees to different chambers and forwarded these names to the appropriate SC chamber representatives, who then vetted the nominees for their eligibility to represent chamber interests. It was only at the January 2000 SC meeting

*Table 4.1*

## Original participants in the FSC-BC steering committee (SC) and standards team (ST)

| Chamber | Steering committee | | Standards team members |
|---|---|---|---|
| | Member | Alternate | |
| Environment | Tamara Stark (*Greenpeace*) | Cheri Burda (*David Suzuki Foundation*) | Tom Green (*ecological economist, Nelson, BC*) |
| | John McInnis (*unaffiliated*) | Merran Smith (*Sierra Club of BC*) | Greg Utzig** (*MS, P.Ag., Kutenai Nature Investigations, Nelson, BC*) |
| Economic | Bill Bourgeois (*Lignum*) | Sandy Lavigne (*Western Forest Product*) | Patrick Armstrong (*Moresby Consulting, Nanaimo, BC*) |
| | John Brink (*Brink Forest Products*) | Reine Kahlke (*Forest Dynamics Management*) | Jim Burbee (*RPF, Venture Forestry Consulting, Prince George, BC*) |
| Social | John Cathro* (*Kootenay Conference on Forest Alternatives*)/ Nicole Rycroft (*Clayoquot Progressive Ventures*) | No alternate appointed | Brian Tuson (*Forest Resource Officer, Pulp and Paper Workers of Canada, Prince George, BC*) |
| | Hans Elias (*Harrop-Procter Watershed Protection Society*) | Rami Rathkop (*Harrop-Procter Watershed Protection Society*) | Jessica Clogg** (*LLB, MES, forestry lawyer, West Coast Environmental Law, Vancouver, BC*) |
| Indigenous peoples | Dave Monture (*Shuswap Nation Tribal Council*) | Shane Wardrobe (*Shuswap Nation Tribal Council*) | Russell Collier (*mapping consultant, Ecotrust Canada, Terrace, BC*) |
| | John Yeltazie (*First Nations Summit Society*) | George Watts (*First Nations Summit Society*) | Dave Mannix (*Forestry and Economic Development Officer, Snuneymuxw Nation, Nanaimo, BC*) |

*Source:* FSC-BC 2002b.

\* In mid-2000, John Cathro resigned from the social chamber to take up a position as paid chair of the SC; Nicole Rycroft, affiliated with a small Vancouver Island-based community economic development organization, filled the vacant position in the social chamber.

\*\* Brian Horejsi, the original environmental chamber appointee to the ST, resigned after its first meeting, resulting in a restructuring. Greg Utzig moved from his position representing the social chamber to fill the vacant environmental chamber ST seat, and Jessica Clogg replaced him as social chamber representative.

that the composition of the ST was finally determined. The appointment process underscores how seriously the participants took the formation of the ST.[4] Through an intensive, iterative process, each prospective candidate was evaluated in detail before the SC finally decided who would make up its eight-member standards team.[5]

The existence of an almost complete D1 was a source of some difficulty for the SC, which could neither ignore nor endorse it. As already noted, an early meeting of the ISC planned to pre-release D1 in March 1999 to First Nations, business, and other stakeholders, but this target date was not met and discussions continued into April, when a lengthy debate about what to do with D1 took place. At this point, SC members agreed that D1 should be considered as a starting point for discussion, and they planned a news conference for early May where "the focus ... would be the launch of the stakeholder process NOT the draft standards, although the standards should be publicly released at the same time but without a lot of fanfare" (FSC-BC ISC April 1999).

The first draft was subsequently released for public discussion in May 1999 and, with the ISC preoccupied with the process for electing a new and representative steering committee, there was little further discussion of the draft. As one of our interviewees noted in regard to D1:

> D1 was basically an ad hoc group of volunteers ... about half a dozen ... who actually showed up for a couple of drafting meetings. At that point industry involvement was negligible, although there were some woodlot owners and RPFs [registered professional foresters] and people sort of aligned with industry there, but there was no official representation ... Once there was actually an elected steering committee that was officially recognized by FSC-Canada, I would say that was where the change was made. And then they officially selected a drafting team. Before then, the drafting team was essentially a group of people who showed up to start on the process. (personal interview 2003)

By the end of 1999, the FSC-BC process was well underway. It had a solid committee structure, was recruiting a standards team, and had managed the release of D1 without alienating key stakeholders, who were now focused on the D2 phase of standard development. But it was not all smooth sailing. At a November public information forum, Kim Pollock, the director of environment and public policy for the Industrial, Wood, and Allied Workers (IWA) union, had launched a personal attack on one of the environmental attendees. The attack was deemed completely unacceptable by others present, and at its November meeting the SC agreed that "the Chair should write a letter to Mr. Pollock expressing the SC's concerns, ... that the SC should invite the IWA to attend a Steering Committee meeting at a mutually agreeable time and location to discuss matters of mutual concern," and "that the Steering Committee host a meeting with all BC forest related labour organizations to promote positive engagement in all aspects of FSC" (FSC-BC ISC November 1999). Subsequently, the regional standards coordinator had a telephone conversation with Pollock to convey the SC's position.

The IWA incident was symptomatic of the tension between the main forest workers union and the FSC at the time. The IWA was not represented on the SC because it was not a member of the FSC-BC chapter, and it was not in a position to put itself forward to the social chamber as a nominee. Another source of tension was the IWA's heavy backing of the Canadian Standards Association (CSA), the scheme competing with FSC in Canada. Kim Pollock had been a participant in the development of the CSA. He was also a vocal critic of the environmental movement, which he accused of threatening workers' jobs (Pollock 1996).[6] At the January SC meeting, therefore, in a follow-up to the public information forum, it was agreed that "the Social Chamber representatives should take the lead in inviting forest labour to become more actively involved in FSC-BC and that the IWA should be encouraged to follow through on their FSC membership application" (FSC-BC ISC January 2000). The SC chair, John Cathro, was also asked to approach the labour unions personally to encourage their involvement.

### Negotiating D2

By early 2000, the process of drafting D2 was underway. The ST held its first meeting in early February and developed a draft workplan for SC approval. The workplan covered the completion of both a second and third draft within an eighteen-month period. D2 was to be completed by the end of 2000 and available for public review in January 2001. At a joint SC/ST meeting in February 2000, there was a lengthy discussion about the inclusion of government representatives on the ST. The ST proposed that "both the government representative and the Alternate should participate in all Standards Team and working group discussions throughout the entire process with an active voice but no vote" (FSC-BC ST 2000). The SC was not initially inclined to allow this. In its review of the draft workplan, the SC reaffirmed "that there is only one government representative on the Standards Team and that the alternate shall only participate when the regular representative is unable to attend." (FSC-BC ST 2000). Following further ST representations, however, and expressions of "deep concerns with the Steering Committee's decision to reject their recommendation," the SC agreed to permit both government representatives to sit in an ex officio (non-voting) capacity.[7]

In addition to determining its final composition, the ST proposed adopting consensus decision making, defined as "the agreement to or willingness to accept any given proposition by all voting Team members." Under this approach, "consensus will not have been achieved if any voting Team member(s) cannot accept the wording of the standard concerned" (FSC-BC ST 2000). It was observed that this consensus requirement placed a strong obligation on ST members to conduct negotiations in good faith, and it established a series of expectations that members would articulate their positions fully and honestly, express their underlying reasons and concerns, respect other members, focus on issues (not personalities), and seek to understand the concerns of other members (FSC-BC ST 2000). The ST's decision-making procedures differed from the policy adopted by the steering committee, which specified that consensus should be used as much as possible but that, in the event of a vote, a motion had to be passed "by 75 % of all members voting as well as by at least 50% + 1 of all members voting in all four chambers" (FSC-BC ISC April 2001).

The ST proposed that the negotiation of standards should occur at three "levels." At the lowest level, indicators relating to a single criterion were to be negotiated and agreed upon. Agreement at the first level, however, did not imply agreement at the mid-level, where a package of indicators was to be negotiated for a single principle. Finally, agreement at the mid-level did not imply agreement at the highest level, where the final package of indicators addressing all ten of the FSC's principles was being reviewed. The rationale for this hierarchy was that "individual standards are contingent on standards in other Principles and ... the opportunity to reconsider previous decisions is necessary because of this contingency" (FSC-BC ISC February 2000). The hierarchy of agreement meant that the package of indicators previously negotiated could be subject to eleventh-hour objections. The SC expressed deep concern that this hierarchical process of decision making "could become a serious obstacle to progress" (FSC-BC ISC February 2000), but the SC was finally persuaded by the ST's rationale and agreed to these decision-making arrangements.

The ongoing cost of developing the FSC-BC regional standard was becoming considerable at this stage. In addition to an ST budget of $207,000 (to cover the cost of honoraria, travel, facilitation, recording, and consultation), there was the cost of running the FSC-BC office, which now had a staff of two: Marty Horswill, who continued to work as coordinator of standards development, and James Rutter, the newly appointed executive director. Although FSC-BC's budget was financed by grants from the Ford and Rockefeller Brothers foundations, money was always in short supply and of constant concern to SC members.[8] In 2000, a financial shortfall led to the almost immediate resignation of the newly appointed executive director, prompting a restructuring of the FSC-BC chapter. John Cathro resigned from the social chamber and took over as the SC's paid chair. In subsequent elections, Cathro's seat on the SC was filled by Nicole Rycroft, a member of a Tofino-based community economic development organization.

With the ST established and meeting regularly, the SC turned its attention to the thorny issue of harmonization. A requirement before any FSC national or regional standard is finally endorsed by FSC-IC is that it be harmonized with relevant neighbouring standards. In defining harmonization, the 1998 FSC-AC "National initiatives manual" states:

The aim of harmonization is for the regional forest stewardship standards to provide a consistent interpretation of the Principles and Criteria world-wide. This is of particular concern where ecological boundaries do not match the socio-political boundaries of national or regional borders. FSC recognizes that regional standards, even those developed for the same forest ecosystem, may vary from region to region for legal, political or other reasons. However, significant variations in ecological indicators and verifiers for similar or identical forest ecosystems would imply inconsistent interpretation of the FSC Principles and Criteria and could lead to downward harmonization of standards. (FSC-AC 1998a, 12.3.3)

The manual goes on to detail the measures that a national initiative must address to satisfy the harmonization requirements. These include identifying relevant neighbouring regions, consulting with members in those regions via inter-regional meetings and/or sharing of staff, and circulating draft standards. The manual also notes that where one regional standard-setting process is more advanced than that in a neighbouring jurisdiction, it is important not to "reinvent the wheel" but, rather, to consider using the more advanced standards "as a starting point" (FSC-AC 1998a, 12.3.2). Regional standard-setting bodies are directed to organize harmonization meetings with delegates from neighbouring regions, the objectives being to "compare procedures, identify gaps, and to identify and resolve controversies and differences that may have negative effects on management or markets" (12.3.2). Any differences between regional standards for neighbouring jurisdictions must be justified in writing.

Harmonization did not figure prominently for the ST at this stage, however, as the workplan timeline provided for harmonization discussions to occur in January and February 2001 following the release of D2. The issue came more sharply into focus for the SC as a consequence of FSC-AC's commission of enquiry into events on the other side of the country, where there had been a failure to harmonize Canada's FSC-Maritimes standard with its US FSC-Northeast equivalent (see the discussion in Chapter 5 on FSC-Maritimes). Another issue that generated pressure over harmonization was FSC-Canada's announcement that it would develop a Canada-wide standard for the boreal forest (FSC-Boreal standard). This decision gave rise to a discussion at FSC-BC on the potential for conflict if both FSC-BC and FSC-Canada simultaneously developed standards for boreal forests. The minutes of the meeting record that

> FSC-Canada has to immediately come up with a process that accommo-
> dates the differences between ecologically based harmonization and the
> legal jurisdictions as defined in Principle 1. It was agreed that FSC-BC
> pro-actively monitor development of boreal standards in other regions of
> Canada with a view to facilitating harmonization or the development of a
> single national standards for boreal forests. It was agreed that FSC-BC pro-
> actively initiate meetings with the US Pacific Coast and Rockies initiatives
> to discuss harmonization, and include ST reps in these discussions. (FSC-BC
> ISC March 2000)

FSC-BC's steering committee was well aware of the importance of harmonization, but it proved extremely difficult to operationalize the concept. First,
although there were two neighbouring jurisdictions where regional standards
were under negotiation – FSC-Rocky Mountain and FSC-Pacific Coast – both
were also at preliminary stages in March 2000. FSC-Rocky Mountain was
producing its own D2,[9] while FSC-Pacific Coast was working on a new draft
of its standard.[10] This raised several difficult questions, among them: Should
FSC-BC take the initiative and seek to harmonize with these neighbouring
standards, or should the neighbouring standards take the initiative and seek
harmonization with FSC-BC? Second, the social and economic conditions
in the US regions were very different from those in British Columbia. The
province was heavily dependent on forestry for profits, foreign exchange,
employment, and tax revenue; the individual states to its south were much
more economically diversified. Third, where 95 percent of British Columbia's
forests were publicly owned, most forests in Washington, Oregon, and California (FSC-Pacific Coast), and in Montana, Idaho, Wyoming, Nevada, the
northern parts of Colorado and Utah, and western South Dakota (FSC-Rocky
Mountain), were in private hands. Finally, whereas British Columbia faced
significant uncertainty around pending Aboriginal rights and title claims,
the US regional initiatives were being negotiated in a context where treaties
and indigenous governments were long established and well settled.

Given these significant differences between British Columbia and its
neighbouring jurisdictions, there was every reason to suspect that standards
would differ. Hence, although the FSC-BC steering committee paid attention
to the issue of harmonization, it was not unduly worried about its ability to
meet its FSC requirements in that respect. According to one interviewee,

> the process of harmonization is meant to say that they should be in keeping
> with meeting the individual criteria, and that the indicators should vary on
> the basis of regional priorities basically. Is public land to be treated differ-

ently than private land in the US? ... I mean it's completely legitimate on an ecological basis, and arguably on an economic level, within Washington State to say, "absolutely no logging of old growth at any time." And, obviously, some people would say that it's legitimate to argue that for BC. We chose not to obviously, and partially based on the industry being so reliant on old growth. (personal interview 2003)

Another factor influencing FSC-BC's perspective on harmonization was the FSC-IC requirement that, where differences were noted between standards in neighbouring jurisdictions, upward not downward harmonization would be required.[11] Hence, most FSC-BC members felt they had little to fear from harmonization. Either their final standard would constitute the baseline for other standards, which would have to harmonize upwards to meet FSC-BC, or FSC-BC would use tougher standards developed by FSC-Pacific Coast or FSC-Rocky Mountain. Either way, the majority of FSC-BC members would achieve their ecological, indigenous, and social objectives. However, it was also recognized that large industry often had the most to gain from harmonization because it could invoke interjurisdictional variations in standards to argue for a "level playing field," which was perceived by some as opening the door to downward harmonization. As one interviewee noted,

it's a key issue and yet it's something that hasn't been picked up on. And even to the point of having a serious discussion about it. Like in Canada, for instance, who do you harmonize with? And one argument is, "Well, you harmonize right across the temperate zone, because that's who you compete with, because it's a global industry." And others say, "Well, no, it's only similar ecological jurisdictions." But I'll tell you, if you were to harmonize with the Rocky Mountains standard, the BC standard would look a hell of an awful lot different (personal interview 2003).

The SC continued to reflect on and discuss harmonization issues over the next two years. It commissioned a paper on the topic from a group of University of British Columbia academics and convened meetings with FSC-BC, FSC-Rocky Mountains and FSC-Pacific Coast leaders. The issue was also pursued at the national level by FSC-Canada and FSC-US via a joint meeting between representatives of the two boards in Arcata, California. In the end, however, practical harmonization was not achieved in British Columbia or, for that matter, most other jurisdictions. The only exception – and it is an important one – were efforts to harmonize FSC-Boreal with relevant standards elsewhere in Canada and internationally (see Chapter 5).

Throughout 2000 and into 2001, the ST met regularly to negotiate D2. In order to manage its work, each ST member was allocated the "lead" on one or, in some cases, two specific principles. For example, Greg Utzig was the lead on Principle 6 (P6), which dealt with ecological values; Pat Armstrong was the lead on Principle 9 (P9) and related issues involving high conservation value forests (HCVFs); and Brian Tucson was the lead on Principle 4 (P4), socio-economic benefits to workers and communities. Through consultations and joint meetings with the SC, the ST slowly assembled D2 using a similar process to that used in international treaty negotiations. Leads were responsible for developing the language for indicators dealing with their assigned principle. Where ST members could not reach agreement on the language, options for the SC to consider were presented in bracketed text. The ST's goal was to produce a final text with as few brackets as possible.

The ST worked hard through 2000 and early 2001 to produce a "bracket-free" text but was unable to achieve that goal. In March 2001, the SC decided that intensive discussions were required to move things forward, and it organized a three-day meeting with the ST. Following an in-depth examination of the indicators and verifiers relating to each criterion, the SC directed the ST to remove all of the bracketed text and present the different wordings as options to be considered. It also deleted the major failures provisions from the addendum and referred to this concept only in general terms.[12] Finally, the minutes record that "all members of the SC would sign a covering letter accompanying the public release of D2 indicating their support in principle for the document and its release for public comment" (FSC-BC ISC March 2001). With plans to release D2 in May, the SC disbanded the ST. Four former ST members, Russell Collier, Jessica Clogg, Greg Utzig, and Patrick Armstrong, were immediately named to a new body, the Technical Advisory Team (TAT), to advise the SC on D3.

**From D2 to D3**
Following the release of D2 in May 2001, work on D3 commenced immediately. The TAT, as its name suggested, functioned differently in these negotiations than the ST had done. The ST had, in effect, drafted D2 – albeit with continuous consultation with the SC. The role of TAT, in contrast, was largely advisory, with the SC taking on the role of lead drafter. The most important output from the TAT was an issues analysis report, which was based on stakeholder submissions on D2 and comments from certifying bodies following field tests conducted in August and September 2001.

Public comment on D2, combined with technical briefings and field tests, produced a large body of data from which the TAT distilled the key issues.[13] Tier 1 issues were the most crucial, and these were grouped into three broad categories: (a) the size, organization, and structure of forest management operations; (b) the usability, auditability, measurability, and consistency of the standard; and (c) indicators developed for specific principles. The latter included such matters as the nature of the "long-term commitment" of forest managers (P1), the specificity of "protocol" agreements between First Nations and forest managers (P3), the definition of the concept of "local" (P4), the application of the concept of "range of natural variability" (P6), the definition of HCVFs in the BC context (P9), and the appropriate level of plantations within a management unit (P10). Principle 8, involving monitoring and assessment, was the only principle for which no issues were identified. Principle 6, which dealt with ecological values, had the most Tier 1 issues.

Two professional facilitators were brought in to manage the negotiations at this point – Daniel Johnston and Alex Gryzbowski. Despite their efforts, however, discussions broke down during the SC's April 2002 negotiations, when large industry, represented by Bill Bourgeois of Lignum Ltd., announced that it was unwilling to approve D3 until it was known what the "impact would be by completing an assessment of the regional standards effectiveness and impact" (FSC-BC ISC April 2002). As a result, large industry did not sign on to D3.

Why did large industry not endorse D3? And why did its objection emerge so late in the negotiation process? Participants in the negotiations have different perspectives on these questions. Some blame large industry directly for negotiating in bad faith and betraying other members of the SC. Others find fault with the process that was adopted, particularly with some of the techniques employed by the facilitators. Still others charge that there was considerable "selective listening" going on throughout the D3 negotiations. Finally, some argue that the breakdown was "inevitable" because what was at stake were participants' differing visions about how rigorous, detailed, and "prescriptive" the standard should be.

Several SC and TAT members experienced a deep sense of betrayal after large industry's announcement in March 2002. In the words of one participant, "It was completely unexpected and an absolute shock. I mean people were – the level of emotion and concern and frustration was enormous" (personal interview 2003). This comment reflects the perspective that large industry had agreed to a negotiating process, engaged willingly in that process, made certain commitments concerning the indicators to be included in the standard, and then, at the eleventh hour, reneged on previously announced agreements and, in effect, betrayed the negotiation process. Although there is little doubt that some SC members were deeply distressed by these events, others close to the negotiations place some responsibility on the facilitators for pushing too hard for a negotiated settlement.[14] The view that the facilitators were partly to blame may be countered by the observation that agreement was only going to be reached in a high-intensity environment where people felt pressured into giving up what they considered vital elements of the final package. And perhaps the facilitators can be forgiven for assuming that participants would signal clearly to them, and to others around the table, any disagreement with specific outcomes. Instead, it appears that the representative of large industry remained silent on at least some key issues toward the end of the process.[15]

One explanation offered for large industry's last-minute rejection of D3 focuses on a fundamental tension that pervaded the FSC-BC standard-development process. Industry saw the process as a choice between a "boutique" and a "general store" standard (see the "Issues in Forest Certification" section of Chapter 2). Environmentalists and social justice advocates tend to reject this framing of the issue. In their view, what they were negotiating was a standard that, in the long run, would achieve the ambitious vision of sustainable forestry embodied in the FSC's principles and criteria. This vision, they contended, could only be realized if a rigorous and credible standard were in place, supported by the pressures imposed by the market. Ultimately, one of the reasons large industry did not sign on to D3 was that it concluded the draft had too many of the hallmarks of a "boutique" standard.[16]

At the same time, it should be acknowledged that large industry was never fully engaged in the FSC-BC standard-development process as a cohesive group. The company Bourgeois represented at the time, Lignum, was one of only a few forest companies to take an interest in the FSC in British Columbia. In contrast, most other large companies supported the CSA scheme and worked behind the scenes to neutralize what they perceived as the FSC threat. As the large-industry "representative" on the SC, Bourgeois was in a quandary because he had no real constituency to report back to or receive instructions from. This severely constrained his ability to represent the interests of large industry. In the absence of any support for his position on the SC, he had to make a judgment about whether his company, and a small group of like-minded companies,[17] would be able to operate under D3. In the end, he opted to vote against D3 because he concluded that it was too "onerous" and "prescriptive" and that further negotiation would be required to render it palatable to large industry in the province.

Despite the contentious end to the three-year negotiations, the SC was entitled, under the prevailing decision-making procedures, to vote on its recommendation that D3 be submitted to FSC-Canada for approval and transmittal to FSC-AC for endorsement. By a seven-to-one margin (which included 50 percent support from each chamber), the SC approved the D3 standard and forwarded it to FSC-Canada in late April 2002, hoping that that national body would approve it at the 24 April board meeting. However, FSC-Canada declined to make a decision and instead agreed "to meet again to further develop its plan for endorsing the BC regional standards and to advise FSC-BC in writing of its plans for this process" (FSC-BC ISC May 2002). Given D3's complexity, it was unreasonable for the SC to expect the FSC-Canada board to approve it so quickly. Moreover, FSC-Canada had established a National Standards Advisory Committee (NSAC) in 2001 specifically to provide advice to the board on regional standards initiatives, and this committee required more time to reflect on D3.[18] Also, the FSC-Canada board had just emerged from an exhausting set of negotiations over the FSC-Maritimes standard and the commission of enquiry set up by FSC-AC (see the discussion of FSC-Maritimes in Chapter 5). In this situation, and given the dissenting vote of the large-industry member on the FSC-BC SC, the FSC-Canada board was committed to scrutinizing the FSC-BC decision-making process carefully before forwarding D3 to FSC-AC.

The FSC-Canada board had been closely monitoring the BC negotiations. Marcelo Levy, FSC-Canada coordinator, had attended several SC meetings in British Columbia, occasionally in the company of Jim McCarthy, FSC-Canada executive director. There had also been at least one full joint meeting of the SC and the FSC-Canada board, and staff members for both bodies were in regular contact. However, the intense negotiations of early 2002 had occurred in relative isolation, and FSC-Canada was keen to clarify the reasons behind large industry's negative vote.[19] This focus by some on the failure of the SC to achieve complete consensus led others to highlight the "exemplary" process adopted by FSC-BC for standards negotiation, the implication being that D3 should be forwarded to FSC-AC without demur. Curiously, if the FSC-Canada board had used the same "consensus" decision-making structure as the SC, it would not have been able to forward D3 to FSC-AC because the required 50 percent support of the economic chamber was not obtained. However, at the time the FSC-Canada by-laws specified that board decisions could be taken by a simple majority; with five in favour, one against, one abstaining, and one absent, a motion to approve D3 was legitimately adopted.[20]

**Preliminary Accreditation**
Once FSC-Canada made its decision to approve D3, the matter moved to the international level, where it was referred, in the first instance, to FSC's Accreditation Business Unit (ABU). The ABU assessed D3 against its five-point checklist covering contents, requirements and attributes, and the processes used for negotiation and adoption. The ABU's evaluation of D3 was scathing, with numerous major difficulties identified. These included:

- insufficient field-testing of innovative approaches (e.g., range of natural variation)
- too much use of procedural- rather than performance-based indicators
- too many major failures
- too long and complex
- vague language used (e.g., "effective," "minimize," "significant")
- insufficient stakeholder support in British Columbia and Canada

- inadequate differentiation between small, medium, and large operations
- inadequacy of FSC-Canada's decision-making procedures to ensure "significant agreement among all stakeholder groups"
- inadequacy of FSC-BC's response to field-test report recommendations
- insufficient consideration of the BC economic chamber's reports.

In the summer of 2002, a draft of the ABU's report on D3 was circulated to a newly established subcommittee of FSC-AC's board – the three-member Standards Committee. The Standards Committee had been set up only six months before to streamline FSC-AC's standards accreditation process. Prior to its establishment, draft standards were forwarded with an accompanying ABU report to the entire FSC-AC board, a procedure that was both onerous and inefficient. The Standards Committee referred D3 to the full international board anyway, in part because it did not feel competent to make a decision of such magnitude and in part because FSC-AC's standing procedures allowed any individual board member to request full board scrutiny and approval of any draft standard, and it was anticipated that such a request would be forthcoming in the BC case.

FSC-AC board members were well aware of the high stakes involved in approving or denying accreditation to D3. Board members were mindful of several issues as they debated their decision. First, they were keenly aware that donors had contributed substantial sums of money to support the D3 process and would pose tough questions about FSC's model if all that resulted from that expenditure was deadlock. In addition, they were being "lobbied" by the forest industry over the unfairness of FSC-Canada's D3 approval process.[21] A third consideration was the desire to recognize the dedication of environmentalists and indigenous peoples in British Columbia, who had laboured for so long to produce a standard that was, in many respects, more rigorous than others previously approved. Other issues at stake included the credibility of the FSC's secretariat if the board were to override the considered advice of its ABU and the credibility of the organization's business model if another "boutique" standard was approved that resulted in only little take up by industry. Therefore, as D3 moved through FSC-AC's approval process, in the shadow of the FSC-Maritimes debacle, actors at different levels began to seek innovative solutions to the problem posed by its potential nonratification.[22]

With the D3 accreditation decision looming, members of the FSC-AC secretariat and board were being lobbied by a variety of constituencies. The chair of the FSC-AC board, David Nahwegahbow, had close contacts with the First Nations chambers in Canada and British Columbia and was being lobbied to protect their interests. Similarly, Asa Tham in the economic chamber held discussions with FSC-Canada economic interests to ensure that her vote represented their views. Grant Rosoman of Greenpeace, an FSC-AC board member and member of its Standards Committee, was being lobbied by the environmental community. Yet, despite this chamber-based politics, there was a surprising array of opinion across levels and within chambers concerning D3. FSC-AC board members in the social and environmental chambers liked much of what they read, but they appreciated that D3 did not have much economic chamber support and viewed the FSC-Canada vote in particular as constituting a serious challenge to the FSC's vision, structure, and processes.

Prior to D3, FSC-AC's decisions on standard accreditation took one of two basic forms: accredit with conditions or with preconditions. The approach was consistent with the process used by certifying bodies (CBs) when certifying a forest management operation. In certifying a forest management unit, CBs normally issue a report with conditions or preconditions attached. When conditions are attached, the CB issues a certificate but monitors the forest manager to ensure that it meets the conditions attached over the designated time period. If the conditions are not met, the certificate will be withdrawn. When the report comes with preconditions, however, the CB does not issue a certificate. The forest manager must first meet the preconditions and then reapply to the CB to have the operation certified.

Up until this juncture, the FSC-AC had followed a similar approach for the "accreditation" (formerly "endorsement") of a national or regional standard. When a new standard was accredited with conditions (such as "do x and y by year z"), it was added to the schedule of approved FSC national and regional standards. This, in turn, obliged all accredited CBs to apply the standard when certifying forests in the jurisdiction to which it applied. In contrast, a standard "accredited" with preconditions was not really accredited at all, in the sense that it was neither incorporated into the FSC's schedule of standards nor binding on FSC certifiers. In effect, when a standard was accredited with preconditions, this meant it was being remitted back to the national and regional body for further work. This was the case, for example, with the FSC-Maritimes standard, which was returned to FSC-Canada and, ultimately, FSC-Maritimes with four preconditions and four conditions attached and with the proviso that "the standards be recognized as formally endorsed by FSC, when the preconditions have been met, to the satisfaction of the Executive Director'" (Boetekees, Moore, and Weber 2000, 5; see also Chapter 5).

Over the summer and into the fall of 2002, rumours circulated within the FSC network that the international board would approve D3 with a series of stringent preconditions. Such preconditions could have included requirements to (a) re-engage with large industry, (b) reduce the number of major failures, (c) rework process indicators into ones that were more performance based, and (d) remove the "vague" language. The general feeling in FSC-Canada and FSC-BC was that this would be a bad outcome because there was no political will or resources in the province to re-engage with the process.

In a series of discussions between Heiko Liedeker (FSC executive director), members of the international board, Jim McCarthy (FSC-Canada executive director), Chris McDonnell (FSC-Canada board chair), Martin von Mirbach (FSC-Canada board vice-chair), and Rod Krimmer and Ananda Tan (co-chairs of the FSC-BC SC), it was agreed that further efforts to mediate the dispute over D3 should be undertaken. This job was assigned to an ad hoc "chairs' committee" composed of the chair and vice-chairs of FSC-Canada and the co-chairs of the BC steering committee, with Jim McCarthy participating in an ex officio capacity.

Through a round of teleconferences, this committee explored ways to resolve the D3 impasse during the fall of 2002, in the run-up to FSC's November General Assembly meeting. Initially, discussions bogged down because the committee had no mandate to negotiate specific indicators, without which industry buy-in could not be achieved. A breakthrough took place, however, when the committee shifted its focus from considering the content of D3 to identifying a process that would allow it to be simultaneously approved and amended. Approval would enable environmental and indigenous peoples' interests to claim victory because they now had their "tough" FSC-BC standard. Amendment would let large industry continue to engage with the standard and mould it to better suit their requirements. According to one participant, "I recall in fact it was Rod Krimmer who in frustration after about three teleconferences said: 'You know, I woke up last night, and I kind of have this vision of where we got to be, and I'm not sure we're going to get there with this negotiated path' that he was hearing. And he said, 'You know the vision kind of says we need to accept the standard but we need to get companies using it, and we need to get environmental groups supporting the companies who use it'" (personal interview 2004).

At this point, discussion began to focus on the possibility of an "extraordinary solution" to the D3 issue. Underlying this emerging strategy was a growing recognition of the desirability of mounting a trial of D3 in a large-scale forest operation, in non-threatening circumstances, under an arrangement that would result in its systematic and programmed modification. The aim was to move debate on D3 from the words on the page to the practices in the forest, enabling members in different chambers to reassess the validity of their concerns. Drawing on his experience at the Canadian Standards Association, McCarthy facilitated acceptance of this novel solution by explaining that provisional approval (with the potential for subsequent revision based on emerging experience) was an approach that many other standard-setting bodies had successfully employed.

Further discussion of this "extraordinary solution" continued within the chair's committee and between McCarthy and Liedeker in the lead-up to the FSC 2002 General Assembly in Oaxaca. A key challenge at this point was the need to persuade the FSC-AC board of the merits of the strategy. Its proponents soon encountered political resistance, as some international board members balked at the idea of adopting a special arrangement to accommodate the failings of a rich, developed country like Canada. As one interviewee observed, "There was a bit of resistance to this preliminary standard idea, because it was giving special treatment to British Columbia, to Canada ... And you know, people were pretty frank about it, about the inability of Canada to get its act together on this standard after all the money they had. And you have a lot of these groups in the South and they have very little funding and they managed to get their standards together" (personal interview 2004).

For strategic reasons, then, British Columbia's "extraordinary solution" had to be presented as a generic tool that could resolve similar difficulties in other countries, especially in the South. Proponents argued that it might facilitate resolution of the FSC Brazil Plantation standard, which had encountered a similar difficulty in that it was strongly supported by some chambers but not by others. Likewise, a version of the Papua New Guinea (PNG) standard had come before the FSC international board for approval and been referred back to FSC-PNG with preconditions, resulting in delays in implementation.[23]

At the General Assembly in Oaxaca, two high-level but informal meetings occurred between members of the FSC-AC board, the FSC-Canada board, FSC-BC's steering committee, and officials from the secretariats of FSC-IC and FSC-Canada. At the first meeting, as the above quotation attests, some international board members expressed concern over the need for an "extraordinary solution" for a developed country standard-setting process that had apparently gone off the rails. In addition, because the "extraordinary solution" was presented as a package at the first meeting, some members were suspicious that the whole thing was a done deal.[24] However, at the second meeting, and following informal discussions in the corridors and further reflection, the idea of adopting an "extraordinary solution" was better received. In a decision that was taken informally and ratified at a subsequent FSC-AC board meeting, the board asked the secretariat to prepare a discussion paper to explore the option in detail.

It was at this point that the idea of an "extraordinary solution" became formally reframed as "preliminary accreditation." Matthew Wenban-Smith at the FSC secretariat contracted James Sullivan, a consultant with FSC-AC experience, to investigate the degree to which the process was used in other standard-setting bodies and how it might be implemented by the FSC. The timeline for this contract was short, but in late December 2002 Sullivan reported favourably on preliminary accreditation, confirming that the proposed procedure was consistent with procedures employed by various standards bodies around the world.

Early in 2003, Wenban-Smith drew up a policy document on preliminary accreditation that was circulated to a number of FSC national initiatives for comment (very few were received) and forwarded to the March meeting of the FSC-AC board. FSC-AC approved the policy document, and preliminary accreditation became established FSC policy from that point, although it took several more weeks to complete the paperwork describing how it would apply in the specific case of British Columbia. In June, FSC-Canada received notification that the D3 standard had been approved under the preliminary accreditation policy, and a letter to that effect was drafted shortly thereafter and forwarded to FSC-BC's SC. Finally, after a year in limbo, the BC regional standard was once again moving forward toward final accreditation.

The new policy of preliminary accreditation was "designed to encourage the development and uptake of national and regional forest standards ... by creating a formal mechanism through which national and regional FSC forest stewardship standards can be introduced in the field with FSC's full endorsement, whilst allowing the many stakeholders with an interest in the standard additional time to gain real field experience in the implementation of the standard, in order to resolve outstanding issues of concern" (FSC-IC 2003a, 1). Under the policy, preliminary accreditation status may be granted to a standard that is consistent with the FSC's basic obligations (i.e., one that has indicators developed for all of FSC's principles and criteria), even though the requirements for full accreditation have not been met (such as a higher level of consensus for standards that are substantially in excess of the basics). In such cases, the FSC-AC board was given discretion to grant preliminary accreditation when a clear rationale is presented, provided that improvement of the specific areas identified by the board can be achieved within eighteen months.

## Implementing Preliminary Accreditation

FSC-Canada informed FSC-BC's SC that D3 had been approved under FSC-AC's new preliminary accreditation procedure in July 2003, and they spent the rest of the year putting arrangements in place to implement the new policy. This was a complex undertaking, not only because FSC-Canada and FSC-BC were testing an approach that had not been implemented elsewhere, but also because of the numerous conditions attached to the preliminary accreditation arrangement. There were fourteen conditions in total, all of which were "to be filled to the satisfaction of the FSC Executive Director" as a requirement of preliminary accreditation. Some of the conditions were procedural, such as a requirement that "within three months of preliminary accreditation ... the FSC-Canada Working Group shall submit to the FSC Accreditation Business Unit a work-plan based on the requirements of the FSC Preliminary Accreditation Policy" (FSC-AC 2003c). Others were substantive, such as a requirement that the evaluation of the preliminary standard "shall include a review of all major failures and limit the number of major failures at principle level to the most important standard elements ... The review shall also provide for each major failure a clear rationale on the intent of the FSC-BC Initiative" (FSC-AC 2003c).

The preliminary accreditation policy also established a set of procedures for the revision of the preliminary standard following its use by a certifying body. These procedures required a CB to inform the national initiative (NI) that it was undertaking an assessment under the preliminary standard and to grant the NI the right to include up to three technical observers on the CB's team during the evaluation. Following the field assessment, the CB was to submit its evaluation report to the NI for comment and, if it decided to certify the applicant, to publish its full assessment report on its website. National initiatives were required to publicly support any certificate issued by a CB under the preliminary accreditation standard "unless and until such time as a formal complaint has been submitted, evaluated and decided upon through the normal FSC dispute resolution and complaints procedures" (FSC-AC 2003c). National initiatives were also required to complete a formal review of the preliminary standard within eighteen months. At the end of the process, the preliminary standard would be replaced with a final standard, or the CB would revert to using its generic standards.

A central goal of these arrangements was to balance a certifier's right to use its professional judgment when assessing forest operations against a preliminary standard with a working group's right to comment on the certifier's assessment (to ensure that the preliminary standard had been appropriately interpreted). In effect, it set up a mechanism to enable a CB, a certification applicant, and a national initiative to negotiate the practical application of a standard in a concrete context. In the eighteen months following D3's approval as a preliminary standard, FSC-Canada and FSC-BC worked together with SmartWood and a large forest company, Tembec, to implement the preliminary accreditation policy in the BC context.

In July 2004, Tembec applied to SmartWood to be certified under FSC-BC's preliminary standard, and the CB appointed a certification team composed of Keith Moore, Cindy Pearce, Tawney Lem, and John Gunn. Interestingly, Keith Moore was already familiar with the preliminary standard, having managed the field trials that were conducted on D2. The certification team was accompanied on its field visit by three FSC members – one from FSC-BC and two from FSC-Canada – and an FSC-Canada standards revisions committee was established, composed of the chair of the National Standards Advisory Committee and two other technical experts nominated by FSC-BC's SC.

SmartWood awarded Tembec FSC certification under the FSC-BC preliminary standard in November 2004, and its report went to the FSC-Canada standards revisions committee, which prepared a report for consideration by FSC-Canada and FSC-BC. The FSC-BC steering committee unanimously approved the proposed changes at its late January meeting; FSC-Canada did likewise at its February meeting. Between then and November 2005, when the FSC-BC preliminary standard was approved as a final standard, some tense discussions occurred between FSC-Canada and FSC-AC over the form of the final standard. As noted, FSC-IC had imposed fourteen conditions on FSC-Canada in relation to preliminary certification, which the international body was determined to see met. The details of the debate between the two bodies is contained in an official FSC-AC document (FSC-AC 2005c), and although it is couched in technical language, it provides interesting insights into the way the final standard was negotiated, the controversy surrounding it, and the bureaucratic relationship between the two FSC levels.

FSC-Canada informed FSC-AC in January 2005 that it would "proactively implement" the provisions of FSC's new policy on standards development (FSC-AC 2004c). This decision had important consequences for the develop-ment of the FSC-BC final standard because the new policy set out more clearly than before how standards were to be developed. Specifically, it replaced the "major failures" terminology of previous policies with the new terminology of minor and major "non-compliance." The implications of adopting the new policy (FSC-AC 2004c) were articulated in FSC-Canada's submission to the ABU, which notes that "FSC-Canada made the decision to implement FSC International Policies expediently ... As a result FSC-STD-20-002 'Struc-ture and Content of FSC Standards' has been adopted. The application of this standard requires that certification assessments ensure that all criteria are met and also defines minor and major non-compliance. It also provides direction concerning the structure and language of indicators and verifiers. As a result, all major failure designations were removed from the standard" (FSC-AC 2005c).

FSC-Canada and FSC-BC negotiated a process for revising the preliminary standard that included collecting feedback from certifiers and other stake-holders who had been involved in certifications and monitoring since the BC standard had been given preliminary accreditation; appointment of a technical revisions committee to consider this input, the ABU report, and other material; and a subsequent negotiation process within FSC-BC based on the technical revisions committee's report. This negotiation process led to a consensus on what package of changes to the FSC-BC preliminary standard should be proposed to FSC-AC.

A major sticking point in the package that was submitted to FSC-AC by FSC-Canada was a recommendation that riparian management within the province should be addressed through the elaboration of a new criterion, C6.5*bis* (see Chapter 9). To justify this recommendation, FSC-Canada submitted numerous documents to the ABU, including consultancy reports on riparian management. [25] Even so, this effort to add an additional criterion to the FSC's principles and criteria was disputed by SmartWood's CEO, Richard Donovan, who registered his "deep concern" and observed: "I do not believe that FSC policy (that allows the addition of a criterion) has been adequately vetted and/or discussed within the FSC system." He further claimed that the new criterion would undermine support from British Columbia's large industry and that riparian protections in the existing standard were adequate (Donovan 2005).

Notwithstanding Donovan's objections, FSC-AC's executive director eventually approved the FSC-BC final standard on 31 October 2005. However, a further fourteen conditions were imposed on FSC-Canada with respect to the FSC-BC final standard, most of which were to be met within a year following the accreditation decision. Many of the conditions were technical and not particularly onerous. Perhaps the most important was one that required FSC-Canada to develop a process for collecting and evaluating feedback from auditors regarding those standards it viewed as containing "vague expressions" in order to revise them.

The end result of the eighteen-month preliminary standard process was that the FSC-BC final standard was considerably revised compared to the D3 standard. Many of the more controversial indicators were modified in an industry-friendly direction. Notably, these included the elimination of all twenty-nine major failure provisions, more flexible arrangements for negotiating with First Nations, more time for the phasing out of pesticides, and more flexibility in planning reserves and riparian management, albeit with a new criterion in this area. In addition, a number of indicators were deleted to eliminate duplication or because they were deemed not auditable, resulting in a shorter, more manageable standard. Critically, a separate standard for small operators was developed for the BC context, which not only clarified the forest management practices expected from this sector but also enabled the removal of much of the "vague" language related to "size and intensity of forest management operations" in D3.

The fact that Tembec was the only company pursuing FSC certification in the province put it in a unique position to influence the outcome of the FSC-BC preliminary standard process, along with SmartWood, its certifying body. Tembec's manager of environmental affairs, Chris McDonnell, was a former chair of FSC-Canada. McDonnell and Troy Hromadnik (Tembec's manager of BC forest operations) represented large industry on FSC-BC's SC. Their knowledge of FSC and Tembec allowed them to play a unique role in bridging the two very different corporate cultures. Tembec's influence was attenuated, however, by the appointment of three FSC-Canada technical experts to monitor the certification assessment, by the requirement that

SmartWood publish the complete certification report on its website, and by the knowledge that all changes to the preliminary standard would be closely scrutinized by FSC-BC members.

Although the accreditation process led to substantial changes to D3, the resulting FSC-BC final standard compares favourably to analogous FSC standards in terms of design, scope, and content (see Part 2), placing significant obligations on forest managers in British Columbia and representing a new benchmark for FSC standards development.

## Conclusion

This chapter has charted the "hard bargaining" that occurred throughout the negotiations over the FSC-BC final standard. A key tension in these negotiations concerned how high FSC-BC should set the certification bar. In large industry's view, setting the bar too high would deter many companies from seeking FSC certification and give rise to an underutilized "niche" or "boutique" standard. Environmental, First Nations, and some social representatives rejected these warnings as self-serving and speculative. In their view, realizing the lofty aspirations of the FSC's principles and criteria in the BC context required a rigorous, innovative standard that was consistent with an ecosystem-based approach to forest management and closely attuned to the province's unique social and legal characteristics, including the challenge of recognizing and respecting Aboriginal rights and title claims.

Finding a compromise between these two positions proved inordinately difficult. The initial bargain struck in D3 left large industry isolated, and although the draft was subsequently recommended by FSC-Canada, the lack of economic chamber support proved highly problematic for FSC-IC. FSC-AC's instinct in such circumstances was to refer D3 back to FSC-Canada and FSC-BC for further negotiations, yet it feared that this would leave the FSC-BC process in indefinite limbo, precipitating a system-wide FSC crisis. A duly constituted, well-funded, and lengthy standard-setting effort in a major jurisdiction would be seen to have failed, calling into question the FSC's basic premise that committed and willing actors could reach agreement on the meaning of sustainable forest management. To head off such a crisis, an innovative and tailor-made process of preliminary accreditation was established, which validated the work done on D3 while creating incentives for the parties to return to the bargaining table.

The difficulties in reaching agreement were compounded by the structure of the FSC-BC negotiation process, in which committed actors accustomed to confrontation were asked to compromise deeply held values. Especially toward the end of the process, after the standards team was disbanded, actors engaged in direct, across-the-table bargaining over the language of the standard assisted only by the technical advisory team. This scenario, although perhaps unavoidable, presented at least two lurking risks. The first was the incentive it created for some negotiators to specify managerial obligations in minute detail to prevent backsliding. The second was that the cumulative effect of compromises reached over individual indicators would render the final standard unworkable. Both of these risks appear to have materialized in the BC case. To mitigate against these risks, other standard-development processes have adopted a more centrally managed process. It is instructive to compare the process adopted to negotiate the FSC-BC final standard with those employed in other jurisdictions, an exercise we undertake in the next chapter.

# 5
# Beyond British Columbia: Standards Development in Other Jurisdictions

In this chapter, we compare the "grassroots" model of standards development that prevailed in British Columbia with analogous processes in our comparator jurisdictions: FSC-Sweden, FSC-Maritimes, FSC-Rocky Mountain, FSC-Pacific Coast, and FSC-Boreal. This analysis highlights the role of leadership in standards development and underscores both the unique features of the BC approach as well as its similarities to some of our comparator processes.

Leadership is critical in two senses. First, and most obviously, a key feature of the Forest Stewardship Council (FSC) model is the role of regional and national initiatives. If there is no leadership at the regional or national level to initiate standards development, none will emerge. Leadership also plays a key role in determining what type of standard-development process ensues and, indeed, what type of standard emerges. All the processes we examine in this chapter were "first-generation" FSC standard initiatives, yet only the FSC-Maritimes process resembles British Columbia's (chronicled in Chapter 4) in terms of its grassroots leadership style. Elsewhere, both in the processes considered here and in most other FSC jurisdictions, standards development has tended to be more centrally led and managed, closely overseen by a "small, secretariat-based core group responsible for orchestrating consultation with interested stakeholders" (Gale 2004). In what follows, we provide an overview of standards development in our comparator jurisdictions and reflect on how they were led and on the contextual features present in each case that help to explain the various modes of standards development deployed. We commence our analysis with the first-ever national standard to be approved by FSC: the FSC-Sweden national standard.

## FSC-Sweden (Approved May 1998)
Efforts to get FSC certification off the ground in Sweden date back to 1992, following a feasibility study by Per Rosenberg of WWF-Sweden.[1] Rosenberg's report confirmed that some key players in the Swedish forest industry, nota-

bly AssiDomän and Stora Enso, were beginning to view forest certification as a means of maintaining their "social licence" to practise forestry. Not only was the Swedish forest industry under increased pressure from ECSOs in the late 1980s and early 1990s, forest companies themselves recognized that the *Swedish Forestry Act* was outdated and required substantial revision to incorporate biodiversity objectives.[2] In certification, these companies saw not only an opportunity to work with (some) ECSOs to stem the tide of criticism and improve forest practices, but also a means of promoting a more ecosystem-based approach to forest management as a domestic and international marketing tool. Thus, by late 1993, several of Sweden's largest forestry companies had committed themselves to actively supporting forest certification, in marked contrast to the approach adopted by large forestry interests in many other FSC jurisdictions.

Given Sweden's close associational system, these companies managed to persuade other members of the Swedish Forest Industry Association (SFIA) to support certification (Cashore, Auld, and Newsom 2004). Although it was a promising first step, SFIA represented just over a third of Sweden's forestland owners: close to half of the country's forestland was in the hands of 354,000 individuals (Elliott 1999, 357), a substantial minority of whom (about 26 percent) belonged to one of eight forestry associations. The largest of these forestry associations was Södra, with about 30,000 members. Unlike SFIA, Södra and other forest owner associations saw little value in forest certification and consequently resisted joining the certification initiative when they were first canvassed in the fall of 1995.

In early 1996, WWF-Sweden and the Swedish Society for Nature Conservation (SSNC) formed a leadership alliance to promote FSC certification in the country.[3] Later that year, an FSC-Sweden working group was established, comprising three representatives each from the environmental, economic, and social chambers, with the latter representing labour unions and the Sami, Sweden's indigenous peoples, who herd reindeer in the northern part of the country. The working group was chaired by Dr. Lars-Erik Liljelund, a senior advisor to the minister of the environment and worked out of SFIA's offices in Stockholm (Elliott 1999, 383). The FSC-Sweden working group operated without any precedents and established its own procedures. Five subcommittees were formed to deal with standard implementation, labelling and marketing, ecology, workers and Sami issues, and economic impacts.[4] Drawing on external technical expertise as needed, each subcommittee identified the key "fields" in which it needed to establish standards and put together a set of indicators for those fields which, taken together, eventually made up the FSC-Sweden national standard.[5]

The process was intensive: ninety-five meetings were held, twenty-one of which were full meetings of the FSC-Sweden working group. Toward the end of the process, two major issues threatened to block agreement. The first was the issue of reindeer grazing rights in northern Sweden, where the Sami were embroiled in a long-standing dispute with small forestry operators. Södra represented members in southern Sweden, where this issue was not a concern, but it recognized that its northern counterpart, Norrskog, was unlikely to agree to the proposed arrangements. Moreover, private forest owners' weak commitment to certification suggests they may also have been searching for a pretext to disengage from the process. The final straw came in April 1997 when the Finnish FSC working group announced a substantially less rigorous draft national standard.[6] At this point, the private forest owners associations abandoned negotiations.[7] In response, FSC-Sweden reconstituted itself from the remaining members and pressed on. Soon, another contentious issue surfaced: the requirements for the planting and harvesting of exotic species. Ultimately, however, when AssiDomän indicated it was ready to sign the standard as written, pressure grew on other stakeholders to do likewise, and a final compromise was worked out.[8]

FSC-Sweden submitted its national standard to FSC-AC in September 1997. It was approved in May 1998 and immediately adopted by several large industrial operators. As of April 2006, more than 10 million hectares of Sweden's forests were FSC-certified, representing just under half of the country's total productive forest base. Major Swedish companies such as AssiDomän, SCA, and Stora Enso are all now FSC-certified. Private forest owners, on the other hand, have established their own Swedish certification scheme under the auspices of the Programme for the Endorsement of Forest Certification (PEFC 2004).[9]

This "compromise" has created major operational difficulties for the FSC in Sweden. Sweden's industrial forest companies depend heavily on non-industrial private forest owners for their wood supply (Cashore, Auld, and Newsom 2004). Moreover, to minimize transportation costs to the mill, companies frequently swap a supply of fibre in one region for a similar supply in another. At the level of the individual company mill, wood from various sources is normally mixed together, but certification makes this problematic. The FSC's requirements with regard to "percentage-based claims" – "the amount of FSC fiber required to permit the use of a label" (191) – necessitates separation of FSC-certified from other-certified and non-certified timber.

The 1998 FSC-Sweden national standard includes provisions that make it "subject to revision every five years ... by a national FSC body comprising representatives of the parties concerned, taking into account scientific research and the experience already gained" (FSC-Sweden 1998). Although this process of revision began in 2001, reaching agreement on a final revised Swedish standard has proved challenging and time consuming. A new draft standard was circulated for consideration and public comment in early 2003, but agreement on a new final standard was not reached until July 2005. There were major difficulties in the process of restructuring the standard to incorporate all ten of FSC's principles and criteria and in addressing industry's perception that the previous standard had delivered insufficient economic benefits due to the operational difficulties at the mill level (described above). During these negotiations, FSC-Sweden was also undergoing an organizational restructuring to refocus its energy on the business of certification and to reform its voting and dispute resolution procedures. Moreover, as was the case with the BC final standard, there have been delays securing FSC-AC approval as a result of the requirement that the revised FSC-Sweden standard harmonize with other standards in the region, notably the Danish and Finnish standards (Estonian Fund for Nature 2005).

In Sweden, nationally based groups negotiated a single, country-wide standard. This is an important contrast to the FSC-BC process, where negotiations were directed to developing a region-specific standard. In Sweden, national-level groups were professionalized, resourced, and connected to the Swedish forest policy network, which had itself opened up during the 1980s to recognize the concerns of conservation biology. In British Columbia, in contrast, many of the groups involved were run by volunteers, had limited resources, and were not part of the traditional BC forest policy network. Another notable difference was the catalytic role played by the Swedish state, which provided an independent chair, Liljelund, to facilitate interparty communication. This, in turn, reflected the Swedish state's historic "corporatist" role in mediating interests among competing groups. In contrast, the BC state was viewed by representatives of key interests as strongly biased in favour of the forest industry, and this view, together with ongoing serious disagreements about the direction and content of BC forest policy, prohibited the state from playing, or being allowed to play, a mediating role.

**FSC-Maritimes (Approved December 1999)**
Like the FSC-BC process, the development of the FSC-Maritimes standard was plagued by a lack of large industry support. The process eventually broke down, with FSC-AC launching a formal commission of enquiry following a complaint by Blake Brundson, the industry representative from J.D. Irving Co. Ltd.[10] The commission conducted its enquiry in early 2001, receiving submissions from over twenty organizations in Nova Scotia and New Brunswick. Its final report, released in May 2001, provides a detailed insight into the structure and operation of the FSC-Maritimes standard-development process (FSC-AC 2001a).[11]

The FSC-Maritimes standard-development process commenced in April 1996 at a meeting in Truro, Nova Scotia, attended by about 175 participants. The assembly divided itself into nine groups and appointed two members from each group to form the Technical Standards Writing Committee (TSWC). The TSWC spent the next two years developing a draft FSC-Maritimes standard and "met almost every month for two or three days" to achieve this objective (FSC-AC 2001a, 2). By August 1997, the committee had completed a first draft of the Maritimes standard (D1-M). A dozen individuals directly involved in forest certification reviewed D1-M, field-tested it on four woodlot operations, and subsequently released it for public comment in November.[12] The TSWC incorporated the public comment into a second draft (D2-M), which was released for review in May 1998. Fifty-five people attended a second public meeting on D2-M, held in Truro in June that year, which was designed to "allow committee members to answer questions about the standards, and provide information about the FSC process at the regional, national and international levels" (3).

The TSWC was assisted in its work by the Falls Brook Centre (FBC), which acted as its secretariat. The director of the centre, Jean Arnold, and other FBC staff took care of the logistical work involved in writing the FSC-Maritimes regional standard. The commission of enquiry noted that "the FSC was also able to raise some much-needed money to support the [TSWC] because it was an independent private organisation, and [its] financial matters were handled directly through the FBC" (FSC-AC 2001a, 33). At the same time, this arrangement created a lack of clarity about the FBC's role: Was it an independent secretariat facilitating the process of FSC standard-setting in the Maritimes region? Or was it a partisan stakeholder seeking to achieve specific outcomes from the process? This ambiguity led the commission to recommend that in future standard-development processes "the provision of good administrative support and financial management should be separated from an individual perspective or committee representative" (33).

At the June 1998 meeting, the TSWC was disbanded and replaced by a new body, the Maritimes Regional Steering Committee (MRSC), almost half of whose members were former members of the TSWC. Although neither the TSWC nor the MRSC was formally balanced in terms of chamber membership, the commission of enquiry concluded that, "even if not formally structured under the four-chamber model, both committees did include at least two representatives from each of the four chambers" (FSC-AC 2001a, 22). However, it also noted confusion among members of the MRSC about "which hat they wore" when negotiating the FSC-Maritimes standard. Some viewed themselves as directly representing specific interests; others considered themselves to be acting on their own behalf.

In addition to confusion within the MRSC over members' roles, the MRSC's own emergence from the TSWC was murky. All nineteen members were acclaimed to the MRSC, leading the commission of enquiry to conclude that "there does not appear to have been any clearly established and agreed process for putting nominations forward or conducting an election" and "no clear rules about who was eligible to be nominated, or how many people were to be elected" (FSC-AC 2001a, 28). Moreover, representation from some potentially important interests was overlooked. In particular, representatives of the

Nova Scotia Forest Products Association, who had in fact been nominated to the MRSC, were not acclaimed and did not take part in the final standard-drafting exercise. Cashore and Lawson note that the structure of the MRSC perpetuated large industry's relative ineffectiveness in influencing standards development, which had been evident from the commencement of the FSC-Maritimes process (2003, 16).

The MRSC moved rapidly to finalize the FSC-Maritimes regional standard. At a June 1998 meeting, a mid-July deadline for final comments on D2-M was set. Although a large number of submissions was made to the newly constituted MRSC, no formal process for reviewing them was put in place. In fact, the commission notes, "the public comments were completely dealt with in the very first meeting of the new committee," which "occurred only five days after the deadline for comment [had] passed and only four days after the committee was established" (FSC-AC 2001a, 30). Moreover, only twelve of the MRSC's nineteen members were present at this meeting, and the industry representative was notably absent. At this meeting, the MRSC considered the formal comments received on D2-M, made some final changes, and forwarded it to FSC-Canada for approval and transmission to FSC-IC.

As of August 1998, the FSC-Canada working group had just been created. Indeed, the board had not even held its first face-to-face meeting, and it decided to postpone a decision on D2-M until its next meeting, in September 1998. At this meeting, the FSC-Canada board decided that the standard required further work "to address the concerns raised by representatives of the forest industry, and to seek harmonization with standards in adjacent regions" (FSC-AC 2001a, 4). The second part of this directive was especially interesting because one of the key concerns dividing environmentalists and industry in the Maritimes was the use of biocides. D2-M called for the immediate elimination of pesticides and herbicides, whereas the standards being developed in the neighbouring US Northeast region (FSC-Northeast, encompassing the states of Maine, New Hampshire, Vermont, Rhode Island, Massachusetts, and Connecticut) were much more permissive.[13]

At a stormy meeting of the MRSC in mid-November 1998, those present, including Blake Brundson from large industry, reached an agreement on the final wording of the FSC-Maritimes standard (D3-M). The consensus proved to be fleeting, however. The next day, Brundson rescinded his agreement, indicating he could no longer support the biocide indicators. After some consideration, the MRSC concluded that it had met its obligations to stakeholders in the Maritimes region and that it had fulfilled FSC-Canada's requirements to further consult parties. It thus forwarded D3-M once again to the FSC-Canada working group.

Upon receiving D3-M, the FSC-Canada board reconsidered the matter and agreed to forward it to FSC-AC for endorsement. This decision prompted a formal complaint from Brundson to FSC-AC, which was referred back to the

FSC-Canada for consideration. The FSC-Canada board appointed a three-person dispute resolution committee (CWG-DRC) to investigate the complaint. It worked through the spring and summer of 1999 and rendered a decision in September. According to the commission of enquiry, "in response to Mr. Brundson's appeal, the CWG-DRC recommended changes in the future structure and operation of the Maritime Regional [Standards] Committee. Nevertheless, it unanimously recommended that the CWG uphold its November 1998 decision to endorse the standards" (FSC-AC 2001a, 6).

While the CWG-DRC was undertaking its work, FSC-AC continued to process D3-M according to its internal procedures. In late 1998, D3-M was submitted to an FSC-IC board meeting with a recommendation for endorsement subject to four "preconditions" and four "conditions." The FSC-AC board took a formal decision to endorse the FSC-Maritimes standard (FSC-M 1999) in January 1999, to come into effect "when the FSC-AC Executive Director, Dr. Timothy Synnott, was satisfied that the CWG had met the listed preconditions" (FSC-AC 2001a, 6). This turned out to be an unfortunate arrangement for both FSC-AC and its executive director.

Upon receiving conditional endorsement from FSC-AC, FSC-Maritimes set about ensuring that each of the preconditions was met. Although most of them ultimately proved negotiable, one precondition and one condition were "at the center of the controversy over the final endorsement of the standards. Both of these requirements affected the biocide standards, which the FSC-AC board felt both exceeded the international criteria and included the use of an unacceptable committee that would authorise exceptions to the general ban on biocide use" (FSC-AC 2001a, 5).

By March 1999, the MRSC concluded that it had met all preconditions and conditions and forwarded the revised standard (D4-M) back to FSC-Canada. The FSC-Canada board unanimously approved D4-M and forwarded it to FSC-AC. However, the board itself imposed two "preconditions," one of which was that the MRSC "spend July making a 'best effort' to consider all the concerns raised by Mr. Brundson and his employer, the J.D Irving Co., Ltd. (JDI), and to try to work out differences directly related to the biocide clause" (FSC-AC 2001a, 6). The MRSC was unable to resolve these issues. Nonetheless, in August FSC-Canada determined that MRSC had met the preconditions and recommended that the FSC-AC executive director accredit D4-M.

FSC-AC's executive director, Tim Synnott, now faced a major dilemma. Although both the CWG-DRC and FSC-Canada were recommending that FSC-AC accredit D4-M, large industry was completely opposed and was strenuously lobbying against endorsement. Synnott took several steps to gauge the situation, including making a field visit to the Maritimes in November 1999; commissioning a confidential report to ascertain whether the preconditions set by FSC-AC had, in fact, been met; and holding discussions with a wide variety of stakeholders.[14] Convinced that he could not approve D4-M as written, Synnott proposed alternative wording for the biocide indicators, suggesting that it be reconsidered within one year. This proposal (D5-M) met with the approval of FSC-Canada's members and with many on the MRSC, and the result of these negotiations was an executive decision to endorse D5-M as the final FSC-Maritimes regional standard.

When the decision was announced on 20 December 1999, it was greeted by a storm of protest from large industry, in part due to a misapprehension that the FSC-Maritimes regional standard (D5-M) had been endorsed with the original, not the modified, biocide clauses. As the accusations continued to fly in 2000, the FSC-IC board called for a commission of enquiry into the handling of the entire FSC-Maritimes regional standard approval process.

There are several parallels between the FSC-Maritimes process and the FSC-BC standard-development process. A small group of volunteers took responsibility for drafting the standard; comments were sought through public release and workshops; revisions were made over the course of several years; and a final standard was achieved that had the support of most stakeholders. In both cases, large-industry concerns about the content of the standard led to serious reverberations throughout the broader FSC network, necessitating intervention from national and international boards and their associated staffs.

Despite the similarities, there were also some notable differences. The FSC-Maritimes standard-development process was poorly funded and largely undertaken by volunteers; FSC-BC, on the other hand, had a budget of almost one million dollars (Canadian) to remunerate members of the standards and technical advisory teams, its board chair, and staff. Furthermore, FSC-BC's steering committee was constituted in a more formal and balanced manner than either the TSWC or the MRSC. Moreover, procedural questions surrounding the FSC-Maritimes process were of a sufficient magnitude to trigger two formal FSC-sponsored inquiries, but no such inquiries were launched into the FSC-BC standard-development process, which was broadly considered to be transparent and procedurally fair.

### FSC-Rocky Mountain (Approved September 2001)
FSC-Rocky Mountain covers a number of border states, including Montana, Wyoming, Nevada, and Idaho, the northern parts of Colorado and Utah, and the western part of South Dakota. FSC-US, which was keen to complete regional standards for the entire country, initiated the standard-setting process. In 1998 it contracted Steve Thompson, a Montana-based natural resource consultant, "to recruit and coordinate a regional working group to develop certification standards for the Rocky Mountain Region" (FSC-RM 2001). Thompson gathered a group of twenty-four individuals from

the affected states to participate in a loosely constituted Rocky Mountain regional initiative.[15]

The FSC-Rocky Mountain working group consisted of individuals with diverse backgrounds from various parts of the region. Table 5.1 categorizes the individuals by chamber, location, and skills based on information provided in versions of the FSC-Rocky Mountain standard (FSC-RM 2000, 2001, 2004). Unsurprisingly perhaps, given Thompson's base of operations, the vast majority of individuals (63 percent) were from Montana. The membership of the working group reflected a fairly broad range of skills and interests, including Native organizations and forestry experts, although ECSOs were arguably under-represented, and the interests of forest workers were not directly represented. Another notable feature of the FSC-Rocky Mountain regional initiative was the large number of government representatives participating in the standard-development process.

The working group met for a total of fourteen days over six occasions, "split about evenly between full committee and subcommittee meetings with some field trips" (FSC-RM 2001). The process for developing the FSC-Rocky Mountain regional standard consisted of Thompson, the FSC-Rocky Mountain regional coordinator, preparing drafts for review, debate, and revision at working group meetings.[16] Complicating the process was the fact that FSC-US was simultaneously developing a set of national indicators to better harmonize standards development across the country (see also the discussion of the FSC-Pacific Coast standard below). This meant Thompson was not only negotiating his own regional initiative, but was also in discussion with the FSC-US Standards Committee, providing input on the emerging FSC-US national standard. In addition to a 1999 field test, "two drafts [of the FSC-Rocky Mountain standard] were submitted for public review and comment in spring 1999 and spring 2001" (FSC-RM 2001, 2). The final draft of the FSC-Rocky Mountain standard was submitted to and approved by the FSC-US board in August 2001. It was endorsed by FSC-AC the following month.

Several features of the Rocky Mountain process stand in direct contrast to that of its FSC-BC counterpart. In the first place, the initiative to commence a standard for the area came from outside, not inside, the region, and the regional standards coordinator was appointed by the national initiative, FSC-US. Second, as a regional entity FSC-Rocky Mountain was less coherent and more artificial than FSC-BC, in part because of its jurisdictional diversity – it spans seven US states. Third, and in part because the initiative originated outside the region, Steve Thompson was able to orchestrate membership of the FSC-Rocky Mountain regional initiative so that it was largely representative of mainstream forestry interests from Montana. This not only minimized the potential for conflict, but also resulted in a relatively painless negotiation of the forest standard because many of those present already

*Table 5.1*

**Persons participating in the FSC-Rocky Mountain regional initiative working group**

| | Economic | Government | Environmental | Social | Indigenous peoples |
|---|---|---|---|---|---|
| Montana | Doug Mote (*Private land resource manager*) Mark Wigen (*Logger and forest management consultant*) | Ed Lieser (*Forest Service silviculturalist*) Ed Monnig (*District ranger, Fortine Ranger District*) Chris Toctell (*Bureau chief, Service Forestry Program, Montana Department of Natural Resources and Conservation*) John Well (*Consulting forester, Montana Forestland Services*) | Jane Adams (*Wildlife biologist*) Steve Barrett (*Research forester*) Peter Geddes (*Forest ecologist, Foundation for Economics and the Environment*) Steve Thompson (*FSC-Rocky Mountain coordinator and natural resource consultant*) Malcolm Thompson (*Audubon Society representative*) | John Anderson (*Balance Technologies*) Prof. Al McQuillan (*Forestry, University of Montana*) | Rolan Becker (*Forester, Confederated Salish and Kootenai Tribes*) Joe Bryan (*Indian Law Resource Center*) |
| Wyoming | | | | Prof. Bill Baker (*Geography, University of Wyoming*) | |
| Idaho | Craig Savidge (*Consulting forester, former resource manager for Louisiana Pacific*) | | | Prof. D. Adams (*Forestry, University of Idaho*) Teresa Catlin (*Forest ecologist*) | Dave Bubser (*Nez Perce Tribal Forestry*) Janel McCurdy (*Forest manager, Coeur D'Alene Tribe*) |
| Northern Utah | David Nimkin (*Resource economist, Confluence Associates*) | David Schen (*Eco-system management coordinator, Utah Division of Forestry, Fire and State Lands*) | Ivan Weber (*Sierra Club activist and environmental/land use planner with Kennecott Utah Copper*) | | |

*Source:* Compiled by the authors from the list of names and descriptions of Rocky Mountain working group members listed in the FSC-Rocky Mountain standard, (FSC-RM 2001, 33-34). Some textual descriptions are more detailed than others, so readers should consider the list indicative, not definitive.

shared an industrial forestry perspective. Finally, as a consequence of the way the process was structured, the Rocky Mountain standard is relatively short, contains no major failure provisions, and places considerable trust in certifying bodies to exercise their professional judgment. Accordingly, in contrast to FSC-BC's D3, the FSC-Rocky Mountain standard moved rapidly through the FSC accreditation process.

### FSC-Pacific Coast (Approved July 2003)

The FSC-Pacific Coast (FSC-PC) regional standard has a long history with its roots in the Pacific Certification Council (PCC), a group of loosely affiliated "ecoforestry" practitioners that formed in 1993 and operated in the "Cascadia" region of the Pacific Northwest.[17] All US members of the PCC (which included the Institute for Sustainable Forestry, the Rogue Institute for Ecology and Economy, and the Ecoforestry Institute) were also members of the FSC. In late 1995, the American members of the PCC, along with a large number of other stakeholders, established the Pacific Coast Working Group (PCWG), which was composed exclusively of existing regional FSC members.

The PCWG was a more fluid body than its FSC-BC and FSC-Rocky Mountain counterparts. Its membership is tabulated by location and chamber in Table 5.2, although it is important to note the proviso in the FSC-PC final standard, which states that "several participants listed above were not active as near-final and final drafts were prepared, and they may not agree with the content of this standard. Listing as a participant does not necessarily imply concurrence with the submission draft" (FSC-PC 2002, 8). While the PCWG was composed of thirty-four organizations and individuals representing a wide range of interests, it was not "balanced" by chamber like FSC-BC. Environmental interests dominated, not only because they constituted nineteen of the thirty-four members, but also because many of the ten economic-chamber members – including Collins Pine, Big Creek Lumber Co., Scientific Certification Systems, Robert Hrubes and Associates, and SmartWood – were already committed to the FSC's vision of certification. As one participant in the process observed, "the environmentalists had the process by the tail, with the social people and the economic people bending over backward to try to placate the environmentalists" (quoted in Cashore, Auld, and Newsom 2004, 109).

The composition of the PCWG reflected the political economy of the FSC's emergence in the United States and particularly the strength of the large-industry group, the American Forest and Paper Association (AF&PA), which had rallied its members around the emerging alternative SFI standard (Cashore, Auld, and Newsom 2004, 88-126). AF&PA worked hard to ensure its members toed the line on forest certification by informing them about the deficiencies of the FSC and offering them a low-cost alternative. The large-industry group was assisted by disagreement within the environmental

Table 5.2

**Organizations participating in the FSC-Pacific Coast regional initiative working group**

| | Economic | Environmental | Social | Indigenous peoples | Total |
|---|---|---|---|---|---|
| Washington state | | Experience International Northwest Natural Resource Group (NNRG) | | Yakama Nation | 3 |
| Oregon | Collins Pine Company Columbia Forest Products The Rogue Institute for Ecology and Economy | Ecoforestry Institute Headwaters Inc. Pacific Rivers Council | | | 6 |
| California | Applied Forest Management Big Creek Lumber Co. Harwood Products Robert Hrubes and Associates Scientific Certification Systems Watershed Research and Training Society | EcoTimber Institute for Sustainable Forestry Forest Soil and Water Pacific Environment and Resources Center (PERC) Pacific Forest Trust Rainforest Action Network Tracy Katelman | Dominique Irvine Humboldt State University | Hoopa Valley Tribe | 16 |
| National and external | SmartWood (a program of the Rainforest Alliance) | American Lands Alliance Environmental Defense Fund Greenpeace National Wildlife Federation Natural Resources Defense Council Wilderness Society World Wildlife Fund – US | United Methodist Church | | 9 |
| Totals | 10 | 19 | 3 | 2 | 34 |

Source: Compiled by the authors from the list of names and descriptions of Pacific Coast working group members listed in the FSC-Pacific Coast Standard (FSC-2003). Some textual descriptions are more detailed than others, so readers should consider the list indicative, not definitive.

community over whether managers of National Forest lands should be allowed to seek FSC certification. Some environmentalists supported certification of National Forest lands, but several did not, including, notably, the Sierra Club. These groups advocated a zero-cut policy in National Forests and made the non-certification of timber harvested from these lands a condition of their support for FSC (96-97). In the end, the FSC-PC accepted this zero-cut principle, adopting a standard that applies only to "private (including American Indian lands) and non-federal public lands" (FSC-PC 2002, 6) and, as such, excludes US Forest Service, Bureau of Land Management, US Fish and Wildlife Service, and National Park Service federal lands. This decision alienated many in rural communities that depend on continuing flows of timber from forests, regardless of their ownership status (Cashore, Auld, and Newsom 2004, 110).

The stark contrast in the composition of the FSC-PC and FSC-Rocky Mountain working groups resulted in very different draft standards, with the former containing substantially tougher environmental provisions than the latter, especially in early drafts. Indeed, the relative rigour of FSC-PC's initial drafts – in comparison to those that emerged from the Rocky Mountain and FSC-Northeast processes – contributed to FSC-US's September 1999 decision to initiate a national harmonization process. Through this process, FSC-US developed a set of national "baseline" standards to ensure greater consistency in the regional interpretation of the FSC's principles and criteria.

The FSC-PC draft standard was completed in June 2002, just as the FSC-US board decided to take over responsibility for the finalization of regional standards. Some changes were made as a consequence of public submissions and field testing, but final approval followed swiftly and the standard was submitted to FSC-AC with a request that it be accredited as the FSC-PC final standard. Accreditation was granted two months later.

### FSC-Boreal (Approved August 2004)

Of the processes examined here, the FSC-Boreal is the most recent and, as such, reflects a significant amount of organizational learning.[18] In designing the Boreal process, members of the FSC-Canada working group were mindful of recent FSC standard-setting experiences in the Maritimes and British Columbia. They were also mindful of the complexities and sensitivities associated with negotiating a single standard for all of Canada's boreal forests: a process that they understood could easily spin out of control.

FSC-Canada took the initiative in establishing the Boreal-standard process and commissioned Gillian McEachern to undertake a study canvassing the various models that might be used (McEachern 2000). Of the five possibilities identified in her report, FSC-Canada selected the "core standard with regional variation" option on the basis of four criteria – consistency, efficiency, learning, and continual improvement.[19] This national approach was intended to ensure that there was minimal variation in the provisions of the standard across the country, from Newfoundland and Labrador in the east to Alberta and British Columbia in the west. It was also intended to ensure that scarce resources of money, personnel, and expertise would be deployed to maximum effect, reducing the consultative burden on affected groups, especially First Nations. Finally, and in contrast to the processes in both the Maritimes and British Columbia, FSC-Canada was committed to crafting a standard that would maximize take up by major industry players. In its words: "rather than try and design a 'perfect' standard the first time around, we will strive to develop a 'good' standard, and to provide for ongoing improvements" (FSC-CAN 2002a, 3).

The challenges faced in developing a single standard for Canada's boreal forests were daunting. These forests span the entire northern region of the country and account for 35 percent of Canada's total land area and 77 percent of its total forested area. Not only are the boreal forests biophysically distinct – characterized by the dominance of a small number of tree species such as black spruce, balsam fir, trembling aspen, white birch, jack pine, white spruce, and larch – they are also home to approximately 80 percent of Canada's indigenous communities. Another key feature of the boreal forests is their socio-economic importance; they account for over 354,000 direct jobs, many associated with large integrated forest companies such as Abitibi Consolidated, Canfor, Domtar, Tolko, and Weyerhaeuser.

To manage this complexity, the FSC-Canada working group adopted a "centre-out" negotiating strategy that involved the preparation of three drafts of the Boreal standard over a period of two years. The process was managed by FSC-Canada, with responsibility devolved to a twelve-person Boreal Coordinating Committee (BCC) representative of FSC-Canada's four chambers – economic, environmental, social, and First Nations. The task of the BCC was to manage the standard-development process, develop drafts, review input, reconcile opposing views, commission expert advice, establish subcommittees as required, ensure effective communications, and meet the established timeline (FSC-Boreal 2004, 16-17). To this end, the BCC mounted a massive consultation exercise that built on the work that World Wildlife Fund-Canada and FSC-Canada had commenced in 1999 to develop an FSC-Ontario boreal standard. Important milestones included the hosting of an indigenous peoples and Forest Stewardship Council conference in Ottawa in August 2001; a major workshop at Cantley, Quebec, in September 2001 to agree on the framework for the negotiation process; and a national boreal forum, held in May 2003, to affirm support and generate final recommendations. In addition, regional committees (representing British Columbia, Alberta, Ontario, Quebec, Yukon, and the Maritimes) participated throughout the process, and certifiers from SmartWood, KPMG, SGS, and Woodmark conducted four field trials of the standard.

Aboriginal interests were closely and extensively involved in the negotiation of the Boreal standard. Their representatives on the BCC were Angus
Dickie (National Aboriginal Forestry Association) and Jim Webb (Little Red
River Cree). From this perspective, it would appear that the emerging standard has been well received. According to Webb, the "standard provides
Indigenous Peoples with an opportunity to cooperatively work with industry
in a manner acceptable to them towards mutually agreed solutions in the
forest. The challenge ahead is to communicate clearly and help Indigenous
communities benefit from this opportunity" (FSC-CAN 2003). Martin von
Mirbach, who chaired the BCC, concurred with this assessment, noting that
the FSC-Boreal standard is "very rigorous and comprehensive" and obligates
companies seeking certification "to reach agreement with each affected Aboriginal community that lives in or uses their management unit" (*Alternatives
Journal* 2004).

Harmonization was an important concern to FSC-Canada and the BCC. It
constituted a key criterion in the decision to select the "core standard with
regional variation" model set out in McEachern's 2000 report.[20] Given the
boreal standard's geographic applicability across Canada and the existence
of boreal forests in Scandinavia, eastern Europe, and Russia, harmonization
at three levels – internal, national, and international – was considered vital.
This was a daunting task and required an explicit harmonization strategy,
which was later set out in FSC-Canada's national boreal standards process
(FSC-CAN 2002a). A key component of the strategy was the preparation
of a detailed report that compared the boreal standard with regional and
national standards in Canada and elsewhere. The comparison, coordinated
by the director of FSC-Canada's standards program, Marcelo Levy, identified
numerous "significant differences" and "potentially significant differences"
between Canada's regional standards and those of Sweden, Finland, and
Russia (Levy, Roberntz, and Hagelberg 2003). Interestingly, and despite the
attention paid to it, harmonization of the boreal standard with other jurisdictions was not achieved. This was because, as Levy, Roberntz, and Hagelberg
note, "Every standard is developed with a given number of trade-offs that
members of the affected Working Groups or writing committees can 'live
with'" (9). For this reason, they argue that harmonization should be looked
upon as a longer-term, cyclical process, with comparative information collected in the first round of standards development feeding into the standard
revision process over time (9).

The Boreal process built on lessons learned from standard setting in other
jurisdictions, including British Columbia, the Maritimes, and the United
States. It differed from the BC and Maritimes processes in its explicit and
early commitment to a standard that would maximize large-industry buy-in,
and in the way it managed the consultation, development, and negotiation
processes. By establishing the BCC, FSC-Canada was able to insulate itself

from day-to-day responsibility for standards negotiation and development. It was therefore able to devote its time to "managing" the process at a macro-level to ensure that it did not become deadlocked. In contrast, in the BC and Maritimes cases, FSC-Canada played a much more limited managerial and facilitative role.

## Conclusion

In all of the standard-development processes we have examined, a key variable is leadership. In each instance, leadership played a particularly critical role in determining how and when standards development was initiated and the nature of the interests that were brought to the negotiating table. When it comes to explaining outcomes, however, in each case this "agency" element interacted with specific structural features in each region, including the region's prevailing ecological context, its history of forest management conflict, and, more broadly, the political economy of its forest sector.

In three of our comparator jurisdictions, the impetus to initiate FSC standard development came from the environmental movement. Thus in Sweden, the SSNC and WWF-Sweden played a leading role in the process that culminated in approval of FSC-Sweden's first national standard. The environmental movement played a similar role in the development of the Maritimes standard and the US Pacific Coast standard. Our remaining comparators depart from this pattern. In both the US Rocky Mountain and the Boreal cases, standard development was initiated and closely managed by FSC national offices.

Given the extraordinary and pioneering challenges confronted in virtually all of the processes examined, levelling procedural criticisms in hindsight at the results of these leadership efforts is hazardous, particularly given the resource constraints under which many laboured. Moreover, even though some of these processes led to serious organizational strains on, and indeed crises for, the broader FSC system, it is important to resist the temptation to judge their success by the "efficiency" with which they completed their mission. It is still too early to gauge these standards by what are, perhaps, the two most important metrics of success: namely, the extent to which the emerging standard accommodates the interests of relevant stakeholders over the longer term and the degree to which they result in well-managed forests in the future.

This said, in our view, the Boreal standard-development process would appear to have much to commend it, combining both the efficiency of the FSC-Rocky Mountain process with much of the substantive rigour and breadth of its BC and Maritimes counterparts. However, three features of the Boreal negotiations should be borne in mind by those inclined to promote it as a "model" process. The first is that it was extremely well funded by a $2.2 million grant from the Pew Charitable Trusts, which allowed it to hire staff, run consultation workshops, and underwrite the cost of a nation-wide process involving a huge number of individuals and organizations. Moreover, it was a process that built on accumulated experience in standards development elsewhere in Canada and beyond. In this regard, one of the important lessons learned from predecessor efforts was the need to distance the standard-setting team from the actual FSC board so that the standard negotiations could occur in a less "politicized" setting. Hence, FSC-Canada established the Boreal Consulting Committee, chaired by a senior environmentalist from the Sierra Club of Canada, Martin von Mirbach. Finally, despite ongoing attention, the FSC-Boreal process came no nearer than FSC-BC to achieving the practical harmonization of its standard with other regional standards in Canada or internationally. Practical harmonization risked upsetting the delicate regional and national balances achieved in existing standards, leading those responsible for it to conclude that practical harmonization is a much longer-term process than it had been conceived originally.

As noted above, it would be unfair to criticize other standard-development processes for not adopting the approach taken in Boreal negotiations since the Boreal process was itself the product of experience derived from its predecessors. At the same time, we would not want to suggest that the structural and process reforms implemented in the Boreal negotiations are the sole reason for its apparent success. Considerable credit must be given to those directly involved in the process and to their capacity to set a standard that satisfied the diversity of interests involved. Notwithstanding the advantages of accumulated experience and stable funding, there remained a significant potential for the process to go awry.

It is notable, too, that the FSC-BC process paralleled its FSC-Boreal counterpart in its early stages insofar as responsibility for standards negotiation was devolved to the standards team. It was only with the release of D2 that the ST ceased operation and the FSC-BC steering committee took over negotiations directly, advised by a smaller Technical Advisory Team. In hindsight, had funding allowed, perhaps the ST could have carried on with a renewed mandate and resolved all the outstanding issues, building on comments and the field-test results.[21] This, of course, did not occur, and what was already a political process became even more politicized, setting the stage for the complex machinations that surrounded approval of the third draft of the BC standard before culminating in the FSC-BC final standard. It is to a comparative analysis of this final standard that we now turn in Part 2.

# Part 2
# Analyzing the FSC Standard

Through its principles and criteria, the Forest Stewardship Council seeks to ensure that forest managers employ a model of forestry that is environmentally sound, socially responsible, and economically viable. To secure certification, managers must also satisfy a host of regionally and nationally developed conditions that aim to protect the integrity of the forest, empower local communities, and respect indigenous peoples' rights. When they aspired to craft an FSC-accredited standard that would translate these lofty principles into on-the-ground realities, the drafters of the FSC-BC final standard faced a number of challenges unique to the practice of forestry in British Columbia.

The four chapters that follow canvass these various challenges and highlight the solutions ultimately reached in order to acquaint the reader with the substance of the FSC-BC final standard and to locate it in its historical, social, political, and environmental context. To this end, Part 2 offers an in-depth analysis of four substantive themes: tenure, use rights, and benefits from the forest (Chapter 6); community and workers' rights (Chapter 7); indigenous peoples' rights (Chapter 8); and environmental values (Chapter 9). In each of these chapters, the BC final standard is assessed on its own merits and in relation to its counterpart standards in five FSC jurisdictions: Sweden, the Canadian Maritimes, the US Rocky Mountains, the US Pacific Coast, and the Canadian boreal region.

For ease of reference, it should be noted that, unless stated otherwise, all citations to FSC principles, criteria, and indicators for British Columbia are to the main FSC-BC final standard that was accredited by FSC-AC in October 2005 (FSC-BC 2005a). In most cases, our comparison draws on material at the criterion or indicator level; consequently, we cite the specific indicator as it appears in the text of the standard (i.e., Indicator 6.3.10). In some cases, however, a citation is to other material contained in the standard – such as an appendix, the glossary, or statements of intent. In this case, we provide the page number (i.e., FSC-BC 2005a, 77). We adopt the same system in the sections devoted to our comparator standards, citing them by the date when they were accredited by FSC-AC. Thus, unless otherwise noted, citations in these respective sections are to the Sweden standard (FSC-Sweden 1998, accredited November 1998), the Maritimes standard (FSC-M 2003, accredited March 2003), the US Rocky Mountain standard (FSC-RM 2001, accredited September 2001), the US Pacific Coast standard (FSC-PC 2003, accredited July 2003), and the Canadian Boreal standard (FSC-Boreal 2004, accredited August 2004).

# 6
# Tenure, Use Rights, and Benefits from the Forest

The Forest Stewardship Council (FSC) principles and criteria seek to promote a wide variety of economic, social, and environmental benefits. Such ambitious objectives are difficult to achieve unless a manager's rights to use forestland and resources are comprehensive, clearly defined, and adequately protected. As well, a prerequisite for FSC certification is that, in undertaking to achieve the required standards, a manager has the legal authority to make such commitments over the long term. The existence and force of this authority depends upon the institutional arrangements surrounding the forestland tenure and use rights in the relevant jurisdiction. Tenure arrangements determine the extent of managerial discretion and may influence such fundamental variables as the mix of forest products produced, management regimes and practices, and how benefits from the forest are distributed.

In British Columbia, approximately 96 percent of forestland is publicly owned, of which only 1 percent, including "lands Reserved for Indians," falls under federal jurisdiction (*Constitution Act, 1867*, s. 91[24]). The management of a majority of provincially owned forestland is delegated to the private sector through various forms of long-term licensing arrangements that are referred to generically as "Crown forest tenures" (Haley and Luckert 1990). Licences of this type account for close to 80 percent of the timber harvested on public land in British Columbia. Consequently, the nature of the BC forest tenure system and the degree and nature of control that the provincial government exercises over forest management are important considerations in the FSC standard-setting process.

The complex and, in many respects, unique attributes of the BC forest tenure system created serious difficulties for those negotiating the FSC standard for the province, and Principle 2 (P2) in the FSC-BC final standard remains controversial. Most problematic is the prevalence of tenure arrangements that do not provide forest use rights within the boundaries of a defined forest area but simply provide managers with a licence to harvest a designated volume of timber within an administrative land unit. Such arrangements, generally referred to as volume-based tenures, are arguably inconsistent with the requirement, in P2, that "long-term tenure and forest use rights *to the land and forest resources* shall be clearly defined, documented and legally established" (emphasis added). Yet failure to allow such tenures to be FSC certified would exclude a majority of the province's forestland base from FSC certification.[1] Another important issue is that most forestland in British Columbia is held under licences that only provide rights to harvest timber. The rights to all other forest resources are retained by the Crown – an arrangement that is contested by many citizens who claim customary rights to a variety of non-timber forest products.

Of further concern is the role played by forest tenure arrangements in determining a manager's ability to control the volume of timber that can be harvested annually or periodically. The rate of harvest is addressed under FSC Criterion 5.6 (C5.6). In British Columbia, all Crown forest tenures have an allowable annual cut (AAC) that is set by the province's chief forester. The rules governing the rate at which timber can be cut are important not only for sustaining forest ecosystems, but also because they affect the health and long-term viability of forest-based communities and the size and vigour of the province's timber economy. For many years, because of their ecological and economic importance, permissible timber-harvesting rates have been the source of major discord between environmental organizations on the one hand and the forest industry and government on the other. Consequently, during the negotiations leading up to the approval of the FSC-BC final standard, provisions governing the rate of harvest were exceptionally contentious.

The focal point of this chapter is forest tenures in British Columbia, the difficulties they present for successful FSC certification, and the strategies adopted by FSC-BC to surmount these challenges. In the next section, we will describe BC public policy with respect to timber allocation and harvest rate determination. This will be followed by an analysis of the FSC-BC indicators and verifiers that deal with forest tenure, use rights, and the rate of harvest of forest products, including their evolution, the issues they raise, and the difficulties that will be faced in their implementation. The chapter concludes with a comparison of the FSC-BC final standard for tenure, use rights, and the rate of harvest under Principles 2 and 5 with those standards in force in our five comparator FSC jurisdictions.

## Standard-Development Context

### Characteristics of British Columbia's Crown Forest Tenure System
In British Columbia, a complex mix of legislation, regulations, contractual agreements, and government policies that make up the forest tenure system are at the heart of the institutional arrangements controlling the governance, allocation, and use of the province's forest resources. The BC *Forest Act* makes

provision for eleven different types of Crown forest tenure under which har-vesting rights are granted; however, only five of these – tree farm licences (TFLs), forest licences, timber sale licences (TSLs), woodlot licences and com-munity forest agreements (CFAs) – are pertinent to the certification debate.[2]

A distinguishing feature of the Crown forest tenure system in British Columbia is the importance of volume-based licences. Such arrangements, which are rare in most other jurisdictions, including other Canadian prov-inces, account for almost half of British Columbia's AAC.[3] These licences, which are not area-specific and merely provide rights to harvest prescribed volumes of timber, include forest licences and renewable timber sale li-cences. Area-based tenures provide exclusive rights to harvest timber from a designated land base and include TFLs, TSLs, woodlot licences, and CFAs. Forest licences, the predominant volume-based tenures, account for about 58 percent of the province's AAC; TFLs, the major long-term area-based tenures, account for approximately 20 percent.

The various licensing agreements under which forest tenures are held in British Columbia differ in detail, but they have many common features. Table 6.1 summarizes the characteristics of the various tenure arrangements that are most significant from an FSC-certification perspective.

*Table 6.1*

**Some Important Characteristics of Crown Forest Tenures in BC**

| Tenure type | Provincial AAC Overall[a] | % | Average size of individual licences[b] | Area or volume allotment | Term (yrs) | Replace-ability (yrs) |
|---|---|---|---|---|---|---|
| Tree farm licence (TFL) | 16,832 | 19.6 | 526 | Area | 25 | 5 to 10 |
| Forest licence | | | | | | |
| Replaceable | 29,500 | 34.4 | 174 | Volume | ≤20 | 5 to 10 |
| Non-replaceable | 19,433 | 22.7 | 98 | Volume | ≤20 | Non-repl. |
| First Nations | 812 | 1 | 45 | Volume | ≤20 | Non-repl. |
| Timber sale licence (BC Timber Sales) | 14,211 | 16.6 | c. 40 | Area | ≤4 | Non-repl. |
| Woodlot licence | 1,100 | 1.3 | 1 | Area | ≤20 | 10 |
| Community forest agreement (CFA) | 510 | 0.6 | 18 | Area | 25-99[c] | 10 |
| Other | 3,324 | 3.8 | | | | |
| Total | 85,722 | 100 | | | | |

*Sources:* British Columbia Forest Act [RSBC 1996] chap. 157; British Columbia Ministry of Forests and Range 2008a, 2008b.
a  Thousands of cubic metres
b  Thousands of cubic metres of AAC
c  After 5-10 years probation

Apart from woodlot licences, neither area- nor volume-based tenures are limited in size, and many of the major industrial tenures tend to be very large indeed. The average size of TFLs, in terms of allocated AAC, is about 526,000 cubic metres, and the average AAC for forest licences is approximately 129,000 cubic metres. In contrast, the average AAC for woodlot licences is 1,000 cubic metres, with the corresponding average for CFAs about 18,000 cubic metres.

Rights granted under British Columbia's forest tenure system are limited in scope. With the exception of CFAs, the only right provided by the various forms of tenure is an exclusive right to harvest designated volumes of Crown timber. In the case of community forests, in addition to the right to Crown timber within the area covered by the agreement, the *Forest Act* also provides rights to "harvest, manage and charge fees for botanical forest products and other prescribed products," leaving the door open to the commercialization of a broader range of forest products.[4]

With the exception of CFAs – which, following a five- or ten-year probationary period, can be granted for a term up to ninety-nine years – no Crown forest tenure in British Columbia has a term exceeding twenty-five years, and most are less. Tree farm licences are granted for twenty-five-year terms, most replaceable forest licences for twenty years, woodlot licences for twenty years, and competitive TSLs for one to four years. Furthermore, no tenure is *renewable* when its term expires. Instead, most can be *replaced,* after a designated number of years, by a new licence that has the same term as the one it replaces. Tree farm licences and forest licences are replaceable five to ten years after the existing licence was granted. Thus, a TFL granted in 2000 with an expiry date in 2025 might come up for replacement in 2010. The replacement TFL would then run until 2035. For woodlot licences and CFAs, ten years must elapse before the tenure is replaceable. If licensees choose not to replace their licences in the designated year, the licence simply runs its term and then expires. At the time of replacement, the government can include in the new contract terms and conditions that it deems to be in the public interest; however, in the case of volume-based licences, the allocated AAC cannot be reduced. In the case of area-based licences, the area described in the original licence cannot be changed. These arrangements provide the government with an opportunity to change the terms of tenure contracts every five to ten years in order to reflect changing public priorities. On the other hand, if the new terms are unacceptable to a licensee, the licence may continue under its original conditions until its term expires.

Under the terms of their contracts, licensees must meet a broad range of operating requirements. Holders of long-term, area-based licences are assigned more management responsibilities than those holding volume-based licences. For example, tree farm licensees are responsible for total resource inventories, silviculture, protection and conservation of a broad range of non-timber resources, forest protection, timber supply analyses, integrated resources management, and road construction, maintenance, and deactivation. Cut controls, which regulate the volume of timber that can be harvested annually and periodically, are enforced for most tenure types. The only unregulated timber harvesting taking place in British Columbia is on private land and under timber licences not included in TFLs. All licensees are required to reforest all areas harvested to designated standards and meet the requirements of the *Forest and Range Practices Act.*[5]

Planning procedures differ depending on the form of tenure. For long-term, area-based tenures, a general management plan is required. For TFLs, this plan must be for at least twenty years, and it is updated and approved every five years. For all tenures, licensees must obtain a cutting permit before harvesting can proceed on a specific block of timber. The period of a cutting permit cannot exceed four years, and these permits are only issued after certain detailed planning requirements, provided for in the *Forest Act* and the *Forest and Range Practices Act,* have been fulfilled.

Day-to-day tenure administration and performance monitoring is carried out by government staff, but an independent body – the Forest Practices Board – is responsible for auditing forest practices of both government and licensees and for investigating complaints lodged by the public. The *Forest and Range Practices Act* contains a comprehensive compliance and enforcement regime. Significant features of this regime include powers of search and seizure and authority to issue administrative orders, such as stop-work and remediation orders. Maximum penalties include fines of up to $1,000,000, imprisonment for up to three years, or both.

### Rate of Timber Harvesting

One of the most important decisions facing forest managers, and certainly one of the most controversial policy issues in British Columbia, is how much timber should be harvested annually or periodically. This is a key factor in the sustainability of forests and, as such, is an important component of FSC-AC's forest-certification standards. However, because the forest industry is such a crucial part of the provincial economy, the Ministry of Forests (MOF) regards the establishment of harvesting rates as a strategic decision with profound implications for the province's prosperity, now and in the future. As we have seen, AACs for Crown forest tenures are set by the MOF and are an important element of each licence agreement.

The environmental movement consistently maintains that the province's forests are being "overcut." A 1999 report sponsored by the David Suzuki Foundation found that "the Allowable Annual Cut (AAC) allocated by the Ministry of Forests to tenure holders is substantially higher than the Ministry itself deems to be environmentally sustainable. By Ministry calculations, there is insufficient timber to sustain the level of logging currently undertaken, even if all other values of the forest are ignored. In the province as a whole, the Ministry estimates that current allocations of timber cutting rights exceed long-term timber supplies by about 20 percent. In some regions, the excess is very much above that level" (Marchak, Aycock, and Hebert 1999, 2). In a similar vein, the environmental organization ForestEthics has accused the provincial government of endangering forests by "operating under a policy of over-cutting" and stated that "current policies require logging levels far higher than the amount BC's Timber Harvesting Land Base can grow each year" (2003, 2).

To a large extent, the dispute over permissible timber-harvesting rates in British Columbia rests on the interpretation of "sustainable." At present, two concepts underlie the determination of the rate at which timber in the province can be harvested. The first, referred to as the "long-term harvesting level," or sometimes the "long-run sustainable yield," is the volume of timber that can be harvested from a management unit annually, or periodically, in perpetuity, taking into account the productivity of the unit's operable land base, the objectives of management, and the strategies and operational approaches that will be used to achieve these objectives.[6]

The second concept is the allowable annual cut (AAC), the officially authorized volume of timber that a manager is permitted to harvest annually from a specified management unit. Although long-term sustainable harvesting levels are principally a function of the productivity of the forestland devoted to timber production, determinations of AACs also take economic, social, and, some would maintain, political considerations into account.

Under the *Forest Act,* the chief forester is required to determine an AAC for Crown land in each timber supply area (TSA) and each TFL in the province every five years. Allowable annual cut determinations for woodlot licences and CFAs are the responsibility of either regional or district managers. The *Forest Act* specifies the factors that must be taken into account in determining AACs. These include the rate of timber production that may be sustained on the area; the short- and long-term implications to British Columbia of alternative rates of timber harvesting; the government's economic and social objectives for the area, the general region, and the province; and any abnormal infestations and devastations of timber in the area.[7]

Frequently, the chief forester's AAC determinations exceed sustainable long-term harvesting levels. For example, an AAC determination for the Arrowsmith TSA on Vancouver Island fixed the AAC at 373,700 cubic metres, 31 percent higher than the projected long-term sustainable harvest level of 286,250 cubic metres. Similarly, the Kamloops TSA in British Columbia's southern Interior has an AAC of 2,686,770 cubic metres, but a long-term harvesting rate of 2,246,000 cubic metres, a difference of 20 percent. The difference between the AAC and the long-term harvesting level is often referred to as the "fall-down effect." In 1998, following the first timber supply review, the difference, province-wide, between the AAC and the long-term harvesting rate was 15.6 percent (Marchak, Aycock, and Hebert 1999).

The apparent contradiction between a policy of sustained yield and allowable harvests that exceed long-term sustainable timber harvesting levels is the result of the age-class distribution of the trees in BC forests. In many management units there is a preponderance of old-growth stands that have achieved a much greater age than the planned harvesting age, or rotation, for subsequent crops. These forests, when harvested, yield a much greater volume of merchantable wood per hectare than will the shorter-rotation forests that are planned to succeed them. For example, a coastal old-growth forest of average site quality may yield 700 to 800 cubic metres per hectare or more, but the stands that succeed it, if harvested at an age of seventy to eighty years, may yield only 400 to 500 cubic metres per hectare.

The fact that primary, or old-growth, stands remain an important component of British Columbia's forest resources offers a range of options in terms of cut control policy. One alternative is to immediately fix the volume harvested at its long-term sustainable level. This option, favoured by environmentalists, has been embraced by the US Department of Agriculture Forest Service for the US National Forests. A second alternative, which has been adopted in British Columbia, is to permit more timber to be harvested now than is sustainable over the long term, allowing volumes harvested to gradually decline over time until a long-run steady state is achieved, where the volume of timber harvested annually will just equal the volume of timber the forest produces each year – i.e., its mean annual incremental growth. For most management areas, this reduction in AAC will be spread over several decades, the actual rate of decline being a policy decision. Policy dictates that no approved AAC will drive long-term future harvests below the rate at which they can be perpetually sustained. Because timber supply reviews are mandatory at five-year intervals, policy makers can adapt to improved information, changing social and economic imperatives, and technological advances as they occur, and adjust allowable annual timber harvests accordingly.

Environmentalists argue that the province's current timber-harvesting policies are an extension of the "liquidation-conversion project" that dominated BC forest policy throughout the latter half of the twentieth century (see Chapter 3). By favouring the accelerated liquidation of old-growth stands in order to create a "normal" second-growth forest, BC policy displays a timber bias and fails to acknowledge the many benefits attributable to the existence of old growth. On the other hand, an immediate reduction in harvests to their long-term sustainable level would cause an abrupt decline in economic activity, resulting in lower provincial revenues, reduced export earnings, job losses, and severe economic impacts in many forest-based communities. In 1999, the reduction in the provincial timber harvest from Crown land as a result of immediately reducing harvests to their long-term sustainable levels would have been about 16 to 18 percent for TSAs and 9 percent for TFLs (Marchak, Aycock, and Hebert 1999). Instead, current provincial government policy, according to the BC MOF, spreads the costs of declining harvests over time – often over periods of several decades – thus providing an opportunity for mitigating strategies to be planned and implemented. The fundamental philosophical differences between these two approaches are at the heart of the controversy surrounding the rate of harvest set out in FSC-BC's final standard.

## Highlights of the Final Standard

### Principle 2: Tenure and Use Rights
In the FSC final standard, Principle 2 and its three criteria are stated as follows:

*Principle 2: Tenure and Use Rights and Responsibility*
Long-term tenure and forest use rights to the land and forest resources shall be clearly defined, documented and legally established.

2.1 Clear long-term tenure and forest use rights to the land (e.g., land title, customary rights, or lease agreements) shall be clearly demonstrated.

2.2 Local communities with legal or customary tenure or use rights shall maintain control, to the extent necessary to protect their rights or resources, over forest operations unless they delegate control with free and informed consent to other agencies.

2.3 Appropriate mechanisms shall be employed to resolve disputes over tenure claims and use rights. The circumstances and status of any outstanding disputes will be explicitly considered in the certification evaluation. Disputes of substantial magnitude involving a significant number of interests will normally disqualify an operation from being certified.

Indicators under C2.1 recognize five alternative circumstances under which a manager in British Columbia may claim a legal right to manage the lands to which certification is sought. The first is the straightforward case of private land where the manager is named on the certificate of title to the area. A second is where the manager claims customary rights (e.g., Aboriginal title) to manage the land. A third arises if the manager has an area-based, replaceable tenure of sufficient duration to achieve the objectives set out in the management plan. The fourth alternative is when the manager holds a replaceable, volume-based tenure; and the fifth is where the licence held is non-replaceable. Where clear title is not held – which includes more than 80 percent of the managed forestland base in British Columbia – the manager is required to demonstrate that the province does not impose constraints that prevent the implementation of either the FSC-BC final standard or the management plan.

Of these three criteria, the first two are especially difficult to satisfy in the BC context. First, there are questions concerning the establishment of "clear and long-term tenures and forest use rights," particularly surrounding volume-based and non-replaceable tenures. Second, the definition of "legal and customary rights" remains a challenge and raises the issue of whether interested, or directly affected, persons should be regarded as having rights holders' status.[8] The third criterion is particularly germane to areas subject to First Nations' land claims.

### Volume-Based Tenures

Apart from issues of Aboriginal rights and title (discussed in Chapter 8), probably the most serious problem confronting FSC certification of Crown forests in British Columbia is the predominance of volume-based as opposed to area-based forest tenures. At first blush, volume-based tenures are inconsistent with P2's basic requirements that long-term tenure and use rights *to the land* be clearly defined, documented, and legally established. Failure to meet this fundamental requirement makes compliance with other FSC principles difficult, if not impossible, because tenure holders have no control over the land base once their harvesting rights have been exercised and their reforestation responsibilities fulfilled.

During the development of the FSC-BC final standard, some commentators argued that volume-based tenures should ordinarily not qualify for certification (FSC-BC TAT 2001). Nevertheless, because of the importance of volume-based tenures in the province, FSC-BC endeavoured to define conditions that would allow the certification of these Crown tenures. The final standard proposes two possible routes to the successful certification of volume-based tenures, both of which involve the cooperation of the province – as the landowner – in the certification process.

First, the manager and the province may jointly apply for certification. Following the release of D2 of the FSC-BC standard, this option was vigorously opposed by the BC MOF, which asserted that a joint application for certification by the Crown and the manager would imply that the government prefers the FSC to other certification bodies and would require legislative changes to the tenure system (Pedersen 2001). Furthermore, if future licensees are required to adhere to the FSC-BC final standard, it was argued that this would render certification compulsory, a requirement that would be contrary to both public policy in British Columbia and the intent of the FSC.

The second route requires the manager to demonstrate that the certification application supports the province and other tenure holders within the TSA; that all permits, licences, and plans issued by the MOF for the management unit adhere to the relevant portions of the FSC-BC final standard and the FSC-approved management and operational plans for the area; and that the chief forester, subsequent to initial certification, has established an AAC specific to the management unit being considered for certification. This option also calls for significant involvement of the MOF and for legislative changes that are unlikely to be imminently approved. For example, significant amendments to the *Forest Act* would be needed if the chief forester were to partition the harvesting rights in a TSA in order to allocate an AAC, calculated according to FSC criteria, to a particular licensee within a designated area.

Both of these proposed approaches to the certification of volume-based tenures assume that a specific area subject to a certification application can be designated within a TSA. In order to accomplish this, the FSC-BC requires that "a map showing the manager's chart/operating area, accompanied by written confirmation from the Province of the manager's rights in this area," is included in the application (Indicator 2.1.2). Although forest licence holders do have so-called chart areas in which they are likely to be granted cutting permits within a TSA, these are informal arrangements between licensees and the Crown and have no legal standing. For the province to confirm a manager's rights in such an area would be to create a de facto area-based tenure that would limit the MOF's discretion in issuing cutting permits and managing the TSA as a whole.

However, in addition to these two alternatives, there are other potential vehicles available for certifying volume-based tenures that FSC-BC may wish to explore in the future; namely, innovative forestry practices agreements (IFPAs) and cooperative management among forest licensees within a TSA.

Holders of volume-based licences may enter into IFPAs to improve the productivity of the forest resource. An IFPA, which cover periods of up to twelve years, gives the licence holder territorial responsibilities for the management of the forest during its term. In order for an IFPA to be granted, there must be an approved area-based management plan that describes the innovative practices to be carried out and explains how these will improve the productivity of the timber resource sufficiently to justify an increase in the AAC of the participant's licence. Innovative practices embrace a number of silvicultural and forest inventory initiatives and "activities that will enhance other resource values, including, but not limited to, water, fisheries, wildlife, biological diversity, soil productivity and stability, forage production, grazing and recreation values."[9] IFPAs may enable certification of volume-based tenures, as they are area-based and allow for forestry practices that could form the basis of a management plan acceptable to FSC certifiers.

An even more promising certification alternative for volume-based tenures is one in which a group of licensees within a TSA agree to cooperatively manage the management unit, or a portion thereof, and jointly apply for certification. There are already instances of cooperative management in the province that involve various volume-based Crown tenure holders working together for the purpose of certification. In both the Kamloops and Merritt TSAs, for example, the licensees (twelve in the Kamloops TSA and fourteen in the Merritt TSA) have worked with local stakeholder groups, First Nations, and the public to develop sustainable forest management plans. Once these plans and environmental management systems are in place, licensees will be able to apply for certification for their individual operating areas. Certification under the Canadian Standards Association is currently the objective, but there seems to be no reason why such cooperative management could not be used as a basis for FSC certification. Although this approach to the certification of forest licences is not explicitly mentioned in the FSC-BC final standard, it was cited as a promising solution by an FSC-BC executive in a interview conducted by Ecotrust for that group's recent review of the certification of forest and range agreements (Ecotrust 2005; and see Chapter 8).

*Limited Rights*

A second serious challenge confronting the certification of forests in British Columbia is the limited nature of the rights conferred on tenure holders. As we have seen, with the exception of CFAs, Crown forest tenures in the province merely provide the right to harvest a specified volume of timber; all other rights in the land and forest resources are retained by the Crown. Tenure holders do not even hold rights to the productivity of the land for timber production and, therefore, have no equity in stands that are established, either naturally or by planting, following harvesting (Haley and Luckert 1998). The FSC-BC final standard requires that the manager have "the legal right to manage the lands and to utilize the forest resources for which certification is sought," so it would seem that BC forest tenure holders can only seek certification for their timber-harvesting operations (Indicator 2.1.1). This constraint appears incompatible with FSC P5, which requires the efficient use of the "forest's multiple products and services to ensure economic viability and a wide range of environmental and social benefits." Crown forest tenure holders may prepare, and present to certifiers, management plans that incorporate strategies designed to "produce a diversity of non-timber forest products compatible with site conditions and local economic objectives," but they do not have the legal authority to manage for such products (Indicator 5.4.2). Nor, under any circumstances, can they benefit financially from their sale.

The BC government has legal authority to exercise its rights in non-timber forest products produced on Crown land at any time, without consideration for existing management plans prepared by Crown forest tenure holders. For example, licences to conduct commercial recreational operations can be, and in fact are, issued on all classes of Crown land. The government could, and probably will at some future date, regulate mushroom harvesters or sell rights to harvest mushrooms or other non-timber botanical products. In other words, overlapping tenures can be established on the same land base. If the government were to recognize integrated management plans prepared for certification purposes, its ability to exercise its legal rights to non-timber forest products and services would be severely constrained. Yet, without such recognition, considerable uncertainty surrounds the certification process. These important constraints on the certification of Crown forest tenures have not been addressed by FSC-BC, although they were raised by the former chief forester in his critique of an earlier draft of the BC standard (Pedersen 2001).

### Non-Replaceable Tenures

A final area of concern in the application of C2.1 is tenure arrangements that are non-replaceable. These include certain forest licences and most timber sale licences.

Non-replaceable forest licences do not make up much of the province's total AAC, but they tend to be held by licensees – including both First Nations and non-First Nation communities – who are concerned about promoting sustainable forest practices and have an interest in becoming certified. However, a licensee holding a licence with a term of fifteen years or less that is neither renewable nor replaceable cannot be a candidate for FSC certification.

Non-replaceable TSLs, on the other hand, currently account for about 15 percent of the province's AAC but are scheduled to rise to more than 20 percent of the AAC as the province's new Market-Based Timber Pricing System is fully implemented. Non-replaceable TSLs are area-based and generally run for terms of less than four years. On termination, the land occupied by the licensee reverts back to the Crown. TSLs are accessed through competitive bidding processes administered by BC Timber Sales, a government agency, and are widely distributed throughout the province. It is clear that licensees cannot certify TSLs, given the short term of the licence. Only the government itself could seek certification of this class of Crown forest tenure, as the FSC-BC final standard recognizes. In 1999, the Forest Enterprise Branch (recently reorganized as BC Timber Sales) explored the readiness of its program for certification under ISO 14001, CSA, and FSC protocols. In 2001, the branch embarked on a series of pilot projects to test these three major certification schemes throughout British Columbia. As of 2006, it has certified all of its operations under ISO 14001, making it the largest government forest program to be so certified.[10]

### Customary Rights

FSC Criterion 2.2 addresses overlapping property rights in an area that is the subject of a certification application. In British Columbia, these might include rights to carry on certain recreational activities, water licences issued under the *Water Act,* grazing licences or permits issued under the *Range Act,* or a registered trapline or guide-outfitter licence issued under the *Wildlife Act*.[11] If the holders of overlapping property rights have legal title or a legal licence or lease agreement, the issue is straightforward. In the case of customary rights, however, the issues are less clear. FSC-AC has defined customary rights as "rights which result from a long series of habitual and customary actions, constantly repeated, which have by such repetition and by uninterrupted acquiescence, acquired the force of a law within a geographical or sociological unit" (FSC-BC 2005a, 62).

In British Columbia, customary rights have two dimensions – those claimed by First Nations and those claimed by non-First Nation communities and individuals. The customary rights of First Nations are recognized in section 35 of Canada's *Constitution Act* and have been upheld by the BC courts and by the Supreme Court of Canada. Reconciling customary Aboriginal rights and the management and use of forest resources is an important component of the FSC certification process that is addressed under P3. These questions are explored in depth in Chapter 8.

The customary rights claimed by non-Aboriginal people in British Columbia are something of a grey area, and there are common misunderstandings concerning the legal status of Crown land and the nature of customary rights. There is a common perception that Crown land is owned by "the people" and, consequently, is freely open for recreational activities and such pursuits as mushroom picking, salal harvesting, and the collection of other botanical products. In fact, formal ownership of all Crown land is vested in the BC government, which retains rights to all products of the land unless they are transferred to a third party by means of a statutory instrument, such as a lease or licence. As mentioned earlier, most Crown forest tenures in the province only provide rights to harvest timber; they do not convey rights to the land itself or to any other products of the land. The government reserves the right to dispose of these products as it sees fit.

It is true that people throughout the province use Crown land for many purposes and have done so for decades, unimpeded and without charge. Such usage in itself does not establish a common or customary right, although it may be used to substantiate a claim to such a right. A successful challenge of this kind has never been mounted (at least not by a non-First Nations plaintiff), but it would not be surprising to see such challenges arise as forest resources become increasingly scarce and demand pressures continue to grow. In the meantime, until such a claim is made, is upheld by the courts, and becomes a part of the common law, it appears that the use of Crown land by those who do not hold a legal permit, licence, or lease is a privilege that can be suspended by the Crown, as owner of the land, at any time.

This grey area is complicated by the FSC-BC final standard, which confuses "customary rights," as defined by FSC-AC, with the interests of those persons who consider themselves to be "directly affected" by forest management operations. The FSC-BC final standard requires the manager, in consultation with local people, to identify, document, and, where appropriate, map "any legal and customary rights in the management unit held by one or more people who reside within or adjacent to it." Thus, where FSC Criterion 2.2 refers to legal or customary rights held by communities, FSC-BC extends this concern to rights held by individuals or groups of unspecified size. While it is acceptable to take into consideration legal rights held by individuals – rights to a trapline or water rights, for example – extending this notion to customary rights is problematic in that it would enable any individual or group to claim a customary right based on habitual and unimpeded usage. This interpretation is reinforced by written comments submitted to FSC-BC by interested parties during the development of the standard (FSC-BC TAT 2001, 19). Several respondents seemed to interpret "customary rights holders" as synonymous with "directly affected persons," who include, as defined by FSC-BC, people or groups who "consider themselves directly affected by the proposed or current operations" (FSC-BC 2005a, 63). For example, one respondent suggested that "it is important that all people that consider themselves 'directly affected' be included in consultation." Another, representing an FSC-accredited certifier, maintained that "'legal or customary rights' should explicitly include recreational users of Crown land (e.g., campers, hikers, etc.)." A labour organization submitted that long-distance workers should be included in the definition of those who hold legal or customary rights (FSC-BC TAT 2001, 19).

Resolving questions surrounding the definition and recognition of customary rights is important to the successful implementation of the FSC-BC final standard. If holders of legal or customary rights remain dissatisfied that current or proposed management fails to adequately protect their rights, the matter must be dealt with through the dispute mechanisms required under Criterion 2.3 of the standard. If substantial disputes remain unresolved, certification may be denied.

Although FSC-BC's indicators for accommodating customary rights are generally acceptable to environmental interests, serious misgivings raised by industry and government have yet to be addressed. Some industry representatives have expressed concern that the definition of customary rights is too broad and could lead to unreasonable and frivolous claims (FSC-BC TAT 2001, 20) and that the dispute mechanism, which demands that firms go beyond the legally stipulated requirements concerning land-use disputes, could potentially add a costly administrative burden (Spalding 2002). The BC government, as one might expect given the nature of forest-resource ownership in the province, has declared that the "only legally established rights are [those] recognized by the government" (Pedersen 2001). It has further asserted that the dispute mechanism essentially usurps the government's responsibility to make "the necessary [land use] choices based upon the social, economic and environmental needs of society" (Pedersen 2001). Such issues, it is argued, often have implications extending far beyond the manager and the local disputant. Ultimately, it is argued that "government ... has the final decision over Crown land" and "must continue to maintain the decision-making power overall" (Pedersen 2001).

### Principle 5: The Rate of Harvest

The rate of harvest is dealt with under FSC-AC Criterion 5.6: "The rate of harvesting of forest products shall not exceed levels which can be permanently sustained." It is bolstered by five indicators. Four of these address the rate at which timber is harvested; the fifth deals with the sustainable harvesting of non-timber forest products.

The final standard provides a detailed list of factors that must be considered in setting the rate of timber harvest (Indicator 5.6.1). These include up-to-date inventories and information on growth and yield; volume reductions to account for the maintenance of ecosystem integrity; non-recoverable losses such as those resulting from fires, insects, and disease; and reductions required to protect non-timber values and forest-dependent economic activities. These prerequisites are similar to the statutory requirements provided for under British Columbia's forestry legislation and regulations.

Moreover, the final standard requires that managers conduct sensitivity analyses of timber harvest rates that take into account the uncertainty surrounding inventory data, management assumptions, and growth and yield projections (Indicator 5.6.2). It also requires that adequate margins of safety be observed when these analyses indicate significant impacts on timber supply. Such procedures are currently standard in the BC MOF's timber supply analyses and are acceptable to all parties.

There are also detailed provisions with respect to determining AACs (C5.6; Indicators 5.6.4, 5.6.5, and 5.6.6). A manager is required to demonstrate that "present and projected annual timber harvests over the next decade, and the average of timber harvests over all subsequent decades, *do not exceed the projected long-term harvest rate,* while meeting the FSC-BC Regional Standards over the long-term" (emphasis added).

This requirement implies that current AACs, if they exceed long-term sustainable levels, must be immediately lowered in order to meet the certification standards. Furthermore, after ten years of FSC certification, a manager must demonstrate that actual average timber harvests since certification have not exceeded projected long-term harvest rates. Some variations in cutting rates are allowed, but the actual rate of harvest in any given year must be no more than 25 percent above the projected long-term harvest rate, unless an equivalent amount below the long-term harvest level has been cut within the last ten years and subsequent to FSC certification, or unless the long-term annual harvest level of the management unit concerned is less than 10,000 cubic metres. Finally, the manager may elevate the rate of harvest to accommodate catastrophic events, such as fires or insect attacks, provided a public consultation has taken place, the manager has addressed the concerns of directly affected persons, and the five-year average cut does not exceed the projected long-term harvest level.

The FSC-BC final standard's requirement for an immediate decline in harvests to sustainable long-term levels is at odds with current public policy in British Columbia. In most management units, this could mean an immediate 15 to 25 percent drop in the AAC, resulting in significant economic and social impacts that would be politically controversial and would likely meet with vigorous opposition from labour, the forest industry, and affected communities. The BC steering committee worked hard to craft C5.6 indicators that would be acceptable to all parties; in fact, the second draft of the standard (FSC-BC 2001) proposed indicators that were generally acceptable. However, these indicators were one of the main causes of the impasse that resulted in the industry representative's refusal to sign off on the preliminary standard. In the final standard, the indicators under C5.6 remain essentially unchanged.

### Regional Comparisons
In this section, we compare the FSC-BC final standard to FSC standards for Sweden, the Maritimes, the US Rocky Mountain, the US Pacific Coast, and the Canadian Boreal, with particular emphasis on the nature of land-use tenures, the recognition and protection of customary rights (see Table 6.2) and the determination of harvesting rates (see Table 6.3).

### The FSC-Sweden Standard
The majority of forestland in Sweden (74 percent) is privately owned, so the question of establishing long-term tenure and use rights is less of an issue than it is in the Canadian context. Leaving aside the vexed issue of indigenous Sami rights to access forest resources (see Chapter 8), the central uncertainties and potential conflicts in Sweden concern forests owned by communities and the church (where title may be held in common), and private landlord-tenant lease agreements.

The current FSC-Sweden standard (1998) does not address the question of long-term tenure and use rights. This gap is remedied, to some degree, by a pending new draft standard, which requires that the right to manage forestland be substantiated by ownership or legally binding contract. Under this draft standard, if a management unit is jointly owned, all the owners

*Table 6.2*

## Comparison of protection of customary rights

### FSC-AC

2.2 Local communities with legal or customary tenure or use rights shall maintain control, to the extent necessary to protect their rights or resources, over forest operations unless they delegate control with free and informed consent to other agencies.

### FSC-BC

- In consultation with local people, the manager identifies any legal or customary tenure or use rights in the management unit held by people who reside within or adjacent to it.
- The manager obtains free and informed consent from local rights holders to any portion of the management plan that affects their rights and resources; if the local rights holders dispute that current or proposed management protects their rights and resources, the manager implements recommendations developed through a Criterion 2.3 dispute resolution process.

### FSC-Sweden

- The forest is accessible to everyone through a broad right of public access.
- Ownership or other disputes with respect to the right to carry on forestry will preclude certification.

### FSC-Maritimes

- Those who have recognized legal or customary tenure, or traditional use rights, are identified.
- The impacts of proposed forest management operations on such uses are evaluated.
- All holders of such rights have access to information about current and proposed management activities that may affect their use rights.
- There is evidence that free and informed consent to forest management activities affecting legal, customary, or traditional use rights has been given by affected groups and individuals and that their interests have been accommodated.

### FSC-Rocky Mountain

- The forest owner or manager allows well-established customary and lawful uses of the forest to the extent that they are consistent with the conservation of the forest resource and the objectives as stated in the management plan.
- Consultation occurs on ownerships where customary use rights and traditional and cultural areas/sites exist.

### FSC-Pacific Coast

- Forest owners or managers allow lawful and customary uses of the forest to the extent they are consistent with the conservation of forest resources and the stated objectives in the management plan and they do not present a legal liability.
- Consultation occurs on ownerships where customary use rights and traditional and cultural areas/sites exist.

▶

◀  *Table 6.2*  ·

---

**FSC-Boreal**
- Customary tenure or resource use rights held by communities are identified and documented. Local communities with legal or customary tenure or use rights either retain control over their forest operations or give free and informed consent to any portion of the management plan that affects their rights and resources.

---

*Table 6.3*

---

**Comparison of rate of harvest**

**FSC-AC**
5.6 The rate of harvest of forest products shall not exceed levels that can be permanently sustained.

**FSC-BC**
- The rate of timber harvest for the management unit is based on an analysis incorporating a prescriptive list of factors.
- The rate of timber harvest is determined in a manner that adequately reflects reliability and uncertainty associated with inventory data, management assumptions, growth and yield projections, and analysis methodologies.
- The manager ensures that the rate of harvest reflects the best available inventory and productivity data, provides for sustainable production, and is adjusted when monitoring indicates over-harvesting.
- The manager demonstrates that the average of the present and projected timber harvests over the next decade, and the averages of projected timber harvests over all subsequent decades, do not exceed the projected long-term harvest rate, while meeting the FSC-BC standard over the long term.

**FSC-Sweden**
- The rate of harvest must be sustainable in the long term.

**FSC-Maritimes**
- The rate of forest product harvest is determined for defined forest areas and is derived from a list of factors. Harvest-level determinations are publicly available. Harvest levels are monitored and accurately reported.
- Rates of harvest of any forest product must be sustainable within ecological limits.
- A pre-harvest inventory or sale area reconnaissance must be implemented.
- An operating/harvesting plan must be written, available, and used in the forest and must include silvicultural guidelines, volume and basal area targets, residual species composition, tree-marking guidelines, and transportation and access guidelines.

**FSC-Rocky Mountain**
- Harvest volumes remain within the range of the periodic allowable cut and the long-run sustainable yield (LRSY) figures, as established in the forest plan. Standing inventory and actual growth rates are sufficient to maintain the LRSY.

▶

◀ *Table 6.3*

---

- If, due to historical circumstances, inventory and/or growth rates are not sufficient to maintain the LRSY, the harvest rate is restricted to that which will produce compliance as soon as reasonably possible.

**FSC-Pacific Coast**
- The level of sustainable harvest is based on clearly documented projections that use growth and regeneration data, site index models, and the classification of soils. The level of documentation is determined by the scale and intensity of the operation.
- Growth rates equal or exceed average harvest rates over rolling periods of no more than 10 years. In cases where owners or managers harvest timber at intervals longer than 10 years, the allowable harvest is determined by the target stocking levels and the volume of regrowth since the previous harvest.
- The rate and methods of harvest lead to well-stocked stands across the forest management unit (FMU). Under-stocked and over-stocked stands are returned to fully stocked levels at the earliest practicable time.

**FSC-Boreal**
- The applicant demonstrates that the calculation of harvest rates of forest products is based on a list of factors.
- The applicant demonstrates that the analysis and calculation of harvest rates of forest products accurately reflects the requirements under other indicators.
- The wood-supply modelling exercise, in which sustainable harvest levels are identified, has been subjected to peer review.
- Actual harvest rates for timber do not exceed planned average levels.

---

must be committed to certification; failure to meet this requirement will normally disqualify a certification application (Indicator 2.1.1).

Under customary Swedish law, any individual is entitled to enter any forestland – public or private – for a variety of reasons, including travel and gathering berries and mushrooms.[12] This right is acknowledged in the 1998 Swedish standard, which states that the "forest is open to one and all by means of the right of public access," provided that the landowner and residents on the property do not have their privacy disturbed nor their economic interests damaged (Indicator 7.1.3). Any legal claims beyond these established customary rights must be acknowledged, and the landowner must consult with the parties concerned about any forest operations that have significant consequences for these legal claims. No formal mechanisms for resolving disputes over tenure claims and use rights are established. Managers must simply try to resolve ongoing disputes through consultation with the parties concerned. They must also document the process. As for rate of harvest, the Swedish standard is succinct if vague, simply requiring that harvesting be "sustainable in the long-term" (Indicator 7.2.3).

## The FSC-Maritimes (FSC-M) Standard

The Maritimes region differs from other regions in Canada in that substantial areas of forestland are privately owned. For the region as a whole, 80 percent of forestland is publicly owned; however, this figure varies greatly between provinces, from 99 percent in Newfoundland, to 50 percent in New Brunswick, 32 percent in Nova Scotia, and 11 percent in Prince Edward Island. In contrast to the practice in British Columbia, timber-harvesting rights to Crown forests in Newfoundland, Nova Scotia, and New Brunswick are mainly granted through long-term, area-based tenures.

As a consequence of the relatively simple tenure arrangements in the Maritimes, the Maritimes standard is more straightforward than its British Columbia counterpart, simply requiring proof of "legal, long-term (or renewable) rights to manage the land and/or utilize forest resources," evidence of "due diligence in establishing clear title," and documentation of First Nations' or others' claims to the land (C2.1 indicators).

The FSC-Maritimes' approach to protecting customary rights is similar to that for the BC and Boreal regions. Legal and customary tenure or use rights must be identified and documented, and the impacts of proposed forest management operations on these rights evaluated (Indicator 2.2). Implementation of a management plan may only proceed with free and informed consent of the affected rights holders. As in the FSC-BC final standard, the Maritimes standard does not restrict consideration of customary rights to those held by communities but explicitly includes individual stakeholders.

Compared to the BC final standard, the FSC-M standard deals with dispute resolution in a general manner, simply requiring documentation of any ongoing disputes and a commitment to their resolution (Indicator 2.3).

With respect to the rate of timber harvesting, the FSC-M standard does not invoke the traditional concept of a sustainable harvest or yield over the long run, but rather requires that the sustainable harvest of all forest products be based on conservative growth and yield data and long-term ecosystem-based planning that "considers ecological factors, community stability, stakeholder input, and a full range of resource values and constraints" (C5.6.1 indicators). Harvest-level determinations must be made publicly available.

## The FSC-Rocky Mountain (FSC-RM) Standard

In the United States, unlike Canada, public forestlands are held and managed by federal, state, county, and municipal governments. Harvesting rights to public timber are not conveyed to the private sector through leasing and licensing arrangements but are sold on a competitive basis in the form of short-term timber sales that impose few, if any, management obligations on purchasers. In the Rocky Mountain region, approximately 80 percent of forestland, including tribal, is in private ownership; only 20 percent is public.

The current FSC-RM standard is directed at private lands, tribal lands, and public lands other than federal.

Compared to the BC standard, the FSC-RM standard's indicators under P2 are concise and general and reflect the importance of private ownership of commercial forestland in the region. The standard simply requires evidence of long-term tenure and use rights and clearly identified boundaries (C2.1). With respect to customary rights, the standard incorporates constraints on their recognition that reflect a stronger commitment to the protection of private rights than is found in Canadian jurisdictions. For example, certifiers are cautioned in a note that the "provisions of this criterion shall not abridge or compromise the legal rights of private property owners" (FSC-RM 2001, 2). Managers are given the discretion to *allow* "well established customary and lawful uses of the forest to the extent that they are consistent with the conservation of the forest resource and the objectives as stated in the management plan" (Indicator 2.2.a). By way of example, the FSC-RM standard notes that hunting, hiking, and fishing may be allowed to take place on *non-posted* property. There is no requirement that the free and informed consent of legal or customary rights holders be obtained if management operations threaten their interests, but simply an obligation to consult with concerned groups during the planning and implementation of forest management activities (Indicator 2.2.b).

The standard does not spell out how to establish and implement dispute mechanisms. Forest managers or owners must initially attempt to resolve disputes through "communication, negotiation and/or mediation" and, if these fail, through "federal, state, local, and/or tribal laws" (Indicator 2.3.a). Unresolved disputes must be reported to the certifying body (Indicator 2.3.b).

Like the FSC-BC final standard, the FSC-RM standard requires that harvest volumes be commensurate with the "long-run sustainable yield (LRSY)" of the management unit (Indicator 5.6.a). However, the methodology used to determine allowable harvests, and criteria to verify that actual harvests conform to sustainable targets, are not prescribed in as much detail.

### The FSC-Pacific Coast (FSC-PC) Standard

Although approximately 60 percent of forestland in the Pacific Coast region is publicly owned and managed by federal, state, county, and municipal governments, the FSC-PC standard, like the FSC-RM standard, is mainly directed at private forestlands that make up the majority of the region's commercial forests. The FSC-PC standard is similar in structure and wording to the FSC-RM standard and reflects the extent to which private property rights are entrenched in US law and tradition. It is also far less detailed and specific than its BC counterpart.

Other than the clear identification of land boundaries, the requirements for establishing long-term use rights are not explicit. As in the FSC-RM standard, under the FSC-PC standard the forest owner or manager should allow "customary and lawful uses of the forest to the extent they are consistent with conservation of the forest resource, forest management objectives, and do not present a legal liability" (Indicator 2.2.b). Where customary use rights exist, consultation with stakeholders is required, but there is no explicit mention of "free and informed consent."

Dispute resolution mechanisms are identical to those in the FSC-RM standard. Initially, disputes must be dealt with directly by the forest owner or manager through negotiation and/or mediation. If this results in failure, the owner/manager must turn to established legal processes (Indicator 2.3.a). Disputes must be reported to the certifying body (Indicator 2.3.b).

As in British Columbia and the Rocky Mountain regions, the rate of timber harvested must not exceed the long-term sustainable harvest rate. Growth rates must generally equal or exceed harvest rates over rolling periods of no more than ten years (Indicator 5.6.b). Methods of determining sustainable harvests are expressed in traditional timber-management terms, with no explicit recognition of non-timber values. Clearly documented yield projections must be based on regeneration and growth data, site index models, and the classification of soils. The rate and method of harvest must lead to "well stocked stands across the forest management unit." Under-stocked and over-stocked stands must be "returned to fully stocked levels at the earliest practicable time" (Indicator 5.6.c).

## The FSC-Boreal Standard

The Canadian Boreal region spans eleven jurisdictions – nine provinces and two territories – each with its own forest laws and regulations. As a result, there were significant challenges to the development of a common standard. The final result is a document that generally applies common performance-based standards across the whole region while reserving broad discretion for certifiers to consider jurisdictional or local requirements. Compared to the FSC-BC final standard, the Boreal standard is shorter and relies less on detailed indicators than on short, explicit indicators supported by more detailed verifiers.

Although forestland in the Boreal region is mainly in public ownership, tenure systems across the region's many jurisdictions are much simpler than those in British Columbia. As in British Columbia, the rights granted under the tenure system are largely confined to timber harvesting rights, but these are mainly held by large forest-products manufacturing companies under long-term, renewable, area-based leasing and licensing arrangements. Volume-based tenures, a hallmark of BC forest policy, are of little importance in the

Boreal region. For example, less than 20 percent of the AAC from BC Crown land is allocated by means of long-term, area-based tenures, compared to 70 percent in Alberta, over 80 percent in Saskatchewan and Manitoba, and 100 percent in Ontario and Quebec. As a result, the Boreal standard has one simple indicator (Indicator 2.1.1), which requires that either land ownership or forest use rights to the land be demonstrated, compared to the BC final standard, which has four indicators addressing the various nuances of the province's unique tenure arrangements.

The approaches taken to protect customary rights in the Boreal standard are similar to those for British Columbia. In each case, customary and legal tenure or use rights in the management unit must be identified and documented, and, insofar as the proposed management plan affects these rights, implementation of the plan may only proceed with the free and informed consent of the affected rights holders. However, whereas the BC standard explicitly includes rights held by groups or individuals who reside in or adjacent to the management unit, the Boreal standard restricts consideration of customary rights to those held by communities.

The Boreal and BC standards deal with the dispute resolution process in a similar manner, except that the Boreal standard does not require disputants to be involved in the development and implementation of the process (C2.3), while the BC standard describes in more detail those factors that must be considered in determining whether an outstanding dispute is substantial enough in terms of its "magnitude and extent" to stall or prevent certification.

In some respects, the BC and Boreal standards take a comparable approach to dealing with the rate of timber harvest, although the BC final standard is more detailed. Each starts with a description of the factors that must be taken into account in determining harvesting rates and then prescribes how analyses should proceed and the criteria that must be used to ensure that the rate of harvest falls within prescribed limits. Unlike the BC standard, however, the Boreal standard does not explicitly require allowable rates of harvest to reflect the "projected long-term harvest rate" for the management unit but simply demands that "harvest rates of forest products accurately reflect the requirements under other Indicators" and that managers submit their calculations of sustainable harvest levels to peer review (Indicators 5.6.2 and 5.6.3).

**Conclusion**

In this chapter we identified some of the difficulties that the BC forest tenure regime presents for successful FSC certification, and we reviewed the contentious issue of rate-of-harvest calculations. Through comparison with five other jurisdictions, we highlighted the challenges faced and the creative solutions employed to bring the FSC-BC final standard into compliance with FSC Principles 2 and 5.

The first criterion under P2 requires that forest managers have long-term use rights to the land in question that are comprehensive, clearly defined, and adequately protected. These requirements present serious obstacles to the certification process in British Columbia. In the case of privately owned forestland, which predominates in the United States and Sweden and is important in the Maritimes region, proof of legal title is generally all that is necessary to confirm that managers have the authority to commit themselves to FSC standards over the long term. However, in most Canadian provinces, including British Columbia, a significant proportion of forestland is publicly owned, and use rights depend on the nature of the licensing, or tenure, arrangements under which public timber is harvested and forestland managed. The Maritimes and Boreal regions have tenure systems that are relatively simple and generally provide for long-term, renewable rights over clearly identified land areas. Such arrangements allow managers to commit themselves to long-term certified management plans and stand in marked contrast to the BC situation, where long-term, replaceable, area-based Crown forest tenures account for less than 20 percent of the provincial AAC.

The most challenging feature of the BC Crown forest tenure system is the importance of "volume-based" licences, which account for about 58 percent of the AAC. These arrangements do not grant forest use rights within the boundaries of a defined area, but simply provide managers with a licence to harvest a designated volume of timber within an administrative forest management unit. The FSC-BC final standard provides two alternative approaches to the certification of volume-based tenures, both of which require the provincial government's participation in the certification application – a role that, to date, it has been reluctant to assume.

Another element of the BC tenure system, unique among Canadian provinces, is the importance of non-renewable, non-replaceable licences. Such arrangements currently account for about 40 percent of the province's AAC. The FSC-BC final standard recognizes that the only vehicle available for the certification of such licenses is a direct application by the provincial government itself.

A second important element of P2, articulated in C2.2, is the protection of groups or individuals holding legal or customary rights that overlap the management area. In all Canadian jurisdictions, the recognition of non-Aboriginal customary rights presents significant questions and challenges. There is a widespread perception in Canada that common rights to various recreational activities and the harvesting of non-timber botanical products exist on Crown forestland. Formally, however, free public use of Crown forests is a privilege rather than a right.

Notwithstanding the legal status of Crown forestland, the BC, Boreal, and Maritimes standards deal with non-Aboriginal customary rights in a manner that explicitly recognizes their de facto existence and requires that they be identified, documented, and evaluated in terms of managerial impacts. The Rocky Mountain and Pacific Coast standards in the United States take a similar approach to customary rights, constraining their recognition in a manner that reflects a stronger commitment to the protection of private rights than is found in Canada. In Sweden, the issue of customary rights has been less controversial (with the notable exception of the indigenous rights of the Sami, which are discussed in Chapter 8) because, by customary law, any individual may enter any forestland – public or private – for a variety of purposes.

Closely related to the question of forest tenure in British Columbia is the highly contentious issue of how to establish a permitted rate of timber harvesting consistent with FSC Principle 5. The requirements set out in the FSC-BC standard are the most detailed of the standards we have examined. The allowable annual harvest for any management unit must not exceed the projected long-term harvest rate. This requirement, in most units, would result in an immediate 15 to 20 percent reduction in current annual harvests. If implemented, such a strategy would not only significantly reduce public revenue but might well result in serious economic consequences for adjacent communities.

By way of comparison, the Swedish standard is more general and permissive regarding harvesting rates, simply requiring that harvesting operations be sustainable in the long term. Within Canada, the BC, Boreal, and Maritimes standards deal similarly with the rate of timber harvest in some respects. However, compared to the BC standard, the standards for the Boreal and the Maritimes regions are more flexible in prescribing required harvesting rates. In the two American regions examined, the indicators for the determination of harvesting rates have a pronounced timber bias. Like the BC standard, both the Rocky Mountain and Pacific Coast standards require that harvest levels be commensurate with the long-run sustainable yield of the management unit. However, it should be noted that, unlike the case in British Columbia, most of the timber being harvested in the western United States is second growth, and current harvests are generally either below or close to their long-run sustainable level.

As is evident from our analysis, the FSC-BC final standard imposes significant requirements on managers with respect to demonstrating long-term rights to forestland and the calculation of the AAC. Although lesser obligations appear to be placed on managers in other jurisdictions, these obligations occur in a policy contexts dominated by the private ownership of forestland (the United States and the Maritimes), the relative absence of old-growth forests (the United States, the Maritimes, and Sweden), and the existence of long-term, area-based licences (the boreal forests). Any comparison of the rigour of the FSC-BC final standard relative to other comparator standards in terms of tenure and AAC arrangements requires a nuanced appreciation of these very different jurisdictional arrangements. A similar observation can be made of the FSC-BC final standard relative to comparator jurisdictions with respect to community and workers rights, which is the topic of the next chapter.

# 7
# Community and Workers' Rights

Sustainable forest management is not simply concerned with the health and sustainability of forest ecosystems. Of equal importance is the mix of goods and services produced from forests and the distribution of economic wealth they generate. The Forest Stewardship Council's principles and criteria are predicated on the belief that forest management should play a major role in enhancing the welfare of rural populations in nearby forests and that, in the interests of social justice, residents of "local communities" should receive a "fair" share of the wealth generated. These concerns embrace those local people who use, or value, the many goods and services produced by forests; community members who benefit, both directly and indirectly, from the economic activity generated by forests; and the workers who are employed in the forests themselves or in downstream processing facilities.

Issues surrounding the welfare of residents of forest-based communities, the economic viability of such communities, and the protection of forest sector workers' rights are primarily dealt with under FSC Principle 4 (P4), which requires that "forest management and operations shall maintain or enhance the long-term social and economic well being of forest workers and local communities." They are also addressed by Principle 5, which focuses on the local processing of forest products and the diversification of local economies.

As described in Chapter 3, the BC economy has been closely aligned with the fortunes of the forest industry since the early years of colonization and remains so today. The welfare of most of the province's citizens is dependent, to some degree, upon the province's forest resources. Indeed, many rural communities and regions depend on forest resources as their principal, or only, economic base (Horne and Powell 1995). Some are old, well-established communities strategically located to serve as communication and service centres; others are instant towns, their sole purpose to provide a labour force to work in forest-products manufacturing plants and the surrounding forests that supply the plants with raw material (Marchak 1983).[1]

During the 1990s, people living in forest-based communities throughout the province became progressively more anxious about their future and disillusioned with public policies. Concerns about the erosion of local jobs and income were aggravated by a growing appreciation of just how little control local communities, under the prevailing forest tenure system (see Chapter 6), have over the resources that provide them with economic prosperity, water, supplemental sources of food, recreation, spiritual inspiration, and desirable environments in which to live. Arguably, no other single issue has polarized the forest policy debate within BC more than the uncertain future facing many forest-based rural regions.

Many observers believe that the resolution of this problem will determine the future character of the province and its people and, consequently, the legacy that will be handed down to future generations. Given the gravity of these concerns and the high stakes at risk for all those involved, it was not surprising that the volatile atmosphere of mistrust that has characterized relations between the government, the forest industry, and rural residents also played a role in the debate on the FSC-BC standards for Principles 4 and 5. This adversarial context had a major impact on the negotiations that took place and on the content and tone of the indicators that ultimately emerged.

This chapter begins with a brief discussion of the socio-economic and political contexts pertinent to the application of Principles 4 and 5 in British Columbia. It then analyzes the provisions of the FSC-BC final standard that address community and workers' rights and the problems still to be resolved in the standard's application. The chapter concludes with a comparison of the FSC-BC standards for local economic development and the protection of workers' rights with those approved in our five comparator jurisdictions: Sweden, the Maritimes, the US Rocky Mountains, the US Pacific Coast, and the Boreal.

## Standard-Development Context

The history of industrial development in BC forests has been one of boom and bust for communities that depend on forest resources for their survival. Before the introduction of sustained-yield forest management (SYFM) in 1948, the forest industry, predominantly on the Coast and in the southern Interior, literally clear-cut its way from watershed to watershed, creating new, frequently temporary, communities as it advanced, leaving declining communities, some to become ghost towns, in its wake. A 1944 study by the BC Bureau of Economics and Statistics identified thirteen small towns in the East Kootenay district alone that had become ghost towns or entered periods of serious economic decline as a result of reduced forest sector activity during the preceding twenty-five years (Mercer 1944).

In the late 1940s, SYFM was implemented as a public policy, partly as a means to stabilize local forest-dependent communities and, thus, improve the welfare of much of the province's rural population (see discussion in Chapter 3). In return for exclusive harvesting rights to public timber over large areas of forestland, forest companies agreed to assume certain forest management responsibilities, mainly directed at ensuring sustainability of timber supplies. They also agreed that they would either operate or provide a supply of wood to designated manufacturing facilities (a policy usually referred to as "mill appurtenancy"). Allowable annual harvest rates were reduced, and penalties were imposed if companies failed to maintain the annual harvest within designated upper and lower limits.

From the 1950s until the early 1970s, as world markets for forest products went through a period of extended postwar growth, the BC forest industry boomed, and many of the province's dependent hinterland communities grew and prospered. New communities emerged and some established communities went into decline as industrial capacity became concentrated in a smaller number of centres, a response to increased industrial consolidation and concentration, changing technology, and improved transportation. However, by the mid-1980s, conditions had changed dramatically. International demand for timber products softened, and new and aggressive competition appeared. British Columbia forest companies – faced with declining timber quality, reduced allowable annual cuts, and rising costs – found it increasingly difficult to compete in export markets. As a result, during the 1980s and '90s in many communities local wood-processing plants were shut down permanently as firms rationalized their operations to meet new resource and market conditions, and appurtenancy requirements were ignored. Moreover, in many parts of the province, these economic difficulties were exacerbated by public policies that took forestland out of timber production in order to increase the area of provincial land dedicated to parks, wilderness areas, and ecological reserves.[2]

These trends prompted vocal and well-organized protests from members of forest-based communities. As a result of this growing social unrest, the provincial government introduced policies to enhance citizen participation in regional land-use planning.[3] It also created opportunities for communities to manage local forest areas, initially through pre-existing woodlot and forest licenses and, later, through a new form of Crown forest tenure – the community forest agreement, or CFA (Haley 2002). Other public policy initiatives designed to address rural unemployment included Forest Renewal BC (FRBC) (see Chapter 3) and the "spectacularly unsuccessful" Jobs and Timber Accord (Cashore et al. 2001a, chap. 8).[4]

The environmentalists argued that the root causes of the economic turbulence faced by forest-dependent communities included industrial concentration; capital intensification; overemphasis on the production of commodity products – lumber, pulp, and board products – as opposed to further manufactured, or "value-added," products; primary, or sole, reliance on timber products at the expense of the many other goods and services forests are capable of producing; and industrial forest practices that failed to maintain the integrity and inherent productivity of forest ecosystems (Hammond 1991; Drushka, Nixon, and Travers 1993; M'Gonigle and Parfitt 1994). They contended that creating a more sustainable forest economy required measures such as increased local ownership of smaller-scale mills; greater emphasis on producing more labour-intensive, value-added products; ecosystem management practices; and a more diverse tenure system to provide for a larger number of smaller, area-based Crown forest tenures held by individuals, First Nations, cooperatives, regional governments, and communities.[5]

Rationalization of the industry, leading to the elimination or relocation of industrial capacity, has accelerated in the new millennium. Local economies have continued to decline and fuel urbanization pressures as people, particularly the young, leave the countryside in search of jobs and a higher level of economic security. The forestry "social contract," eroded during the 1990s, was finally dismantled in 2003 when amendments to the *Forest Act* – the Forest Revitalization Program – repealed appurtenancy statutes and regulations, removed lower limits on the amount of timber a Crown forest tenure holder is required to harvest annually, and made all major Crown forest tenures freely divisible and transferable.

By the late 1990s, disillusioned by the BC government's apparent inability to reverse the decline in their local economies, forest-based community activists began to look to forest certification as a means to protect their rural way of life. Many of these activists joined the FSC-BC social chamber and played a major role in crafting the indicators and verifiers that were aimed at elaborating and implementing the criteria under Principles 4 and 5. The province's labour movement, on the other hand, remained divided in its support for the FSC process (see Chapter 3). The Industrial, Wood, and Allied Workers was skeptical about the FSC's motives, regarding it as a "green" initiative led by environmentalists that would reduce the number of jobs available to its members across the province. In contrast, from its early days, the FSC enjoyed strong support from the Pulp, Paper and Woodworkers of Canada, a union that was an active participant in the FSC-BC social chamber alongside unions and other organizations representing silvicultural workers.

### Highlights of the Final Standard
FSC Principle 4 (P4) of the FSC-BC final standard focuses on community relations and workers' rights and states: "Forest management operations shall maintain or enhance the long-term social and economic well being of

forest workers and local communities." It contains five associated criteria that address issues of concern to communities, employees, and directly affected persons and groups (see Appendix).

FSC Principle 5 (P5), addresses "benefits from the forest," providing that "forest management operations shall encourage the efficient use of the forest's multiple products and services to ensure economic viability and a wide range of environmental and social benefits" (FSC-IC 2004b). Principle 5 has two criteria relevant to the economic health of local communities. The first requires that forest management and marketing operations "encourage the optimal use and local processing of the forests' diversity of products" (C5.2); the second exhorts forest managers "to strengthen and diversify local economies, avoiding dependence on a single forest product" (C5.4).

Several important issues arise out of the application of P4 and P5 in the British Columbia context. First, because these principles address local employment, local processing, the viability and economic health of local communities, and the impact of forest management on the welfare of local people, the definition of "local" assumes a central role in the certification process. Negotiating this definition in a way that was acceptable to the broad spectrum of FSC-BC stakeholders proved to be a particularly daunting task.

Second, the provisions to encourage local employment were particularly controversial because so many BC communities depend on forest sector employment and because so much local manufacturing capacity had been lost in recent years. Particularly contentious was the extent to which measures to enhance local employment should be prescribed, and the use of quantitative indices to monitor a manager's performance.

Finally, because P4 is concerned with the social impact of forest management operations on people "directly affected" by them, recognition as a "directly affected" party in the certification process is important to a broad spectrum of groups and individuals – some local and others from farther afield. Decisions about defining who should be included in this category were vigorously debated during the development of the final standard, and this remains a contentious issue.

### The Definition of "Local"

There is no single, authoritative definition of community. Communities can be described in a variety of ways; indeed, most people identify with more than one community. Communities are often defined in terms of administrative or political boundaries – a regional district, municipality, city, or province. Alternatively, they may be defined geographically in terms of landscapes – an island, river valley, watershed, or flood plain. A community may have neither a political nor geographical basis but may simply refer to a group of individuals with common goals or a common culture. For example, we speak of the "gay community," the "Muslim community," and the "Aboriginal community." Institutions may also provide foci for communities. For example, those attached to a particular school, place of worship, or society may be regarded, and indeed frequently define themselves, as communities.

While the concept of "community" is therefore critical to certification, this term is not defined in the FSC-BC final standard. Instead, an arguably more pragmatic approach has been adopted in which community is deemed to be made up of those individuals who live close enough to the forest management area to be considered "local." Thus, under the final standard, "People are considered local where they permanently reside within daily commuting distance by car or boat from the management unit or where they are part of a First Nation whose lands and territories contain or are contained within the management unit" (FSC-BC 2005a, 72). This definition attracted a considerable amount of written comment from stakeholders during the development of the FSC-BC standard and was a major topic of discussion at the workshops held by the social chamber working group in July 2001 and February 2002 (FSC-BC 2002a). It was also identified as an issue of special concern by the Technical Advisory Team (FSC-BC TAT 2001).[6] Although this definition is defensible in terms of the philosophy and objectives of certification, it does create some anomalies.

First, it risks excluding some people who actually work on the forest management unit, harvesting timber or performing silvicultural operations. The paradigm that sees wood flowing to local manufacturing plants may be realistic in some instances, but in British Columbia there are many exceptions. Many forestry operations, particularly on the Coast, are served by a labour force working out of semi-permanent camps. It is common for forestry workers living in cities such as Vancouver, Victoria, Nanaimo, and Campbell River to "commute" to work, often by aircraft, on a one- to two-week turnaround basis. Also, BC silvicultural workers are not generally "local" residents. Forest companies do not usually have their own permanent, silvicultural work forces but award contracts for this work competitively. Silvicultural contractors may employ people residing in the vicinity of the management unit, but the seasonal and itinerant nature of the work often requires the employment of part-time workers such as students and other mobile young people.

Second, the FSC-BC definition of "local" includes mill workers living within commuting distance by road or water of the management unit and excludes workers employed in distant mills, sometimes many hundreds of kilometres away, that obtain their wood supply from the management area. This is a major issue in British Columbia, where, as a result of transportation infrastructure, technology, and the economies of location and scale, logs from many forests travel great distances to the mills they support.

Unionized forest sector workers typically argue that the term "local worker" should include both those working in the forest management area and those working in manufacturing plants dependent, or partially dependent, upon the wood from the management area. Nevertheless, a persuasive argument can be made that one of the purposes of certification is to ensure that local residents derive benefits from the forest resources located in their region, and that this is inconsistent with a broader definition of "local" that includes distant workers and communities. In the end, the FSC-BC Standards Committee was not swayed by the distance workers' arguments, and the definition of "local" remained virtually unchanged throughout the negotiations that culminated in the final standard.

A third anomaly in the existing definition of "local" lies in the difference between communities that are widely dispersed and those that are located in highly populated parts of the province. Where communities are widely dispersed, particularly on parts of the BC Coast and in the northern Interior, the definition of "local" adopted by FSC-BC is more appropriate. However, the definition makes less sense in the more heavily populated parts of the province with well-developed systems of high-speed highways. For example, when a certifier is considering provisions for local employment and the procurement of local services, should it include residents of Prince George as part of the community of Quesnel, which is within easy commuting distance by car? Should local residents of Kelowna include everyone residing in the Penticton-Vernon corridor?

Clearly, defining "local" is not an easy task, especially in the BC context. Determining who should be excluded or included for the purposes of applying a particular indicator is a practical decision that depends upon the nature of a particular case. It is also a sensitive decision that, although demanding some general guidelines, should be left to the discretion of certifiers.

**Provisions to Encourage Employment in Local Communities**
Over the lengthy period it took to develop the FSC-BC final standard, the Standards Committee had great difficulty crafting indicators and verifiers that assured residents of forest-based communities that criteria designed to maintain and enhance local employment were being adequately met. What emerged in the preliminary standard (FSC-BC 2003) was a set of highly specific indicators that relied on detailed quantitative indices, which were intended to measure both the proportion and composition of local people employed in certified operations. These indices measured local employees as a percentage of total employees; the percentage of contracted activities captured by local companies or individuals; the person-days of employment on the management unit per 1,000 cubic metres of wood harvested; the ratio of permanent employees to part-time employees and contractors; the

wages and benefits paid as compared to the BC industry average; and other measures identified through a public participation process. Not only would it have been difficult and costly to collect these statistics, it was not apparent how they would be used, because no target values were provided. For example, is 10 percent local employment a reasonable goal or should it be 25 or 50 percent? Nevertheless, in the absence of such targets, managers were required to demonstrate "improvement over time" in the areas measured by the indices, regardless, presumably, of their absolute or relative values.

This appetite for specific, quantitative, performance indicators reflects a perception – held by at least some of those involved in drafting the FSC-BC standard – that industrial forest companies operating within the province could not be trusted to promote local economic and social objectives. In the final standard, however, most of these quantitative performance requirements were eliminated or significantly modified. Required outcomes have thus been simplified, and in many cases they now appear as verifiers, such as the existence of written policies to encourage local hiring and to monitor the implementation of such policies, the percentage of total workers who are "local," and the percentage of total contracted activities captured by local companies or individuals (Verifiers 4.1.1[i][ii][iii]).

### The Definition of "Directly Affected Persons"

One component of FSC's Criterion 4.4 requires that "consultations shall be maintained with people and groups *directly affected* by management operations" (emphasis added). Given its importance under FSC certification, many diverse individuals and interests seek recognition as "directly affected" parties. These include local people who use the area for various recreational pursuits or are concerned about their viewscapes or water supplies, people who gather non-timber botanical products for their personal use or to sell, local workers and businesses that are concerned for their livelihoods if employment in the area is reduced by certification, and local conservation and environmental protection groups whose agendas include preservation of the management area. Others who could plausibly claim to be "directly affected" might include those who work in distant mills that draw their wood supplies from the management area being considered for certification and prominent provincial, national, or even international environmental organizations that claim to represent a broader public interest, both domestic and global, in the impacts of forest management operations.

This raises the question: Who, under FSC rules, should be considered "directly affected" by forest management operations? Which of these many interests should be acknowledged in the certification process and which should be rejected? How can trivial claims be separated from substantial claims? These are exceedingly important questions in a region such as British Columbia, where a high proportion of the population has an interest in the way forests are managed and where opinions are so polarized on many key resource management and social issues.

Clearly, those who reside in communities within or adjacent to the management area should be included if forest management activities significantly affect their health and general welfare. The obvious effects of this kind, which are particularly prevalent in British Columbia, are those that impact the quality of a community's domestic water supply, increase the threat of damage from floods, or have serious consequences for residents' viewscapes. Likewise, those who have statutory or customary legal rights in the management area may be "directly affected," although these individuals are also protected under Principle 2 (see Chapter 6). Beyond these more obvious categories, the delineation of "directly affected persons" becomes a less straightforward matter. At the local level, for example, should those who use the forest for recreational activities or who gather botanical products, such as mushrooms or greenery, be included? If an important recreational lake or scenic trail is damaged by forest management activities, it is likely to be a community matter and the answer is probably yes. But what of individuals who use the forest for casual activities, such as walking their dog or an occasional hiking, hunting, or fishing trip? If "directly affected persons" is defined too broadly, the likelihood of designing a management plan that satisfies all interests is greatly reduced and may result in the failure of the certification process.

The FSC-BC final standard takes an expansive approach to "directly affected persons," defining them as "people or groups *who consider themselves directly affected* by the proposed and current operations; reside in communities within or adjacent to the management unit" or "have legal or customary rights in the management unit" (FSC-BC 2005a, 63, emphasis added).

When this definition appeared in draft versions of the FSC-BC standard, it provoked considerable stakeholder comment and was identified by the FSC-BC Technical Advisory Team as an issue of special concern (FSC-BC TAT 2001). Many commentators expressed strong support for expanding the definition to include "all interested parties," not just those "directly affected." Indeed, some respondents felt that national and international environmental groups should be included in the definition – on the grounds that it was these organizations "who, after all, have made FSC possible" (24). On the other hand, some industry respondents argued that people should have to demonstrate why they considered themselves "directly affected." The definition was also criticized as being so broad as to "include anyone who hikes or visits the area." Certainly, if all such individual users could register as "directly affected persons," the consultation process could become unwieldy and excessively costly, and the chance of successful certification remote.

A second component of Criterion 4.4 specifies that "management planning and operations incorporate the results of evaluations of social impact." In regard to social impact assessment, the manager is required to provide "directly affected persons" with information that allows them to assess the potential impacts of forest management plans on their rights or interests. The manager is also required to take steps to protect these rights or interests through a process of public participation and put in place a mutually agreed-upon dispute resolution mechanism, to be used in the event that the manager and directly affected persons fail to reach agreement.

The final standard also provides that grievances must be investigated, and fully documented, using a mutually agreeable process (Indicator 4.5.1). In the event that a grievance is resolved in favour of the griever, the manager must either refrain from carrying out the offending activity or put acceptable mitigating strategies in place (Indicator 4.5.2). When actual loss or damage to a griever's "rights, property, resources or livelihood" is proven, compensation must be provided that places the griever in the position that he or she would have been in if not for the activities of the manager (Indicator 4.5.5).

### Regional Comparisons

In this section, we compare the provisions of Principles 4 and 5 in the FSC-BC final standard with those of our comparator regions, with particular emphasis on the definition of "local," stipulations relating to local employment (see Table 7.1), and the interpretation of "directly affected persons" (see Table 7.2).

### FSC-Sweden Standard

With respect to the issues under this theme, the FSC-Sweden standard (1998) is at best relatively cryptic if not silent. No attempt is made to define or describe which communities, groups, or individuals should be considered "local." Nor is the promotion of local employment mentioned, apart from a general requirement that forest managers practise responsible, long-term forest management to "maintain or enhance the long-term social and economic well being of forest workers and local communities" (Indicator 4.3). Likewise, there are no specific obligations to promote local processing. The 2005 draft of the FSC-Sweden standard remedies this somewhat by encouraging landowners to adopt forest management, processing, and marketing strategies that "guarantee timber sale, timber deliveries and service systems throughout the country, including sparsely populated areas with long distances for transport" (Indicator 5.2.1).

Nor does the 1998 Swedish standard define or provide any interpretive advice on the question of "directly affected persons." The 2005 draft addresses this deficiency in a rather modest fashion by requiring landowners

*Table 7.1*

## Comparison of provisions related to encouraging local employment

*FSC-AC*

4.1 The communities within, or adjacent to, the forest management area
   should be given the opportunity for employment, training, and other
   services.

*FSC-BC*

- Local forest workers are employed on the management unit and paid
  wages and benefits that are consistent with regional BC industry averages.
- Managers and contractors provide training opportunities and/or collaborate with local training providers and institutions so that local people
  receive enhanced employment qualifications; forest workers receive the
  training needed to comply with the FSC-BC regional standard and legal
  requirements applicable to their responsibilities; and employees receive
  skills upgrading to facilitate advancement within the manager's operations.
- Managers help displaced employees make the transition to new work.
- Managers use local goods and services.

*FSC-Sweden*

- Forest management promotes the long-term well-being of forest workers
  and local communities. The landowner performs responsible long-term
  forest management on his property.
- The potential for forest management in montane forest regions is used to
  promote the employment and livelihood of local people.

*FSC-Maritimes*

- Well-established traditional, non-timber, environmentally appropriate
  uses of the forest by local people or the public are sustained.
- The landowner must provide evidence of support for the local community.

*FSC-Rocky Mountain*

- Employment conditions for non-local workers are equivalent for local
  workers doing the same work, and forest owners or managers use and give
  preference to qualified local workers. Remuneration and hiring practices
  are comparable to local norms for similar work.
- Forest managers and contractors attempt to procure goods and services
  locally and participate in community development and civic activities.
- Forest owners or managers contribute to public education about forestry
  practices and provide and/or support training opportunities for workers
  to improve their skills.

*FSC-Pacific Coast*

- Forest owners or managers use and give preference to qualified local
  workers. Forest work is offered in ways that create a high-quality work
  environment, and conditions of employment are as good for non-local
  workers as they are for local workers. Remuneration and hiring practices
  are comparable to local norms for similar work, and forest owners or
  managers comply with the letter and intent of applicable state and
  federal laws and regulations.

▶

◄ *Table 7.1*

---

- Forest owners or managers demonstrate a preference for the local procurement of goods and services.
- Forest owners and managers contribute to public education about forest ecosystems and their management.

### FSC-Boreal

- Employment preference is given to workers from affected communities; non-resident forest workers are encouraged to reside in local communities while working on the forest. Remuneration is comparable with local industry standards, and employees are treated in a fair and equitable manner through adherence to labour, employment, workplace, and human rights standards. Training is an integral and proactive part of the operation so employees can continually upgrade their skills.
- Applicants, according to their means, contribute to affected communities in a manner that builds capacity and enhances quality of life.
- Applicants emphasize the procurement of goods and services from local suppliers and attempt to mitigate the impacts of technology investment on their employees.

---

*Table 7.2*

---

**Social impact on directly affected persons**

### FSC-AC

4.4 Management planning and operations shall incorporate the results of evaluations of social impact. Consultations shall be maintained with people and groups directly affected by management operations.

### FSC-BC

- The manager develops and implements a plan for ongoing public participation that accommodates directly affected persons. Directly affected persons are provided with information used in making management decisions in a manner that allows them to understand potential impacts on their rights or interests.
- Steps sufficient to protect the rights and interests of directly affected persons are developed and agreed to through the public participation process and implemented by the manager.

### FSC-Sweden

- The landowner takes into account any views on forest management and attends to them.
- Notice of application for a harvesting contract is a public document.

### FSC-Maritimes

- Employees must be given opportunities to participate in, and give feedback on, major management decisions and policy formulation.

---

►

◀  *Table 7.2*

- Local communities and community organizations directly affected by forestry activities must be given an opportunity to participate in setting forest management goals and in forest management planning. Quantitative notice requirements may apply.
- Owner/manager(s) must demonstrate their cooperation with, support of, or assistance to other sustainable management initiatives within the region.
- When planning harvest operations and road design, owner/manager(s) must consider impacts on visual and sound quality in the vicinity of high-use areas.
- The presence of logging operations must not jeopardize the existence of nearby communities/community initiatives within the region.

*FSC-Rocky Mountain*
- Forest owners or managers contribute to designing and achieving goals for forest and natural resource use and protection as articulated in local plans.
- Forest owners or managers provide opportunities for people affected by management operations to provide input to planning.
- Comments and concerns are solicited from people affected by management operations.
- Significant social sites are designated as "social management zones" or otherwise protected during harvest operations.

*FSC-Pacific Coast*
- Forest owners or managers of large-scale operations provide opportunities for people affected by forestry operations to offer input into management planning.
- Comments and concerns are solicited from people affected by management operations. Such concerns are addressed in management plans and operations.
- Significant social sites are designated as "social management zones" or otherwise protected during harvest operations.

*FSC-Boreal*
- Local communities, community NGOs, and the interested public affected by forest management are provided with opportunities to participate in forest management planning. The applicant must demonstrate that efforts were made to contact and work with indigenous forest users without prejudice to Aboriginal rights.
- The public participation process uses clearly defined rules and must contain a list of elements.
- Forest workers are encouraged to report any management activities that threaten the environment or cultural values, or any instances of non-compliance with laws and regulations, and will not be penalized for reporting.
- The applicant shall complete a socio-economic impact assessment and use it to assist with the selection of the desired management option during planning.

to establish "routines for receiving views regarding forestry and the way in which these views are handled within the organization" and by recommending, "in the case of disputes that are difficult to solve," that assistance from a neutral third party be sought (Indicator 4.4.5).

### The FSC-Maritimes Standard

In contrast to the 1998 FSC-Sweden standard, the FSC-Maritimes (FSC-M) standard glossary provides a complex definition of "local community" that embodies concepts of place, political boundaries, and social interactions (FSC-M 2003, 33). "Local community" is defined as "a group of people with similar interests living under and exerting some influence over the same government in a shared locality, having a common attachment to their place of residence where they have some degree of autonomy. People in the community share social interactions with one another, with organizations beyond government, with the larger society, and with local government, moulding the landscape within which it rests and being moulded by it" (33). However, this definition provides no guidance on the question of how close a community must be to the management area in order for its residents to be considered local, a determination that is left to the discretion of the certifier.

With respect to encouraging employment in local communities, the Maritimes standard is brief. Landowners are simply required to "provide evidence of support for the local community" through such activities as "supporting local processing and value added manufacturers, supporting local businesses, supporting local hiring, education and training" (Criterion 4.1.2 indicators). Unlike the BC and Boreal standards, the FSC-M standard makes no explicit references to hiring local contractors or ensuring contractors adhere to the standard, nor does it say anything about the wages and benefits paid to workers. However, managers must provide evidence that "active efforts are made to develop markets for, and optimize use and local processing of, and added value to timber and non-timber products" (Criterion 5.2 indicator). And, to diversify local economies, the manager must produce evidence that "rational and sustainable use of non-timber forest products is encouraged and over-exploitation controlled" (Criterion 5.4 indicators).

The FSC-M standard does not define "directly affected persons." However, it does state that employees of the applicant, local communities and community organizations, and neighbouring landowners must be given the opportunity to participate in planning and/or operational activities. Logging operations "must not jeopardise the existence of nearby communities/community initiatives within the region" (Criterion 4.4.5 indicators), and the impacts of harvesting operations on "visual and sound quality in the vicinity of high use areas" must be considered (Criterion 4.4.6 indicators).

No formal procedures for resolving grievances and determining compensation are prescribed. Managers are simply required to document and employ procedures for resolving grievances and exercise due diligence to prevent loss or damage (Criterion 4.5 indicators).

### The FSC-Rocky Mountain Standard

The FSC-Rocky Mountain (FSC-RM) standard does not provide a glossary definition of "local" or "local community," and interpretations of these terms are not explicit in the text of the standard. Consequently, determination of which communities and persons should be considered "local" is left to the discretion of certifiers, according to local circumstances.

Provisions to encourage local employment are comprehensive but of a general nature. Forest owners or managers must give preference to local contactors, and both managers and contractors are directed to give hiring preference to qualified local workers (Indicator 4.1b). Non-local and local workers must also be treated similarly in terms of remuneration, benefits, working conditions, and training. Compensation for all employees must meet or exceed the prevailing local norms for work requiring equivalent education, skills, and experience. An attempt must be made to ensure that goods and services are procured locally. Forest owners or managers must provide and/or support training opportunities for workers to increase their skills and must contribute to public education about forest practices in conjunction with educational institutions (Indicator 4.1e; 4.1g).

Under P5, preference must be given to local, financially competitive, value-added processing and manufacturing facilities, and owners or managers must adopt sales practices that allow small businesses to compete successfully (Criterion 5.2). They must also explore new markets for less-used species, grades of lumber, and an expanded diversity of forest products (C5.2).

The FSC-RM standard does not provide a definition of "directly affected persons," but an "applicability note" accompanying the text of P4 lists them as "employees and contractors of the land owner, neighbours, fishers and hunters, recreationalists, local water users and forest products processors" (FSC-RM 2001, 5). Forest owners and managers must provide opportunities for such groups and individuals to participate in management planning and must apprise them of the environmental and aesthetic effects of proposed forestry activities in order to solicit their concerns, which must then be addressed in management plans.

No special arrangements for resolving grievances are prescribed, but forest owners or managers must attempt to settle differences and mitigate damage resulting from management activities through open communication and negotiation prior to legal action (Criterion 4.5).

### The FSC-Pacific Coast Standard

With respect to the interpretation of "local community" and provisions related to encouraging local employment and the treatment of "directly affected persons," the FSC-Pacific Coast (FSC-PC) standard is similar to its Rocky Mountain counterpart. Definitions of "local" or "local community" are not provided, and the text of the standard provides no guidance to certifiers on the interpretation of these terms.

Provisions to encourage local employment are succinct and general. Forest owners or managers are required to package and offer forest work "in ways that create a high-quality work environment for employees, contractors, and their employees" (Indicator 4.1a). Conditions of employment must be as "good for non-local workers as they are for local workers doing the same job," and employee compensation and hiring practices must "meet or exceed standards for comparable forest workers within the region" (Indicators 4.1b and 4.1c). Forest managers and their contractors should give preference to qualified local workers when they hire, and they must demonstrate a preference for the local procurement of goods and services (Indicators 4.1d and 4.1e). Applicants are not required to provide worker training or skills upgrading, but they must "contribute to public education about forest ecosystems and their management" (Indicator 4.1g).

Principle 5's requirements to promote local processing and economic diversification include sales strategies that give preference to local, financially competitive, value-added processes and other smaller businesses, and the development of markets for a wider diversity of forest products, both timber and non-timber.

Although the FSC-PC standard does not provide a definition of "directly affected persons," it describes them in a note accompanying the text of P4 as potentially including "employees and contractors of the land owner; neighbours; fishers and hunters, as well as other recreational users; local water users and processors of forest products; and representatives of local and regional organizations concerned with social impacts" (Criterion 4.4). Persons falling into this category must be provided with the opportunity to offer input into management planning and to be apprised of proposed forestry activities and their associated environmental and aesthetic effects so they can present their comments and concerns (Indicator 4.4.b). These concerns must be documented and addressed in management plans and operations.

No mechanisms for resolving grievances are prescribed but, as in the Rocky Mountain standard, forest owners or managers must attempt to settle differences and mitigate damage resulting from management activities through open communication and negotiation prior to legal action (Indicator 4.5a).

### The FSC-Boreal Standard
The Boreal region of Canada is not unlike British Columbia in that legitimate community interests may extend well beyond the immediate boundaries of the management unit being considered for certification. In some instances,

wood from a particular unit may travel considerable distances to processing facilities, and forest workers may not dwell permanently in the vicinity of the management unit but may either commute from other communities or be members of an itinerant work force. Nevertheless, the Boreal standard takes a much narrower view of "local" than the one used in British Columbia, defining it as "any (human) community that is on or adjacent to the forest that is being certified" (FSC-Boreal 2004, 45). In circumstances where "no communities meet this criterion," the note directs that the scope of "local" should be expanded to include "communities within a reasonable commuting distance from the boundary of the forest being certified" (45).

Provisions for the encouragement of local employment are similar in the Boreal region to those in the BC standard, although they are, overall, less specific. Managers are expected to provide employment opportunities to workers and contractors from both local and affected communities, but in contrast to the requirements of the BC standard, no quantitative measures of compliance are required as a means of verification. Non-resident forest workers are encouraged to reside in and contribute to local communities while working in the forest, and the manager must endeavour to build capacity in local communities and enhance the quality of life (Indicators 4.1.4 and 4.1.5). The manager must place emphasis on procuring goods and services from local suppliers using a fair and open process that considers price and the time frame for delivery (Indicator 4.1.6). Under P5, managers must encourage local processing of timber products "where economically viable" and explore the financial and operational feasibility of producing a range of products so as to contribute to the diversification of the local economy (Indicators 5.2.2 and 5.4.1).

Monitoring the social impact of forest management operations is a significant and fairly detailed component of the Boreal standard. Unlike the BC standard, the Boreal standard does not provide a glossary definition of "directly affected persons," although the standard appears to adopt a broad view of those who should be considered, including "local communities, community and non-government organizations, forest workers and the interested public" (Indicator 4.4.1). These groups and individuals must be provided with the opportunity to contribute to forest management planning in a "meaningful" manner (I4.4.1). Managers are required to take a proactive role in contacting indigenous forest users and communities in order to identify issues that affect them, and on Crown land managers must openly seek representation from a range of interested parties so that they might contribute to a public participation process (FSC-Boreal 2004, 46). Whistle-blower protection must be provided for forest workers who report management activities that threaten environmental or cultural values. Like FSC-BC's final standard, the Boreal standard requires that a "socio-economic impact assessment" be completed, made available, and used in management planning.

The elements of a dispute resolution procedure are not prescribed in detail; the manager is simply required to have "a process in place for fairly resolving disputes, including loss or damages" that may result from forest operations (Indicator 4.5.3).

## Conclusion

FSC certification aims to promote a broad cluster of values relating to communities and workers' rights. These include ensuring that forests provide local communities with a range of benefits in terms of jobs, income, and general economic well-being; that forest workers enjoy high standards of health and safety and have the right to organize and freely negotiate with their employers; that the economic health and stability of local communities play a key role in forest management strategies; and that the social impacts of forest operations are addressed directly in management plans (FSC Principles 4 and 5).

The economic welfare of forest-dependent communities is a crucial issue in British Columbia. A high proportion of the province's rural residents depend directly on forest resources for their livelihoods and the general quality of life they enjoy. In recent years, job losses due to the relocation or closure of mills, increasing cyclical volatility in the forest sector, and a general perception that public forest policies promote industrial forest practices that are unsustainable have resulted in growing tensions within forest-based communities. Dealings between local residents, the provincial government, and the forest industry have often been pervaded by an atmosphere of mistrust that has occasionally erupted into open hostility. It is not surprising, in this adversarial environment, that negotiation of the regional standards for Principles 4 and 5 was highly contentious and that the indicators that emerged remain divisive. In particular, three controversial issues arose: the definition of "local"; provisions for the enhancement of local employment; and the definition of "directly affected persons" for the purpose of social impact evaluations.

The definition of "local" adopted in the FSC-BC final standard, considers people local if they "permanently reside within daily commuting distance by car or boat from the management unit" or if "they are part of a First Nation whose lands and territories contain or are contained within the management unit." Communities are considered to be local if they are "within or adjacent to the management area." As we have discussed, this definition, if strictly applied, may create anomalies. First, it may exclude employees who actually work on the management unit, including loggers who commute great distances on a weekly or bi-weekly basis and silviculture workers who are part of an itinerant work force. Second, the definition excludes workers in mills that rely on the management unit for their log supply, but which are located in non-local communities. Finally, in areas where transportation networks are well developed, people considered to be local may include residents living in communities that are within commuting distance, but which have little or no dependence on the management unit and which, in fact, may be larger and more economically viable than the local communities themselves.

In all but one of the other FSC standards examined, the task of defining "local" is left to the discretion of the certifier. The exception, the Boreal standard, defines "local" as "any (human) community that is on or adjacent to the forest that is being certified" or, in the absence of such communities, any "communities within a reasonable commuting distance from the boundary of the forest being certified." The Maritimes standard provides a broad definition of "community" but no guidance on how close a community must be to the management area in order for its residents to be considered local. The Rocky Mountain, Pacific Coast, and Swedish standards provide neither a definition nor any guidance in the text.

Pressure from community activists during the early stages of development of the FC-BC standard resulted in highly detailed and quantitative indicators designed to encourage local employment. This was an important factor in FSC-AC's decision not to approve the FCS-BC D3 unconditionally. In the final standard, many of the overly specific provisions have been eliminated or modified, or they appear as verifiers rather than indicators. Nevertheless, FSC-BC's approach to the stimulation of local employment remains more demanding than that of the comparator regions.

An important component of P4 is the requirement that management planning for certified forests incorporate an evaluation of the social impact of forest operations that is based on consultations with people "directly affected" by management operations. Recognition as a "directly affected" person or group is of crucial importance to those with an interest in the management unit, and the definition of this term proved to be a contentious issue during the development of the BC standard, as disparate groups argued to be included under its rubric.

In the FSC-BC final standard, the term "directly affected persons" is restricted "to those who reside in communities within or adjacent to the management unit" or "have legal or customary rights in the management unit." However, within these constraints, all "people or groups who consider themselves directly affected by the proposed and current operations" have a legitimate claim to be heard in the management planning process.

With the exception of the Swedish standard, the standards for the comparator regions also take a broad approach to the recognition of "directly affected" persons. In fact, the definitions adopted by the Boreal and the two American regions are more inclusive than their BC counterpart. In contrast, the Swedish standard is silent with respect to the definition of "directly affected persons." Even the new draft Swedish standard leaves this determination largely to certifier discretion, although it does propose to require a manager to take into account any views on the management of the forest and attend to them according to a publicly available procedure.

In summary, our assessment of the process culminating in the FSC-BC final standard suggests that values relating to community and workers were a concern that prompted considerably more interest and debate than was the case in our comparator jurisdictions, with the possible exception of Canadian counterpart standards in the Boreal region and the Maritimes. That this was the case is likely due to the political and economic context of forestry in BC and the predominance of public forests both in BC and other Canadian jurisdictions. This said, depending on how certifiers interpret the relevant BC provisions, particularly as they relate to the definition of "local," it is likely that the debate among FSC-BC stakeholders about the language and application of the final standard will continue.

# 8
# Indigenous Peoples' Rights

The interrelated challenges of providing enhanced protection for the rights of indigenous people while promoting more sustainable use of the world's forest resources are receiving ever-increasing and much-deserved international recognition (World Bank 2001). It is a nexus that forms a key component of the FSC certification regime, and one that is accorded special recognition in FSC's Principle 3 and its corollary criteria. Principle 3 (P3) obliges forest managers to recognize and respect "the legal and customary rights of indigenous peoples to own, use and manage their lands, territories and resources." However, as with other international initiatives aimed at protecting indigenous rights, the task of translating this laudable aspiration into measures that will be effective on the ground looms large. This is particularly so in jurisdictions where legal recognition of indigenous rights is inchoate and where a strong domestic judicial and political tradition of guaranteeing such rights is lacking. Can the FSC approach, with its reliance on moral suasion and voluntary compliance, offer a model of forest development that is sustainable and protects indigenous rights, even when efforts to achieve these goals through more conventional political and legal channels have failed? The approach is now being tested in British Columbia, a jurisdiction where conflicts among government, industry, and First Nations over forest resources have been a depressingly familiar feature of the political landscape for decades.

Although judicial decisions rendered since Aboriginal rights and title secured constitutional protection in 1982 have given BC First Nations hope for a lasting resolution of their long-standing rights claims, until recently these judgments have been slow to translate into substantive changes in government policy.[1] Negotiations at treaty tables, established to give contemporary definition to Aboriginal rights and title, have also moved slowly. Since 1992, provincial and federal governments have engaged in treaty negotiations with dozens of First Nations from across British Columbia. Yet, these efforts did not culminate in new treaties until 2007.[2] Meanwhile, in response to the Supreme Court of Canada's decisions in the companion cases of *Haida Nation* and *Taku River* (2004),[3] the BC government has launched several new initiatives to accommodate Aboriginal rights and promote First Nation forestry opportunities. However, the nature and extent of the government's commitment to accommodating First Nations (both at the treaty table and via interim arrangements), remains unclear. As a result, the uncertainty, conflict, and litigation around Aboriginal rights and title – which many would consider to be the province's most pressing public policy issue – seems destined to continue.

In this chapter we assess the nature, implications, and future prospects of P3 and its application under the FSC-BC final standard. In doing so, we seek to identify the lessons that can be learned from the process by which the standard was developed. We begin by setting out the context for First Nations issues in British Columbia. The chapter then reviews and analyzes the relevant aspects of the FSC-BC final standard. It concludes by considering how British Columbia's P3 standard, characterized by some First Nations as one of the most "rigorous" of its kind, compares to standards in other FSC jurisdictions.

## Standard-Development Context

To understand the development of the FSC-BC P3 standard and to reflect on its future prospects and implications, context is critical. In this section we hope to convey the richness of this context by touching on some of the features that made developing a made-in-BC P3 standard such a unique challenge.

In Chapter 3, we discussed how the Supreme Court of Canada's 1973 decision in *Calder*,[4] and the subsequent entrenchment of protection for "existing Aboriginal rights and title" in the Canadian Constitution in 1982,[5] marked a definitive break from the then-prevailing approach to First Nations issues in British Columbia. In *Calder*, the Supreme Court of Canada rejected the notion that Aboriginal title derived exclusively from the Crown in the form of legislative or executive action, leaving the door open for First Nations to establish the existence of inherent title based on proof of continuing use or occupation. Outside British Columbia, treaty negotiations intensified in the period following *Calder* and began to produce significant results. The federal government established a claims commission and gave it the task of negotiating comprehensive land claim treaties for lands that, until then, had remained un-treatied. From the late 1970s to the mid-1990s, the process led to the settlement of seven outstanding land claims in Quebec, Labrador, and Canada's north (Howlett 2001, 123). In contrast to the progress achieved elsewhere in Canada, this federal initiative had little impact in British Columbia, due in large part to the provincial government's continuing refusal to recognize Aboriginal title.

The BC government was finally forced to reverse its non-recognition policy after protection for "existing Aboriginal rights and title" was entrenched in the Canadian Constitution in 1982 (McKee 2000, 29). Still, it was another ten years before the Harcourt government established the BC Treaty Commission in 1992 (33). Labouring under an admittedly heavy workload, the commission's progress has been slow. Although almost fifty BC First Nations have participated in talks at over forty distinct treaty tables, only two final treaties have been concluded in the ensuing years, both in 2007.[6] From its inception, the commission has been dogged by controversy. During its early years, several First Nations withdrew or threatened to withdraw from the treaty talks, citing concerns about the time and expense associated with the negotiation process. The process has also come under continuing criticism from Aboriginal groups for failing to implement adequate "interim measures," which would protect the lands and resources that are at stake during the process, and for requiring, as a condition of participating in the negotiation process, that First Nations refrain from seeking to enforce their rights through litigation.

The BC treaty process was dealt another serious blow when the newly elected Liberal government decided to stage a province-wide referendum on treaty-related issues in 2002. The referendum was uniformly decried by First Nations as an attempt to rewrite the rules of the process midstream. It also threatened to have the scope of constitutionally guaranteed minority rights determined at the ballot box. When their legal challenges to the referendum plan failed, Aboriginal groups urged voters to boycott the referendum. In the end, only about 30 percent of eligible BC voters cast ballots; those who did vote registered strong support for the government to take a harder, more bottom-line approach in treaty talks.

The referendum represented a low point in relations between the provincial government and BC First Nations. However, recent events suggest that a turnaround may be underway. As discussed in Chapter 3, in the spring of 2005, the BC government unveiled an ambitious partnership agreement that it had concluded with the province's three leading Aboriginal organizations: the First Nations Summit, the Union of BC Indian Chiefs, and the Assembly of First Nations (British Columbia Ministry of Forests 2004b). Among other things, this New Relationship agreement commits the parties to work together to develop a new framework for dealing with Aboriginal rights with a view to enhancing economic opportunities on the province's land base. In the forestry context, the New Relationship has led to reform of the Ministry of Forest's Forest and Range Agreement (FRA) program.

The original intention of the FRA program, initiated in 2004, was to achieve "workable accommodations" with First Nations in situations where MOF tenure decisions had the potential to affect Aboriginal rights or title.[7] In return for grants of volume-based forest tenures (typically five years in duration) and a share of provincial forest revenues (calculated on the basis of the population of the First Nation involved), First Nations signatories were required to affirm that the economic component of their Aboriginal rights and title interests had been accommodated for the term of the agreement and to refrain from challenging the adequacy of the accommodation in court or disrupting ministry-licensed forest activities within their territory. Despite First Nations criticisms of the program, including complaints about the population-based revenue-sharing formula and the inadequacy of the volumes allocated, over a hundred FRAs have now been concluded.[8] Further, as a step toward implementing the New Relationship, the Ministry of Forests has recently replaced the FRA program with the Forest and Range Opportunity (FRO) program, an initiative designed to respond to First Nations concerns with its FRA predecessor.[9]

These forest policy initiatives and, more generally, the New Relationship agreement reflect the provincial government's growing appreciation of the need to enhance relations with First Nations in order to create a more stable business climate for resource development (British Columbia Ministry of Forests 2004b). This new pragmatism has, in turn, been driven by a growing body of judicial precedent on Aboriginal rights and title issues, including a series of landmark decisions in the Supreme Court of Canada: *Sparrow* (1990), *Van der Peet* (1996), *Delgamuukw* (1997), and *Haida Nation* (2004).[10] These decisions confirm that neither the province nor the federal government can unilaterally extinguish (through regulation or other means) constitutionally protected Aboriginal rights; that governmental infringements of such rights must meet strict justificatory criteria that include a duty to consult, accommodate, and usually compensate affected First Nations; and that Aboriginal title exists as a distinct "inalienable, communally held" species of Aboriginal rights.

Of these decisions, *Haida Nation* has had the most significant impact on provincial forest policy. In this case, the Haida challenged an MOF decision to renew a forest licence in their traditional territory, arguing that the consultation process was inadequate. The BC government contended that, because the Haida had not legally established their Aboriginal rights or title to the area in question, a duty to consult did not arise. The Supreme Court unequivocally rejected the government's argument. According to the court, the honour of the Crown obliges provincial and federal governments to consult with First Nations where it has knowledge of a potential rights or title claim that could be adversely affected by government action; the scope and nature of this duty will vary depending on a variety of circumstances, including the strength of the claim and the nature of the impact of the contemplated action on the Aboriginal interest being asserted.

Thus, in recent years, BC First Nations have made significant advances, at the treaty table, through litigation, and in provincial forest policy, in securing recognition and respect for their rights in relation to forest resources. However, the costs of this progress – in terms of time, money, and other resources – have been enormous, and the benefits, as yet, difficult to quantify.[11] Although the BC treaty process has given First Nations a venue in which to articulate and advocate their agenda in a comprehensive way, significant concerns remain about the adequacy of the protection the process provides for the forest resources that are at stake in the negotiations. In a recent case, for example, the MOF was judicially criticized for its inflexible employment of the FRA interim measures program to accommodate First Nation's interests.[12] And, although First Nations have scored important victories in the courts, these decisions have often yielded outcomes so narrow in their application (or so conceptually broad) that their impact in terms of government policy has been modest relative to the costs incurred.[13] Moreover, the courts have rejected several key arguments mounted by First Nations, perhaps most notably the contention that the duty to consult should extend to corporations operating under Crown resource licence agreements.[14]

For these reasons, First Nations across Canada, and particularly in British Columbia, have begun to look to new strategies and approaches to secure recognition and respect for their rights. Can the FSC-BC standard succeed where negotiation and litigation have failed? Does it offer a means by which First Nations can exercise control over forest developments on their traditional lands in a way that will provide immediate protection to, and benefits from, those resources without compromising the rights they continue to pursue at the treaty table and through the courts?

For First Nations, the threshold challenge confronted in the FSC-BC regional process was to design a P3 standard that ensured proactive protection for Aboriginal rights and title. This aspiration is reflected in a FSC-BC commentary on P3 in the second draft of the standard (D2): "Before anything else, Principle 3, its four Criteria, and all its Indicators and Verifiers must unequivocally ensure protection of Aboriginal rights if BC's First Nations' participation in FSC processes is to become a reality. The Standards that deal with Aboriginal rights must confirm those rights. And ... the Standards must give First Nations the power to permit or deny certification based on the quality, the depth and the sincerity of that protection" (FSC-BC 2001, 127). By securing a standard that met this threshold challenge, BC First Nations hoped to avoid many of the uncertainties and expenses associated with seeking recognition and respect for Aboriginal rights and title through legal and political channels. Moreover, many were optimistic that such a standard would ensure First Nations communities derived significant new benefits in terms of jobs and revenues from the forest resource. There was also a broad recognition that, in order to achieve these goals, First Nations needed a standard that would allow them to develop new and mutually beneficial relationships with the province's forest companies.

The debate within FSC-BC over P3 was therefore less about ends (on which there was general agreement) than means (i.e., what form of standard best advances these agreed-upon goals?). For example, First Nations participants in the process held the view that a prerequisite to relationship building was development of a standard that dramatically enhanced their ability to negotiate with forest operators and participate in forest management; not all agreed on how this end should be pursued. Some argued that the legacy of mistrust between First Nations and the forest industry favoured a standard that defined the nature and parameters of this relationship in detail. Others argued for flexibility, contending that a standard would only work if it eschewed legalisms and mandatory requirements and allowed parties to exercise discretion to deal with on-the-ground conditions. This, in turn, would allow new relationships to evolve and adapt to local conditions, interests, and needs.

First Nations were also divided on the question of whether, and to what extent, the standard should specify requirements governing how approval for proposed forest management plans was secured within and beyond the First Nation community directly involved. For some, a paramount concern was to avoid certifying operations that faced substantial internal opposition within a First Nation community or that did not have the support of all First Nations that shared jurisdiction over the territory to be certified. (The latter scenario is commonly referred to as a "shared territory" situation.) Some First Nations participants argued for a standard that specifically addressed and sought to minimize these risks. Others, however, contended that such an approach would undermine Aboriginal self-government by requiring FSC certifiers to evaluate the merits of First Nations' decision-making processes. It was also argued that incorporating such requirements into the standard would, in many instances, hamper, if not thwart, the certification process by conferring a veto power on disaffected groups or individuals.

Forest operators involved in developing the standard also had significant interests at stake. Heavily reliant on Crown timber that is the subject of pending Aboriginal claims, BC forest operators have become acutely aware of the costs associated with the continuing uncertainty created by the impasse in resolving such claims. Many operators, particularly those with markets in Europe and the United States, felt it was a priority to develop a standard that would reduce or eliminate the occurrence of Aboriginal-sponsored blockades, protests, and litigation, events that have become a regular feature of business in the province. To achieve this goal, forest operators were cognizant of the need for a standard that fostered the development of new, constructive relationships with First Nations communities. They were also generally supportive of a standard that enhanced the ability of First Nations to participate more meaningfully in forest planning and management.

Like some First Nations, however, most forest operators favoured a standard that established general parameters and goals over one that imposed specific requirements. This preference flowed from two closely related concerns: viability and cost. Of particular concern on the former front was the prospect of a standard that imposed requirements with respect to internal community support and shared territory situations. The viability of a highly detailed and specific standard was also an issue for some small operators concerned about their capacity to meet such a standard. In addition, a frequently voiced concern was whether, on a cost-benefit analysis, a detailed standard could create adequate incentives for forest operators to seek certification.

### Highlights of the Final Standard

The FSC's Principle 3 states that "the legal and customary rights of indigenous peoples to own, use, and manage their lands, territories, and resources shall be recognized and respected." P3 is associated with four criteria that deal with free and informed consent, non-diminishment of resource and tenure rights, identification of sites of special significance, and compensation (see Appendix).

The drafters of the FSC-BC P3 standard faced several daunting challenges. One of the first was to develop language that defined the nature of these rights in such a way that it provided sufficient guidance for them to be readily applied in the field. A second challenge arose from the fact that judicial and political definition of these rights is still at an early stage and is the subject of considerable controversy. As a result, the drafters sought to define the relevant rights in a manner that referred to current "minimum legal requirements" but, at the same time, made it clear that managers were expected to commit themselves to exceeding these requirements. Third, there was the question of who would ultimately control how these rights were defined. As we have seen, until recently the BC government took the position that, in the absence of a signed treaty or a judicial ruling to the contrary, the existence of First Nations' rights and title should not be presumed. As we shall see, the BC final standard reverses this presumption. It gives "relevant First Nations" (a terminology employed in the standard to identify First Nations whose legal or customary rights may be affected by forest operations) the ability to define the scope and nature of their relevant rights, an approach that is consistent with (but goes considerably beyond) the Supreme Court of Canada's 2004 decision in *Haida Nation*.

In this section of the chapter, we will examine the procedural obligations that the standard imposes on managers and then consider how the standard addresses the four substantive criteria associated with P3.

### Overview of the P3 Process under the Final Standard

Under the FSC-BC final standard, implementation of P3 occurs in a three-stage chronological process:

1  Complete a protocol agreement with any relevant First Nations (if requested by a First Nation).
2  Develop a management plan (either through a process of consultation or joint management agreement with relevant First Nations).
3  Document the consent of relevant First Nations that their legal and customary rights over the lands in question are recognized and respected by the proposed forest operations.

*Stage 1: Protocol Agreement*
When a relevant First Nation requires it, the final standard mandates that the manager enter into a protocol agreement aimed at defining in broad terms the "nature of the relationship between the Parties." Such an agreement must include the following components: provisions that address how the parties will establish and conduct their relationship; descriptions of their respective roles, responsibilities, and interests; provisions that identify the "appropriate decision-making authorities for all parties"; and a "framework for subsequent agreements necessary to give effect to the protocol" (Indicator 3.1.2).

*Stage 2: Development of the Management Plan*
Forest operators have two options when developing their management plan: they can develop such a plan through a process of consultation with affected First Nations or under the auspices of a joint management agreement (JMA) with an interested First Nation. Because of the complexities of the latter option, it is expected that JMAs will primarily be employed by larger operators.

The expectations for a forest manager who pursues the consultation route are spelled out in the glossary definition of "consulting with First Nations" (FSC-BC 2005a, 61). This definition provides a list of what it describes as "characteristics of a good consultation process." Many of these are elements one would anticipate forming part of any effective consultative process (i.e., that the process is designed by and agreed upon by the parties; that it provides an adequate schedule for consultation; that concerns raised are appropriately recorded and reflected in the plan emerging from the consultation). Many of the other elements relate to developing strategies that will protect First Nations' resources, tenure rights, special sites, and traditional knowledge recognized in Criteria 3.2 to 3.4. A final characteristic of a "good consultation process" is that "financial, technical or logistical capacity building support, in proportion to the scale and intensity of operations," is made available to affected First Nations to allow them to participate in the consultation process (61).

In contrast, a forest manager who decides to develop the management plan jointly with First Nations must negotiate a JMA. Under a JMA, the parties embark on what the glossary characterizes as an "enhanced form of consultation" that engages the parties in a process of "jointly setting goals, objectives, strategies, implementation, restoration and monitoring of the forest within the management unit" (FSC-BC 2005a, 69). The standard suggests that JMAs will take a variety of forms, ranging from agreements that deal with "a relatively few areas of common interest" to those that entail a "thorough integration of industry and First Nations' ideas throughout the whole Management Plan" (69).

Mandatory components of a JMA include measures to protect First Nations' resources, tenure rights, special sites, and traditional knowledge described in Criteria 3.2 to 3.4; processes to collaboratively develop objectives and strategies on matters of importance to the First Nation and to develop "all or a part of the Management Plan"; and assurances that appropriate consultation is undertaken with respect to any part of the management plan that is not collaboratively developed. A "good" JMA should be approved by the decision-making bodies identified in the relevant protocol agreement, and it should make "financial, technical or logistical capacity building support, in proportion to the scale and intensity of operations," available to First Nations so they can meaningfully participate in the JMA's development.

*Stage 3: Securing and Documenting Consent(s) from Affected First Nations*
Once a management plan has been developed, the final stage of the process involves securing the necessary consent from the First Nations involved. The final standard describes two types of consent that a manager must secure. The first is the "free and informed consent" of the First Nations to the management plan (Indicator 3.1.4). According to the glossary, such consent must be given both freely (i.e., without manipulation, undue influence, or coercion) and knowledgeably (i.e., based on all relevant and necessary information). To determine whether such consent has been validly secured, the standard requires the certifier to ensure that the First Nation possessed the "financial, technical and logistical capacity" adequate to participate on an informed basis in planning and decision making (Indicator 3.1.4, Verifier [ii]).

A forest manager must also convince the certifier that a second form of consent is present. This is determined based on two requirements. The first is that relevant First Nations provide oral or written affirmation that their legal and customary rights over the affected lands have been recognized and respected; the second requires the certifier to conclude objectively that First Nations' "interests and concerns are incorporated in the Management Plan" (Indicator 3.1.1, Verifiers [i] and [ii]).

Two additional consent-related issues merit attention: the role of "consent" after certification and the necessity of consent in shared territory or overlapping territory situations. The concept of consent is not to be construed as a one-time, "static" event. This is evidenced by the requirement that a management plan record "the conditions under which consent has been given and under which it may be withdrawn" (Criterion 3.1.5). As for situations where more than one First Nation is affected by proposed forestry operations, the standard stipulates that the consent of each affected Nation is "ordinarily required" (Criterion 3.1.6). This principle applies in situations where two or more First Nations have adjacent traditional territory within the proposed forest management unit (C3.1.6) and in situations where such territories overlap or are shared; it is clarified in a detailed statement of the intent of Criterion 3.1 (FSC-BC 2005a, 13-14). In either of these situations, the standard clearly stipulates that if one of the affected First Nations is dissatisfied with proposed forestry activities, "under no circumstances should certification proceed" (13-14). However, the absence of consent from an affected First Nation does not necessarily bar certification. In cases where the First Nation has declined to participate in the certification process due to lack of interest or capacity, certifiers will be called upon to exercise their judgment. The same applies in situations where the fate of a certification application is being affected by a dispute between First Nations over overlapping territory (13-14).

### Criterion 3.1: Legal and Customary Rights to Lands, Territories, and Resources

FSC-AC has approved a generally applicable international definition for the subject matter of this criterion. In the FSC-BC final standard, the term "lands, territories and resources" thus refers to "the total environment of the lands, air, water, sea, sea-ice, flora and fauna, and other resources which Indigenous peoples have traditionally occupied or used," an internationally approved definition borrowed from Article 26 of the "Draft United Nations declaration on the rights of indigenous peoples" (United Nations 1993). P3 requires forest managers to recognize and respect the "legal and customary rights" of First Nations with regard to this broad subject matter.

A key task for the FSC-BC drafters under this criterion was to define, in the BC context, the meaning of the term "legal and customary rights." The approach adopted in the standard is to equate "legal and customary rights" with the concept of "Aboriginal rights and title." The glossary defines "Aboriginal rights" in a two-sentence paragraph that summarizes the ruling of the Supreme Court of Canada in *Van der Peet*.[15] Similarly, it defines "Aboriginal title" in a paragraph that summarizes the court's reasons in *Delgamuukw*.[16]

The final standard is rather cryptic about how these broad-brush glossary definitions are to be applied on the ground. The absence of guidance in this regard is of less consequence where the scope and nature of Aboriginal rights are defined by treaty. In this situation, the terms of the treaty will define the applicable Aboriginal rights or title. However, where treaties do not exist, as is the case throughout most of British Columbia, the governing interpretation of "Aboriginal rights and title" will be that of the affected First

Nations themselves, at least insofar as geographical breadth of the applicable traditional territory is concerned.

## Criteria 3.2 (Maintenance of Resources and Tenure Right) and 3.3 (Protection of Special Sites)

Criteria 3.2 and 3.3 impose particular requirements on managers with respect to Aboriginal resources and tenure rights (RTRs) and sites of special cultural, ecological, economic, or religious significance to First Nations (special sites). Here again the standard incorporates a made-in-British Columbia definition. The glossary definition of RTRs is extremely broad, articulated in the form of a non-exhaustive illustrative list that includes references to various natural resources (water, fisheries, timber and forests, non-timber forest products, and subsurface mineral and oil deposits), site- or area-based amenities (access routes to resources, hunting and gathering areas, tribal heritage parks, and traplines), and activities (guiding operations and cultural tourism). Similarly, special sites are defined in the glossary through a lengthy catalogue of illustrative sites,[17] including sites related to traditional oral history, supernatural beings, recreation (gathering places, games, or competition places), and education.

No attempt has been made to tie these definitions to rights currently recognized in Canadian law. Instead, the standard vests in affected First Nations the responsibility for identifying what RTRs and special sites they wish to have protected.[18] Once these are identified, the manager is obliged to devise means to maintain and protect RTRs and special sites from being impaired by forestry activities (Indicators 3.2.1 and 3.3.1). Where such activities lead to the loss or diminishment of RTRs or special sites, the First Nation must be satisfied with the mitigative or compensatory measures that a manager proposes.

## Criterion 3.4: Compensation for the Application of Traditional Knowledge

As it does for RTRs and special sites, the final standard provides a glossary definition of "traditional knowledge" that takes the form of a non-exhaustive, illustrative list. This list refers to a variety of forms of indigenous knowledge potentially relevant to forest management, including knowledge concerning local fish, wildlife, and plant life and cycles; climatic changes and cycles; and local ecosystem information, including responses to natural or human disturbances, as well as qualitative information about the human uses of medicinal or edible resource plants.

Once again, the final standard contemplates that First Nations will control the process of determining how this definition is to be applied. Where a First Nation and a forest manager agree, the standard provides that traditional knowledge will be incorporated into the management plan. Moreover, it stipulates that First Nations are to retain control over their traditional knowledge and be fairly compensated, to their satisfaction, when that knowledge is used by a manager.

### Regional Comparisons

As noted earlier, First Nations proponents of the FSC-BC P3 final standard contend that it is one of the most rigorous of its kind in the world (personal interviews 2004 and 2005). In the following section, we will evaluate this claim by assessing the BC standard against the comparators in our study (see Table 8.1 on pp. 180-81).

### The FSC-Sweden Standard

In Sweden, the primary indigenous group is the Sami, a nomadic people who populate northern Scandinavia. There are forty-four Sami communities in Sweden. Like many other minority indigenous populations, the Sami have been engaged in a long struggle for recognition of their traditional rights. Currently, the primary vehicle through which their rights are protected is the *Reindeer Husbandry Law* of 1971, which governs Sami land and water access rights in connection with reindeer ranching. Although a Sami parliament was established by the state of Sweden in 1993, hopes that this would enhance Sami self-government and protection of Sami culture and language have not been realized to date.

The FSC-Sweden standard calls on landowners to comply with Sweden's *Forestry Act* and "accept and give consideration to" reindeer ranching by Sami people in areas designated by the Swedish government as reindeer husbandry areas (Indicators 4.2.1 and 4.2.2). The revised 2005 FSC-Sweden draft standard enhances the Sami's consultation rights. It provides that managers shall "accept and respect" Sami reindeer husbandry on their lands and requires them to provide a map of planned logging areas to the Sami village affected in advance, "if possible." However, the revised standard still contemplates that once consultation has been completed, forest operations may proceed, even if there are objections or "divergent views," as long as there is a proper record of the consultations made and confirmed by the parties (FSC-Sweden 2005, Indicator 3.1.2).

The main indigenous resource right identified in both the current and draft Swedish standards is arboreal lichens on which reindeer graze. Under the current standard, landowners are required "to consider" protection of lichen in forest planning (Indicator 4.2.3). The 2005 draft standard imposes more specific management obligations, requiring landowners to identify the location of lichen and leave riparian buffers and tree patches that are necessary to allow the lichen to reproduce. Input from Sami villages describing areas of particular significance for lichen growing is to be "taken into special account" (FSC-Sweden 2005, Indicator 3.2.1).

Both the current and draft Swedish standards also address protection of sites of special significance to Sami (including old settlements, cultural remains, migration routes, gathering places, overnight resting areas, lichen areas, work paddocks, calving areas, and culturally important paths and sacrificial places) (FSC-Sweden 1998, Indicator 4.2.4; FSC-Sweden 2005, Indicator 3.3.1). Here again, however, landowners' obligations are relatively modest. Unlike comparator standards, they are not obliged to implement affirmative protective measures beyond taking into account and respecting – in cooperation with Sami people – sites falling within this description. Nor is there any requirement to attempt to mitigate or provide compensation where such sites are adversely affected by forest operations. Finally, the 2005 draft standard expresses the conclusion that the FSC criterion relating to traditional knowledge is "not applicable to present Swedish conditions" (FSC-Sweden 2005, Criterion 3.4).

### The FSC-Maritimes Standard

There are approximately thirty-three Aboriginal bands in the Maritime region (Canada, Ministry of Indian and Northern Affairs 2005), and the total Aboriginal population is currently approximately 54,000 (Statistics Canada 2001). During the colonial era, primarily the sixteenth and seventeenth centuries, ancestors of modern-day First Nations in the Maritimes entered into treaties with new arrivals from Europe. Disputes over the validity and interpretation of these treaties persist, however. In some cases, as in British Columbia, these disputes relate to rights to forest resources.

Like many of the standards assessed in this section, the FSC-M P3 standard is relatively brief and uncomplicated, devoid of references to legal principles or definitions. Indeed it is the least specific of the standards in force in our Canadian comparator jurisdictions.

The approach adopted in the Maritimes standard is to exhort managers to consider the interests of First Nations, encourage First Nations participation in forest management, commit to resolving outstanding disputes involving First Nations, and generally comply with existing legal obligations to First Nations. On balance, however, the standard contains few rights that are likely to assist First Nations in negotiating new relationships with forest companies or deriving additional benefits from forest resources.

With respect to the key issue of the right of First Nations to control forest management on their lands and territories, the indicator language is relatively vague and weak, failing to address the question of how such rights are to be defined and indicating only that First Nations rights shall be "recognized" and given "fair accommodation." The related verifiers reveal the limited nature of what "fair accommodation" in this context entails: among them are requirements that the manager make efforts "to get First Nation participation in forest management decision-making"; that the manager have "a

*Table 8.1*

## Comparison of criteria related to indigenous people's rights

| FSC region | Extent of indigenous people's (IP's) control of their lands (C3.1) | Protection of resource and tenure rights (RTR) (C3.2) | Protection of special sites (C3.3) | Compensation for use of traditional knowledge (C3.4) |
|---|---|---|---|---|
| FSC-BC | IP control – manager must obtain consent for the management plan on indigenous land; affected First Nations' geographical interpretation of their "lands, territories, and resources" prevails | Forest activities are carried out to maintain RTR unless IP sign off | Forest activities are carried out to protect special sites unless IP sign off | Traditional knowledge is used where mutually agreed and IP receive fair compensation for such use |
| FSC-Sweden | Landowners must consider Sami reindeer husbandry occurring on their land | Plans must show consideration for fruticose arboreal lichens and specific soil types | Landowner takes into account and respects special sites | Not applicable; no rights |
| FSC-Maritimes | IP knowledge, practices, and insights are considered in planning and operations | RTR not defined Indicators link RTR criteria to Aboriginal and treaty rights, planning involvement, and employment | "Culturally sensitive sites" are identified and incorporated into plans Economic, ecological, or religious significance is not addressed | IP are compensated for the use of traditional knowledge; IP must agree on compensation |

| | | | | |
|---|---|---|---|---|
| FSC-Rocky Mountain | IP control – forest managers or owners must secure consent prior to any forest management (subject to the degree of occupation and use of lands and territories) | IP are invited to jointly plan operations affecting their resources | IP participate in identifying special sites IP and managers jointly develop measures to protect special sites | Agreement to protect IP's traditional knowledge required prior to commercialization |
| FSC-Pacific | IP control – managers must secure consent | IP are invited to jointly plan operations affecting their resources | IP participate in identifying special sites IP and managers jointly develop plans to protect special sites | Agreements to protect IP's intellectual property made prior to commercialization |
| FSC-Boreal | Applicant obtain agreement from IP that their interests and concerns are incorporated into the management plan | Applicant develops management activities in the management plan to ensure indigenous resources are not threatened or diminished Tenure rights are assessed but not included in the protection granted to resources | IP develop "indigenous areas of concern" protection agreement If IP indicate that forest activities seriously threaten significant sites, operations cease pending a resolution process | IP are compensated for the use of traditional knowledge if its use leads to certain prescribed ends |

program/procedure for consulting with local First Nations"; and that the "local First Nations have not challenged the management plan in court" (Indicator 3.1.1).

The Maritimes standard deals in a similar manner with the requirement that forest management not diminish or threaten Aboriginal resources or tenures. A manager must "consider and meet obligations" with respect to Aboriginal and treaty rights, but this requirement only applies to those rights that have been "duly established" (Indicator 3.2). Although the term "duly established" is not defined, it appears to refer only to those Aboriginal rights that have received formal judicial recognition or state approval (in the form of clear treaty language, for example).

The Maritimes standard is also relatively weak in terms of its protection of special sites. In this regard, the standard distinguishes between "culturally sensitive" and "culturally significant" areas. The former definition is broader and appears to encompass all sites that have a special cultural, ecological, economic, or religious significance to First Nations. With respect to areas within this definition, a manager is simply obliged to incorporate protective measures into the management plan. Only if the site is one of "cultural significance" (defined as being an area of spiritual or religious significance or an area subject to a pending land claim) is the manager required to secure the consent of the affected First Nation prior to embarking on forest operations (Indicator 3.3.2).

Finally, a distinguishing feature of the Maritimes standard is the concept of "major problem." If an applicant fails to meet a requirement that is designated a major problem, remedial steps must be taken within a reasonable time if the applicant is to maintain certified status. There are two criteria and one indicator in the Maritimes standard that are so designated: the informed consent provision (C3.1); the requirement that forest management not threaten or diminish indigenous resource or tenure rights (C3.2); and the requirement for consent to operations in "culturally significant" areas (Indicator 3.3.2).

### US Jurisdictions

Indian tribes that are recognized by the US government have a special legal and political relationship with government agencies. The federal government has the authority to regulate commerce with Indian tribes; it also regulates the management of Indian lands, on which many tribes are engaged in forestry activity. However, recognized tribes enjoy a variety of self-government powers and have a government-to-government relationship with the US government. As in Canada, the US government owes a fiduciary duty to Indians. Unlike the situation in British Columbia and parts of the Boreal region, indigenous rights to land and resources throughout much of the United States are addressed by treaties, mostly dating back to the colonial era.

There are nine FSC regions in the United States. As several of these regions began to draft standards, significant variations between regions led FSC-US to establish a standards committee, made up of experts in environmental, social, forestry, and economic issues, that was responsible for developing a model set of national indicators. The stated purpose of the national indicators is "to foster a sense of trust and transparency by representing a common baseline" for the development of FSC-US regional standards (FSC-US 2001a, viii). Of particular note is the fact that the national indicators grant an exception to the general prohibition on commercially harvesting timber in old-growth, high conservation value forests (HCVFs) (xiv). This exemption recognizes the fact that Indian tribes own or control a disproportionately high percentage of HCVF-designated areas, and applying the general prohibition would severely undermine the viability of many tribal forest operations. In order to be certified, however, exempted tribes must set aside a representative portion of such forests as reserves.

## The FSC-Rocky Mountain Standard

The FSC-Rocky Mountain (FSC-RM) standard was accredited in 2001. With respect to indigenous people's rights it is identical in its terms to the draft US national indicators. The FSC-RM standard defines "indigenous peoples" in broad terms to include "Native American or Indian tribes, nations, communities or bands, and their members," including those not formally recognized by the federal government (FSC-RM 2001, 28). The issue of consent is addressed by unequivocally providing that "forest management on tribal lands takes place only after securing the informed consent of tribes or individuals ... whose forest is being considered for management" (Indicator 3.1.b). Tribal landowners are also given the right to insist that a manager use "tribal experience, knowledge, practices and insights" when carrying out forestry on tribal lands, and the right to have input into forest management planning "in accordance with their laws and customs." However, the standard fails to define "tribal lands," which may create uncertainty as to the scope and nature of these rights in practice. This is compounded by a caveat to Indicator 3.1, which states that "the degree of consultation or informed consent required for tribal traditional territories is related to the degree of occupation and/or use of those lands and territories" (3).

In other areas, the FSC-RM standard contemplates an even less rights-based approach. For instance, with respect to C3.2 (that forest management not threaten or diminish the "resources or tenure rights of indigenous peoples"), the standard requires managers to identify and contact Indian groups that may potentially be affected and invite them to participate in joint forest-management planning processes (Indicator 3.2.a). However, the term "resource or tenure rights of indigenous peoples" is not defined. Similarly, managers carrying out forest operations within watersheds that affect tribal lands are asked to take steps to ensure that these activities do not adversely affect tribal resources (Indicator 3.2.b). As it is currently framed, this obligation likely falls short of what managers in this situation would be required to do under the common law. A similar approach is adopted with respect to the protection of "special sites." Here again, managers are exhorted to invite tribal representatives to participate in a process of identifying sites of this kind (Indicator 3.3.a). However, in contrast to the requirements for potentially affected indigenous resources and tenure rights, where special sites are involved the standard requires only that managers and tribal representatives "jointly develop measures to protect or enhance areas of special significance" (Indicator 3.3.b).

Finally, with respect to traditional knowledge, the standard obliges managers "to respect the confidentiality of tribal knowledge ... assist in the protection of tribal intellectual property rights" (Indicator 3.4.a), and enter into written agreements with individuals and/or tribes before traditional knowledge is "commercially exploited" (Indicator 3.4.b).

### The FSC-Pacific Coast Standard

In most respects the FSC-Pacific Coast (FSC-PC) standard tracks the language of the FSC-RM standard, reflecting the general trend toward regional uniformity among US standards. Its definition of indigenous peoples is quite similar to its Rocky Mountain counterpart's, as is its requirement that managers not carry out forest activities on tribal lands until the informed consent of affected tribes or tribal members has been secured (Indicator 3.1.a). The FSC-PC standard goes somewhat further, however, by incorporating a requirement that a manager delineate areas of restricted access in consultation with affected tribal members and in accordance with tribal laws and customs (Indicator 3.1.c).

With respect to obligations not to threaten or diminish indigenous resources or tenures, the FSC-PC standard is closely analogous to the FSC-RM standard (Indicator 3.2.b). Provisions with respect to the protection of special sites are identical in both standards.

Finally, provisions regarding use of traditional knowledge are identical save for an additional requirement in the FSC-PC standard that, when a manager requests use of traditional ecological knowledge, protocols are jointly developed to protect tribal intellectual property rights (Indicator 3.4.c).

### The FSC-Boreal Standard

Indigenous rights protection is a key issue in the Boreal region, which is home to 80 percent of Canada's First Nations people (FSC-Boreal 2004, 22).

Attention to these issues in the standard-development process produced a standard that has been applauded for its recognition of and respect for indigenous rights.

Unlike its BC counterpart, the Boreal standard does not oblige a manager to enter into a protocol agreement with affected indigenous communities, nor are there specific requirements relating to the process by which the management plan is to be developed. Instead, the drafters of the Boreal standard have chosen to establish various performance-based requirements that a manager must meet to ensure that the proposed management plan satisfies the interests, and receives the consent, of affected indigenous communities. To this end, the Boreal standard obliges a manager to secure agreement of all affected indigenous communities to the management plan; to demonstrate within the plan a good working knowledge of indigenous communities and their interests in the forest management area; to support indigenous communities' capacity "to participate in all aspects of forest management and development"; and to jointly establish, with indigenous communities, "opportunities for long-term economic benefits" and a process for resolving disputes (Indicators 3.1.1 to 3.1.5).

Although the informed consent requirement stipulates that the manager must obtain agreement to the management plan from each affected indigenous community, the standard recognizes that there may be "exceptional circumstances that may influence whether or how consent is achieved" (FSC-Boreal 2004, 39). In these circumstances, the manager is obliged to "make best efforts to obtain positive acceptance" by ensuring that all affected communities clearly understand the plan (30). In this regard, the Boreal standard appears to depart from the BC standard in that the latter prevents certification if affected First Nations are dissatisfied with proposed forest management activities.

Another distinction between the BC and Boreal standards concerns their treatment of indigenous resource and tenure rights. Unlike its BC counterpart, the Boreal standard does not define these rights. Another potential weakness of the Boreal standard is that it allows a manager "to make use of an existing assessment" of the rights as a baseline for developing its management plan (Indicators 3.2.1 and 3.2.2). Moreover, it contains no specific provisions that compel a manager to suspend operations or provide compensation when indigenous resource or tenure rights have been threatened or diminished.

However, the Boreal standard takes a more rigorous and detailed approach in its protection of sites of special cultural, ecological, economic, or religious significance. It designates sites within this category as being "Indigenous areas of concern" and provides a glossary definition that largely tracks the one used in the BC standard. In relation to these sites, under the Boreal standard a manager must support indigenous communities in the development of a joint protection agreement; support indigenous monitoring of forestry impacts on such sites in accordance with the agreement; and suspend forestry activity when it poses a "serious threat" to such sites (Indicators 3.3.1, 3.3.2, and 3.3.3).

The Boreal standard also differs somewhat from the BC final standard with respect to the definition and use of traditional knowledge. The Boreal standard does not define the term "traditional knowledge"; instead, it uses the term "traditional ecological knowledge" (FSC-Boreal 2004, 144). Despite this difference in terminology, the definitions in both standards appear to be analogous. The principal differences between them is that the Boreal standard obliges a manager to enter into agreements with affected indigenous communities to address compensation for the use of traditional knowledge that contributes to commercial use of forest species or improved management planning or operations. In contrast, the BC final standard does not require managers to secure these agreements in advance and does not impose restrictions on when compensation for the use of such knowledge should be granted.

## Conclusion

That FSC-BC was able to develop and unanimously endorse the P3 standard is a significant accomplishment that speaks to the commitment of the parties involved to tackle the challenge of finding a workable, new approach to reconciling the legitimate aspirations of First Nations with the interests of those who will rely on the province's forest resources now and in the future.

The FSC-BC final standard is notably less specific and legalistic than predecessor drafts, but by virtue of its strong "consent" requirements it appears to meet the threshold concern of most First Nations: the need to unequivocally ensure protection of Aboriginal rights. At the same time, it is a standard that provides a viable foundation for the development of new relationships between First Nations and the province's forest companies. It accomplishes this by carefully meeting the need to create a framework within which these relationships can develop without hampering the process by unnecessary prescription.

This said, implementation of the FSC-BC P3 standard will doubtless present significant challenges. Many of these challenges are likely to flow from the complex politics surrounding First Nations issues in British Columbia. A key challenge arises from the fact that the scope and nature of Aboriginal rights and title within the province remains under negotiation and, in some cases, before the courts. This does not preclude interim accommodation arrangements from being used as a basis for FSC certification, but the interaction between such interim arrangements and FSC standards is far from seamless. As a result, in some situations, FSC certification may not be a realistic option for indigenous tenure holders. Particularly challenging issues arise with respect to forest licences granted to First Nations under the FRA/FRO program. The short duration of these licences is likely to make it difficult for a First Nation seeking FSC certification to satisfy P2 requirements regarding long-term tenure and forest use (see Chapter 6). Because such licences are volume-based, it is also likely that the FSC would require the province to be a joint certification applicant. Given these and other complexities in the application of Principles 2 and 3, the question of whether a licence awarded under the FRA/FRO program could be FSC certified is uncertain (Ecotrust 2005).

Additionally, unlike the Rocky Mountain standard, the BC final standard does not define "First Nation" for the purposes of the standard. This is perhaps not surprising given the complex anthropological and political realities of BC First Nations and, in particular, the distinction between traditional and elected First Nations leaderships. Similarly, the drafters of the standard have chosen, at the apparent urging of First Nations themselves, not to address the question of what type of community approval must exist within a First Nations community as a prerequisite to certification.

Another looming challenge concerns the financial, technical, and logistical capacity of First Nations to participate effectively in the certification process. Throughout the standard-development process, this concern was raised repeatedly. In earlier drafts, it was addressed by a formal requirement that First Nations be provided with such support without specifying the source from which this support was to come (FSC-BC 2001, Indicator 3.1.7). In the final standard, the presence of such support has developed into a Verifier that certifiers must consider in deciding whether the manager has secured the informed consent of the affected First Nation (Indicator 3.1.4, Verifier [ii]).

In all of the areas mentioned, it appears that the FSC-BC final standard, as it has evolved, eschews legal formalism in the hope that the good will of the parties involved, and the good judgment of certifiers, will prevail. Whether this leap of faith will herald a new beginning – in a realm where our political and legal institutions and processes have, in many ways, failed – has yet to be fully tested.

# 9
# Environmental Values

Provisions for safeguarding environmental values are critical components of any forest management standard. In the past, industrial management approaches have overemphasized timber production, devaluing the contribution of forests to biodiversity, soil fertility, hydrological flows, and non-timber forest products. It is a new emphasis on ecological values that distinguishes sustainable approaches to forest management from more traditional ones. All forest-certification schemes claim to result in sustainable management of forestlands, but there are substantial differences in the degree to which they incorporate environmental objectives in their principles, criteria, and indicators. Of the currently available forest-certification schemes, the FSC scheme is widely recognized, particularly in global civil society circles, as the most rigorous in protecting environmental values (FERN 2004). Consequently, many firms opt for FSC certification because of its perceived credibility in recognizing and rewarding environmental stewardship (Cashore, Auld, and Newsom 2004).

A significant reason the FSC scheme is seen as more ecologically rigorous than its competitors lies in its theoretical roots in the relatively young discipline of conservation biology.[1] Conservation biology is organized around the central idea of the ecosystem, a concept that highlights the interdependence of natural processes and the emergence and endurance of identifiable plant and animal communities (Grumbine 1994, 1999). Linked to this shift from a reductionist, timber-focused approach to one that is more holistic and based on ecosystem health and integrity, is the recognition that simply protecting species and species habitat, while normal commercial forestry carries on in adjacent areas, does not necessarily safeguard environmental values and processes.[2] Consequently, environmentalists and conservation biologists argue for the need to expand substantially the size and representativeness of parks and protected areas and to calibrate management practices on the remaining land base to ensure that viable populations of native plants and animals endure through space and over time.

Grumbine's (1994) groundbreaking work on ecosystem-based management (EBM) elucidates the range of ecological values that environmentalists often advocate should be incorporated into forest and land management practices.[3] Precisely how EBM translates into best practices on the ground continues to be the subject of considerable debate. Nonetheless, however it is interpreted, it is a management model that poses fundamental challenges to the traditional, sustained-yield forest management (SYFM) paradigm and the modern industrial forest economy that prevails in British Columbia.[4]

Building a consensus for the implementation of EBM can be scientifically and politically challenging, and it can be profoundly influenced by institutional context. Political, administrative, and private property boundaries often divide the landscape into ownership and management jurisdictions that make little ecological sense (Grumbine 1994). This is most assuredly the case in British Columbia, where implementing EBM will require forging new partnerships and enhanced interagency cooperation to ensure that management takes place within appropriate ecologically defined spaces.

The FSC-BC standard-development process became a forum where policy actors operating in different networks and with contrasting approaches to forest management confronted the challenge of negotiating the meaning of EBM in an ecologically diverse, regional context. In this chapter we focus on the outcome of these negotiations with respect to the environmental values embedded in the FSC-BC final standard, which we then compare to the regional and national standards in our five comparator jurisdictions.

## Standard-Development Context

In British Columbia, growing public realization that forests must be managed on an ecosystem basis for a broad range of values set the stage for a head-on clash between interest groups emphasizing the economic values of the forests and others defending environmental processes, biodiversity, and non-timber values. This struggle began to take centre stage in the late 1980s, famously pitting supporters of conventional SYFM on one side against environmentalists, First Nations, and non-timber interests on the other (see Chapter 3).

This "war in the woods" was fought on two main fronts (Wilson 1998). The first was geographic, involving battles over whether particular areas of the province should be designated as off-limits to industrial logging. These battles often took the form of "valley-by-valley" skirmishes, including iconic struggles over the Stein Valley (1980s), Clayoquot Sound (1990s), and, more recently, the Great Bear Rainforest (see Wilson 1998; Drushka 1999; Hayter 2000). For industry and environmentalists alike, the stakes of these often bitter, watershed-based battles were high due to the significant concentrations of timber and biodiversity values within these natural boundaries. On this front, the environmental movement has made important gains.

Over the past thirty years, as a result of commitments made by the federal and provincial governments, British Columbia's protected area network has tripled from around 4 percent of the total land area to almost 14 percent (BC Parks 2007). Despite these initiatives, in some regions of the province less than 5 percent of total land area is protected, and significant gaps remain in the structure of ecosystem representation (WCELA 1997; Sierra Club of Canada 2007).[5]

Battles on the second front have targeted forest management practices, particularly monocyclical silviculture systems that rely on clear-cut logging to produce even-aged forest stands. The theoretical premise of postwar forest policy in British Columbia was SYFM, an approach that viewed old-growth forests with low rates of growth as "unproductive" (see Chapter 3; Dellert 1998; Hessing and Howlett 1999). During the 1980s, as the adverse implications of SYFM for wildlife populations, biodiversity, and environmental services (such as soil stabilization and water purity) were documented, the provincial government came under growing pressure to revisit its commitment to this model of forest management. In response, in the mid-1990s, the New Democratic Party government introduced the *Forest Practices Code,* which reaffirmed a commitment to the SYFM concept but required managers to meet a range of additional management goals for fisheries, watershed, biodiversity, terrain, riparian, and visual quality values. This new, bureaucratized approach to SYFM provoked intense criticism by the forest industry throughout the 1990s due, in large measure, to its high compliance costs.[6] When the Liberal government took office in 2001, this led to experiments with more "results-based" approaches to forest management, culminating in the *Forest and Range Practices Act* of 2004 (Hoberg 2002; see Chapter 3).

For the past two decades, these struggles over forest policy and practices have tended to entrench the value gulf between environmentalists and industry, echoing the divide between public and productive interests in larger Canadian policy circles (Hessing and Howlett 1999). Moreover, in the wake of the Clayoquot conflict, ECSOs began to appreciate the full potential of international market campaigns as a means to leverage social and political change, and they began to revise their tactics accordingly (Walter 2003).

At its outset, the FSC-BC standard-development process was seen as offering a timely and unique venue to promote a broader and more creative dialogue about the future sustainable forest management in British Columbia in the context of the global marketplace. Predictably, however, this process inherited many points of conflict, some of which had the potential to become "showstoppers" in the ensuing negotiations over the FSC-BC standard. These conflicts were particularly notable in five areas: old-growth preservation; clear-cut logging and silvaculture practices; riparian zone protection; use of chemically based pesticides, herbicides, fertilizers, and biological control agents; and the conversion of natural forests to plantations. In the remainder of this chapter, we propose to employ these defining issues for British Columbia's forest policy community, which also figure prominently in industry-environmentalist debates in many other jurisdictions, as a framework for assessing the FSC-BC final standard against standards negotiated in our five comparator jurisdictions.

**Highlights of the Final Standard**

FSC Principles 6, 9, and 10, which deal, respectively, with environmental impacts, maintenance of high conservation value forests (HCVFs), and plantations, are specifically designed to protect ecological values (see Appendix). Principle 6 (P6) states that "forest management shall conserve biological diversity and its associated values, water resources, soils, and unique and fragile ecosystems and landscapes, and, by so doing, maintain the ecological functions and the integrity of the forest." Its ten associated criteria specify actions managers should take to achieve outcomes consistent with the principle. Among these are requirements that managers undertake an environmental impact assessment prior to any site-disturbing operations (Criterion 6.1), and that they establish conservation zones and protection areas within the forest management unit (FMU) to "protect rare, threatened and endangered species and their habitats" (Criterion 6.2). The other eight criteria elaborate on silviculture obligations designed to promote conservation and forest health.

Principle 9 states that "management activities in High Conservation Value Forests shall maintain or enhance the attributes, which define such forests.[7] Decisions regarding High Conservation Value Forests shall always be considered in the context of a precautionary approach."[8] There are four associated criteria. These require managers to conduct assessments to determine the presence of high conservation attributes on their forestlands; to consult with stakeholders and other relevant groups with respect to high conservation forest attributes; to develop management plans and implement measures to maintain or enhance the high conservation attributes; and to implement monitoring to assess the effectiveness of these management measures (see Appendix).

Principle 10 addresses the contrasting phenomenon: plantations.[9] It states that "plantations shall be planned and managed in accordance with Principles and Criteria 1-9, and Principle 10 and its Criteria. While plantations can provide an array of social and economic benefits, and can contribute to satisfying the world's needs for forest products, they should complement the management of, reduce pressures on, and promote the restoration and conservation of natural forests." The nine criteria for P10 address a range of issues, including specification of management objectives, the design and layout of plantations, tree species selection, and the proportion of overall forest management area to be restored to natural forest cover. Criterion 10.9

states that natural forests converted to plantations after November 1994 will "not normally be certified" (see Appendix).

Together with the associated criteria, these three principles outline FSC's generic provisions for landscape planning and forest management. In British Columbia, the process of regional standard development involved negotiating environmental indicators, and in some cases verifiers, for these generic principles and criteria. Given the history of conflict over forest values in the province, it was clear that the success of these efforts would be judged in terms of the balance struck between the economic values of forests to industry and the ecological values of forests to the environment and environmentalists. That balance is set out in FSC-BC's final standard.

We now turn to a detailed analysis of the final standard in order to assess its application in the five core issue areas set out above: old-growth and biodiversity protection, clear-cut logging and silviculture practices, riparian management, chemical use, and plantation establishment.

**Old-Growth and Biodiversity Protection**
Invariably, old-growth and biodiversity protection is one of the most contentious issues when forest certification standards are negotiated. This proved especially true in a jurisdiction as large and diverse as British Columbia, a province which has been described as "more variable physically and biologically than any comparable region in Canada" (Pojar and Meidinger 1991, 40), with more than 1,100 species of vertebrates alone (British Columbia Ministry of Water, Land and Air Protection 2002).[10] Although many of these species are associated with old-growth forest ecosystems – and a modest number of vertebrates depend on them – serious deficiencies also attend inventories of old growth and scientific understanding of the relationship between old growth and species diversity (MacKinnon 1998). In negotiating the BC standard, therefore, key questions included the following: How should the concept of "old growth" be defined? How should old-growth forests (redefined as HCVFs for protecting biodiversity) be identified and mapped? How much old-growth forest should be protected? And what management practices should be permitted in unprotected old-growth forests?

The final standard aims to secure old-growth and biodiversity protection by means of indicators developed under P6 and P9 that address such issues as inventory arrangements and risk/impact assessments (Indicators 6.1.1 to 6.1.10); rare, endangered, and threatened species (Indicators 6.2.1 to 6.2.5); protected reserves (Indicators 6.2.1 to 6.2.5); conversion of natural forest to plantations (Indicators 6.10.1 to 6.10.3); and HCVF (P9 indicators).[11]

Despite the ubiquity of the term, however, the FSC-BC standards negotiation process was not framed around the concept of "old growth." Early on, FSC-AC had concluded that, as it was originally articulated, P9 was inadequate for wrestling with this and related terms, such as "primary" or "virgin" forests. Difficulties included lack of definitional clarity, conceptual confusion, ambiguous usage by certifying bodies, and the inability to translate the concept into Spanish, FSC's second official language. With respect to the latter, one of our interviewees noted that "FSC operates in two official languages, Spanish and English – so the concept that people were really pushing for was wilderness, or intact undeveloped forest. There is no [equivalent] word for 'wilderness' in Spanish ... so we couldn't use *that* word. Second, we couldn't use 'undeveloped,' because people from the South were offended by that term" (personal interview 2004).

Consequently, based on the work of a task force struck in 1998, the FSC elected to adopt an alternative terminology, namely, "High Conservation Value Forest" (HCVF). From an HCVF perspective, some forests embody important ecological values in the form of mature stands that provide crucial habitat for old-growth-dependent species; others are valuable because they provide vital services for a local community or contribute to a sense of cultural identity. The HCVF approach therefore is intended to shift attention from protecting a forest because of its age, structure, and composition, per se, to identifying conservation values that the forest provides and determining how best to protect them.

The glossary to the FSC-BC final standard provides an extensive elaboration of the HCVF concept (2005a, 67).[12] The first part of the definition is the most important, characterizing HCVFs as forested areas possessing one or more of a list of ecological attributes, including viable populations of most if not all naturally occurring species that exist in natural patterns of distribution and abundance;[13] critical habitats of globally, nationally, or provincially threatened species; critical habitats of endemic species; unusually high naturally occurring species diversity, migratory concentrations of species or individuals, or other rare ecological or evolutionary phenomena; and forest areas associated with high-value fish habitat and other critical aquatic habitat.[14] The definition also covers forests that provide natural services in critical situations, and forest areas fundamentally linked to the basic needs and traditional cultural identity of local communities.

What this detailed and comprehensive definition does not predetermine is the geographic location of HCVFs. Instead, it is left to managers to identify the presence of HCVFs, which, in turn, requires them to undertake an assessment appropriate to the scale and intensity of forest management according to six general HCVF categories.[15] The assessment must be carried out by "qualified specialists" and must include "consultation with directly affected persons and relevant interests (e.g., First Nations, regulatory agencies, local communities, conservation organisations) (Indicator 9.1.5)."[16]

Once the appropriate HCVF assessment has been completed, management strategies must maintain or restore identified HCVF attributes and include a risk assessment and an appropriate monitoring program to facilitate adaptive management (Indicator 9.1.4). Management strategies and selected main-tenance/restoration measures must also be consistent with the precaution-ary approach, with the onus placed on the manager to demonstrate a high probability of long-term ecosystem maintenance or restoration (Indicator 9.3.2). In the event that management strategies and measures for main-taining or restoring an HCVF are disputed by directly affected persons or qualified specialists, "the onus is on the manager to prove that HCVFs and their associated conservation attributes have been adequately identified and assessed, and will be maintained under the proposed management strategies" (Indicator 9.1.6).[17]

Several P6 indicators play an important role in ecological conservation and biodiversity protection (Criteria 6.1, 6.2, 6.4, and 6.10). With regard to inventory and risk assessment, the manager is obliged to take the following ac-tions: collect environmental impact assessment data (Indicator 6.1.1)[18]; build ecosystem inventories, assessments, and information databases (Indicator 6.1.3); and complete pre-harvest cutblock inventories (Indicator 6.1.6).[19]

Furthermore, managers and certifiers are explicitly directed to specific sources to ascertain the existence of rare, threatened, and endangered spe-cies (Indicators 6.2.1 and 6.2.2). Where these are identified on the FMU, the manager must take effective measures to "minimize risk to the long-term persistence of those species and/or plant communities" by establishing a protected reserve network, avoiding habitat alteration that increases risk of species loss, and restoring previously altered habitat to a suitable condition. In order to ensure compliance with Criterion 6.4, certifiers are to check that "a network of protected reserves is established at multiple scales and managed within" the FMU. That network is to be "delineated on maps," there should be "written objectives for each reserve area related to that area's contribution to maintaining or restoring ecological integrity," and the network must meet "the applicable minimum percentage area for ecosystem representation by [biogeoclimatic ecosystem classification] variant within the FMU" (Indicator 6.4.1).[20]

Maintenance and protection of HCVFs and biodiversity is ensured by proscribing large-scale conversion to plantations or other non-forest uses (Indicator 6.10.1). Areas of new conversion must not exceed 5 percent of the FMU's timber-harvesting land base (THLB),[21] and their location must be prioritized to "previously harvested poorly managed forest."[22] Any conver-sion of natural forest to plantations must not "directly result in the area of old growth forest falling below the estimated mean area of old growth forest determined by the description of the *range of natural variability* completed under Indicator 6.1.7" (emphasis added).

The concept of "range of natural variability" (RONV),[23] requires that managers prepare "a written description of the estimated range of natural variability including reference to ecosystem conditions and ecosystem functioning" to serve "as an environmental base case (i.e., benchmark or reference ecosystem conditions) against which to measure potential environmental changes or impacts resulting from proposed management activities" (Indicator 6.1.7). RONV was central to the debate over "what is natural?" in the FSC-BC standard-development process and is a foundational concept that informs many other components of the final standard. The purpose of conducting a RONV analysis is to develop a "base case" and benchmark for forest management planning, which allows managers to determine the relative degree of risk in forest management strategies. Further discussion of RONV and its application is found in the section that follows.

## Clear-Cut Logging and Silvicultural Practices

Although practised in British Columbia since the 1930s, silviculture was initially limited to "encouraging reforestation of logged-off lands" and "after-the-fact activities such as planting seedlings on lands clearcut or burned, and providing for the regeneration of trees by modifying harvesting methods" (Drushka 1999, 210). A more systematic approach evolved around an even-aged approach to forest management, which reached its zenith in 1996-97, when more than 90 percent of the timber cut in British Columbia was harvested using clear-cut logging methods (212).

The belief that clear-cutting does no damage – and the corollary argument that clear-cuts mimic natural disturbances, such as wildfires – was challenged by advocates of ecosystem-based forest management. Although many industrial operators continue to defend clear-cut logging, it is increasingly accepted that the practice is only appropriate for specific forest types.[24] In British Columbia, the debate took a dramatic turn in 1998 when MacMillan Bloedel (MacBlo), a company that had long defended the practice of clear-cut logging and was embroiled in an international public relations struggle with environmentalists over its forest practices, suddenly announced that it would phase out clear-cut logging and move to a regime of variable-retention logging (Hunter 1998). The announcement stunned government officials and politicians and upstaged other industry officials who had until this point vigorously defended the practice. The provincial government followed MacBlo's lead, embracing a more ecosystem-based approach to silviculture and phasing out its blanket endorsement of clear-cut logging. Nonetheless, clear-cutting remains a predominant method of forest harvesting in British Columbia today.

In an effort to mediate the values conflict between industry and environmentalists over clear-cut logging, the FSC-BC final standard binds silvicultural practices tightly to the concept of RONV. The significance of RONV lies in the information it provides the manager about the forest's natural condition. It obliges managers to continually evaluate the forest conditions to ensure that these are maintained within fundamental ecological parameters. As a consequence, without explicitly proscribing clear-cut logging, the strict application of RONV curtails its use, especially on the Coast. When RONV is applied as a "rule" rather than simply as a "tool," approaches such as clear-cut logging become feasible only in certain types of forest ecosystems.[25]

The key elements of the final standard's approach to clear-cut logging and silvicultural practices – constraining forest management by RONV to ensure outcomes consistent with nature, in conjunction with quantitative measures for basal area and green tree/snag retention – were established early on in the FSC-BC standard-development process. The final standard therefore represents negotiated quantitative minimum levels of retention that are calibrated for different natural disturbance types (NDTs) and based on the number of stems/snags per hectare, rather than on a percentage of basal area.[26]

## Riparian Management

British Columbia has a vast network of rivers of all sizes and lengths, myriad lakes, numerous wetlands that provide critical habitat for bird populations, and a long Pacific coastline dotted with islands. Managing this complex water system requires a detailed understanding of terrestrial-riparian ecosystem interactions. Riparian areas, which include these rivers, lakes, wetlands, and the marine coast, are vital from not only an ecological but also an economic perspective. They offer vitally important habitat and ecosystem services to a wide range of species and support a range of social values, including potable drinking water and opportunities for fishing, tourism, and recreation, for residents and visitors alike.

Although the detrimental impacts of logging in riparian areas have long been a source of concern for ECSOs and other public policy actors in British Columbia, little scientific analysis on this topic was available prior to the late 1990s (Voller 1998). Subsequent research has shown that industrial timber production contributes large inputs of organic debris into river systems; that herbicides and pesticides used in silvicultural operations can threaten the health of various species that inhabit riparian areas; and that road building can result in landslides and silt accumulation in neighbouring water bodies that undermine salmon spawning (Voller 1998). As a result of this emerging research, there have been growing calls for new management arrangements to protect riparian areas from the adverse effects of logging in general, and clear-cutting in particular. However, these arrangements can have significant impacts on the volume of available timber and on forestry costs. From the early stages of the FSC-BC standard-development process, therefore, it was clear that negotiations over riparian management practices would be difficult.

The final standard contains several indicators dealing with riparian management. Some deal with the effects of roads and machines (Indicators 6.5.6

and 6.5.7). Others are set out in a controversial new Criterion 6.5*bis*, which reads: "Riparian ecosystems and all their functions shall be maintained or restored."[27]

The newly created criterion includes the most important provisions for riparian management, requiring managers to conduct an integrated riparian assessment (IRA). Managers have the option of doing this according to a detailed framework set out in Appendix B; if they choose not to use that framework, however, then they are permitted to employ an alternative approach providing it "meets the intent and addresses all the issues raised in the framework" (Indicator 6.5*bis*.1a).[28] Appendix B specifies a six-step process for the conduct of an IRA and obliges managers to adapt the riparian management regime to different categories of streams, lakes, wetlands, and marine shorelines.[29] The approach, which is sensitive to the scale and intensity of the operation, establishes minimum requirements for a riparian system and combines riparian reserve zones (RRZ), within which no logging can occur, with riparian management zones (RMZ), for which appropriate levels of variable-retention harvesting are recommended based on stream size and importance.[30] Similar provisions are made for lakes and marine shorelines.

## Pesticide Use

Forests everywhere are subject to various types of disease and pest infestations. In British Columbia, the primary concern in this regard is damage by bark beetles, most notably the mountain pine beetle (*Dedroctonus ponderosae*) and defoliators.[31] The mountain pine beetle (MPB) represents "the largest threat to BC's forests in terms of timber loss, environmental concerns and negative economic impacts," with infestations "causing extensive tree mortality throughout the range of pine in the province" (British Columbia Ministry of Forests 2003a, 8; see also Chapter 3).[32]

To control pest outbreaks in BC forests, managers conventionally apply pesticides, predominantly Foray 48B®, via aerial spraying. The active ingredient in Foray 48B® is *Bacillus thuringiensis var. Kurstaki,* abbreviated to Btk. Just over 2 percent of this naturally occurring bacillus is mixed with a variety of "inert" ingredients to create the commercial compound. Because of commercial confidentiality laws, those concerned about exposure to Foray 48B® are unable, even using the *Freedom of Information Act,* to obtain detailed information about all the chemical compounds in such products, which has prompted considerable public anxiety and political controversy (British Columbia Ministry of Forests 2003a). Public outcry against aerial spraying, coupled with the growing awareness of the consequences for non-target species in the food chain (i.e., birds) has thus raised serious questions about the ministry's policy in this area (Pearce, Behie, and Chappell 2002, 20).[33]

The FSC's recently revised pesticides policy aims to "minimise the negative environmental and social impacts of pesticide use whilst promoting economically viable management of the world's forests" (FSC-IC 2005a, 3). Pests are defined very widely as "organisms, which are harmful or perceived as harmful and as prejudicing the achievement of management goals" and include "animal pests, plant weeds, pathogenic fungi and other micro-organisms." The policy distinguishes two categories of pesticide: highly hazardous and non-highly hazardous (5-6). Highly hazardous pesticides, based on active ingredients, are listed in an annex to FSC-IC's *FSC pesticides policy: Guidance on implementation.*

In the FSC-BC final standard, managers must not use any of the prohibited chemical pesticides identified in this annex and must also commit to phasing out use of permitted pesticides within five years of certification. Managers are also obliged to provide "evidence of consistent effort to meet plans for their phase out, including the use of integrated pest management, with emphasis on prevention strategies" during the five-year phase-out period (Indicator 6.6.2.).[34]

Pesticides like Btk that do not appear on FSC's list of highly hazardous chemicals are broken into two categories: native and exotic (C6.8). Exotic biological control agents may only be used "as part of a pest management strategy for the control of exotic species of plants, pathogens, insects or other animals when other non-chemical pest control methods are, or can reasonably be expected to be, ineffective" (Indicator 6.8.10). Their use is contingent, furthermore, "on peer-reviewed scientific evidence that the agents in question are non-invasive and are safe for indigenous species" (I6.8.10).[35] In accordance with FSC guidelines and global policy, the use of genetically modified organisms is explicitly banned (Indicator 6.8.3). The final standard is also consistent with the position on chemical use outlined by the FSC-Canada National Standards Advisory Council, which permits a phase-out over a reasonable period of time, temporary exemptions (derogations), and exceptional use of chemical pesticides and biological control agents not included in the list of prohibited pesticides, in accordance with FSC policy and the "Mandate for FSC Board Committee on Chemical Pesticides" (FSC-AC 2002d).

## Plantation Establishment
A relatively small group of countries currently account for the vast majority of all plantations established worldwide (Brown 2000).[36] In most cases, the plantation managers plant fast-growing hybrid tree species, mainly eucalypts (*Eucaplyptus* spp.), acacias (*Acacia* spp.), and pines (*Pinus* spp., especially *Pinus radiata*) (Brown 2000). Notably, Canada is omitted from lists of plantation countries because it does not distinguish between natural and planted forests in its official forest statistics. This, in turn, reflects ambiguities

in the definition of "plantation" and in the degree of management intensity and the regularity of tree spacing required to qualify (Brown 2000). Most managed forests in Canada do not strictly conform to the definition because reforestation and afforestation generally involve broadcast seeding of natural species.[37]

In 2001, the Canadian Council of Forest Ministers (CCFM) opened for debate the idea of revisiting Canada's designation as a non-plantation country via its Forest 2020 dialogue (Arseneau and Chiu 2003). This was due to the potential of plantation-style forest development to deliver significantly more wood from a substantially smaller land base.[38] From an ecosystem-based perspective, however, plantations that replace natural forest ecosystems or require large-scale application of artificial inputs to keep them productive are problematic. Use of chemical fertilizers, pesticides, and herbicides create externalities in the form of deteriorating water quality, pollution, and biodiversity loss, which should be, but rarely are, factored in as costs of plantation establishment. On the other hand, if plantations are established on previously degraded land, they have the potential to yield net positive environmental outcomes and reduce pressure on the natural forests to provide fibre. In an Australian context, for example, Ajani (2007) argues that plantations could obviate the need for any commercial harvesting of native forests.

During the FSC-BC standard-development process, a central challenge was to negotiate an agreement on plantations that would satisfy both ECSOs and large producers. The former argued that "no plantations (new or old) should be allowed under FSC certification in BC," a stand that was countered by large industry, which "favoured setting no specific thresholds for old or new plantations" (FSC-BC TAT 2001, 31). The final standard contains a compromise solution that replaces the plantation-natural forest dichotomy with a more complex typology consisting of five different forest types: (i) plantations; (ii) former plantations undergoing restoration; (iii) poorly managed natural forest; (iv) natural forest; and (v) land (forested or not) slated for possible future plantation conversion (FSC-BC 2005a, 76). The plantation definition was also elaborated significantly for the BC context, and both the total size and proportion of new conversions was restricted.[39]

The expanded clarification of the plantation concept provided in the glossary uses a process approach that defines plantation management regimes by the following characteristics:

- a *set of stand characteristics* that are present and observable on site, as a result of past and/or current practices
- a long-term *management regime* to maintain or intensify those stand characteristics
- an *intent* to manage for economic or tree growth objectives to the exclusion of others. (FSC-CAN 2005, 75; original emphases)

Managers and certifiers can determine whether a plantation management regime exists by ascertaining whether the following three criteria occur together:

- At least two ecosystem components are eliminated or maintained in a highly altered state. [40]
- At least three intensive land management practices/treatments are used to maintain highly altered ecosystems over the long-term. [41]
- Management objectives (a) emphasize timber/wood fibre as primary, and (b) do not address maintenance/restoration of stand structural characteristics compatible with RONV. (FSC-CAN 2005, 75)

Managers are prevented from converting more than 5 percent of their THLB on the FMU to plantation, and any conversion must target previously harvested, poorly managed land (Indicator 6.10.1). [42] HCVF cannot be converted to plantation under any circumstances. The total area of existing plantations and new conversions to plantations must not exceed 10 percent of the THLB (Indicator 10.5.1). [43] Where this has occurred in the past, managers must undertake restoration forestry to reduce the total area of plantation on the FMU to required levels (Indicator 10.5.2). In addition to meeting quantitative limits on area and prioritizing the most degraded sites for plantation activity, the manager must also consider broader landscape-level biodiversity objectives when locating plantations under the final standard (Indicator 10.2.1).

The FSC-BC definition of "plantation" is designed to distinguish forest areas managed outside the RONV exclusively for timber production on a continuous basis from forest areas managed for timber production within the RONV that also achieve a range of other values. Because conversion to even-aged forests via clear-cut logging is not designed to produce plantations as defined in the standard and because, in most cases, BC forest management is still intended to provide additional non-timber values on Crown land, relatively little of the province's forestland is likely to be classified as plantation under the final standard.

### Regional Comparisons

In this section we compare the provisions of the FSC-BC final standard to regional standards from five other FSC jurisdictions – Sweden, the Canadian Maritimes, the US Pacific Coast, the US Rocky Mountains, and the Canadian Boreal.

### The FSC-Sweden Standard

The first Swedish standard is the only one among our comparators that does not require extensive data collection for the purpose of an environmental assessment on the FMU (FSC-Sweden 1998). In this early standard, inventory requirements are minimal, obliging the forest manager to give only "proper consideration" to red-listed species in areas "where they are known to occur." It does, however, employ a "mapping" approach in which HCVF is largely predetermined by location, especially in montane forests.[44] Timber harvesting is explicitly proscribed on "virgin-type" montane forests,[45] and the standard requires that all montane forests be subject to special management arrangements on the basis of high biodiversity value. The FSC-Sweden standard's quantitative approach to reserve establishment is similar to that used in the FSC-BC final standard (although forest reserve set-aside requirements are smaller).[46] General provisions for forest reserves are less specific, and there are few consultative provisions.

The FSC-Sweden standard contains relatively few silvicultural practice provisions, which places it at the opposite end of the spectrum from the BC final standard. Although there is a preference for natural regeneration methods, "mother tree" retention, and variable-retention logging in this standard, the requirements are relatively lenient regarding clear-cut logging, which is still practised extensively in the Swedish forest industry.[47] The FSC-Sweden standard's provisions dealing with riparian management are the least specific of all the comparator jurisdictions; most importantly, no thresholds are identified, leaving considerable discretion for both forest managers and certifiers to determine whether management activities adequately protect bodies of water.[48] Similar to the FSC-Boreal standard, the FSC-Sweden standard also has relatively lenient provisions concerning pesticide use, permitting it where properly established procedures exist.[49]

The original Swedish standard was developed before the final version of the FSC's generic P10, on plantations, was adopted, and this principle and its related criteria are not directly addressed.[50] Although the standard expresses a general preference for natural regeneration, there appear to be relatively few constraints on plantation management (see, for example, Criterion 6.5). The revised FSC-Sweden standard specifically addresses this lacuna, drawing a distinction between "managed forests" and "plantations" (FSC-Sweden 2005, 38). It notes that "the Swedish forests have been managed for centuries," producing "the regenerated forests that now prevail in the Swedish forest landscape"(38). This regenerated forest "often holds certain elements of natural complexity, diversity, and structure," but not "to the same extent as a natural forest." Thus these "managed forests" are distinguished from both natural forests and plantations, the latter being separately and rigorously defined (38).[51]

## The FSC-Maritimes Standard

The FSC-M standard was developed for ecosystems and landscapes that had already been substantially transformed. Hence, a major objective of this standard is to promote restoration forestry that will move the industry "toward sustainability – ecologically, economically, and socially" (3). Another difference between British Columbia and the Maritimes is that much of the forested land in the Maritimes is privately owned – either by large companies like J.D. Irving or by smaller woodlotters. The standard does not differentiate between public and private landholders in terms of requirements, however, in contrast to the various standards developed for the United States.

The FSC-Maritimes standard requires that no old-growth stands be harvested (Indicator 6.2.2). It makes reference to developing benchmarks "against which to measure potential environmental change," but it does not provide details on how this should be undertaken. Managers also have limited consultative obligations regarding HCVFs, other than conducting an HCVF assessment. There are no provisions regarding the level of consultation required under P9, although managers are required to consult with provincial and World Wildlife Fund representatives.[52] In contrast to the FSC-BC final standard's complex arrangement for determining the size of forest reserves,[53] the FSC-M standard uses peer-reviewed gap-analysis studies to determine the presence of important ecosystems requiring protection.

Under the FSC-M standard, "clear cutting or other aspects of even-aged management may be appropriate" when "used as the best tools to restore the natural forest type (including non-timber forest values), appropriate to the site" (FSC-M 2003, 14).[54] In terms of riparian management, the standard adopts an approach similar to that of the BC final standard, requiring maintenance of riparian buffer zones adjacent to all bodies of water and watercourses.[55] As regards plantation establishment, the FSC-M standard takes a straightforward approach: "Conversion of natural forest to plantation is prohibited,"[56] and plantation management must aim to restore the natural forest for the district and soil site (Criterion 10.1.1 indicators).[57]

The most controversial feature of the FSC-M standard is its prohibition on pesticides. It states that "management is explicitly committed to using no biocides in its forestry practices, and has developed an integrated silvicultural strategy that supports a no use approach " (Criterion 6.6.1).[58] As is outlined in Chapter 5, the emphasis placed on this issue in the Maritimes reflects the public outcry against heavy use of aerial spraying of Foray 48B® to control a gypsy moth *(Lymantria dispar)* infestation. This unconditional prohibition of what the FSC-M standard terms "biocides"contrasts with the FSC-BC approach, which distinguishes between prohibited and permitted pesticides and encourages the phasing out of the latter over a five-year period, but recognizes that emergency situations may warrant their continued use.[59]

## The FSC-Rocky Mountain Standard
A key contextual feature of the FSC-Rocky Mountain (FSC-RM) standard is that it was developed primarily for application on private lands. It requires inventory data collection for comparison against historical forest conditions,

but as with the FSC-Pacific Coast (FSC-PC) standard, there is no reference to an overarching concept like RONV. "Old-growth forests" are defined by their structural characteristics in the appended glossary, a definition supported by citations of regionally developed and widely accepted descriptions of old growth developed for three sub-regions (Green et al. 1992; Mehl 1992; Hamilton 1993).[60] It also recommends normally designating such forests as HCVFs "due to the scarcity of old-growth forests in the lower 48 states" (FSC-RM 2001, 9), while "appropriate consultations with local and regional scientists and stakeholders" must be undertaken in conjunction with "public review of proposed HCVF attributes and areas" (Indicator 9.1). Provisions for forest reserves are generally less specific. Managers may conduct a gap analysis and ensure that the "size and extent of representative samples on public lands being considered for certification is determined through a transparent planning process that is accessible and responsive to the public" (Indicator 6.4.c).

The FSC-RM standard explicitly permits clear cutting either as a restorative measure or as part of an "even-aged management" regime.[61] It is allowed where it is "ecologically appropriate to the forest type" or where it is the only remedy for an imbalance in the natural disturbance regime created by human activity (Indicator 6.3.a.4). With respect to riparian issues, the standard takes "a conservative approach that puts aquatic and riparian protection above timber production," employing a variation on the IRA approach to riparian management that provides for streamside management zones adjacent to all bodies of water and watercourses (Indicator 6.5.o).[62]

The FSC-Rocky Mountain standard states a preference for integrated pest management but permits use of chemicals that are not highly hazardous on the condition that there is a written prescription that "fully describes the risks and benefits of their use and the precautions that workers employ" (Indicator 6.6.c). Like the FSC-BC final standard, the standard allows for exotic species to be controlled with exotic predators or biological control agents when peer-reviewed scientific evidence demonstrates that the species in question are non-invasive and do not diminish biodiversity.

With respect to plantation establishment, the FSC-RM standard requires that a proportion of previously converted forestlands be restored to natural and semi-natural forest cover,[63] and it sets a forty-acre upper limit for average opening size, although larger openings that "can be justified by scientifically credible analyses" are permitted (Indicator 10.2.a).[64]

### The FSC-Pacific Coast Standard
Of the standards considered in this section, the one developed for the Pacific Coast governs forests that are closest in type to British Columbia's temperate coastal rainforests. Unlike those in British Columbia, however, the majority of forests covered by the FSC-PC standard are not publicly

owned and, as with its Rocky Mountain counterpart, the Pacific standard was explicitly developed for owners and managers of private lands. To this end, the standard underscores the need for more public consultation than is currently required for management practices on private lands to ensure adequate representation of stakeholders and consideration of community values in order to incorporate conservation and restoration requirements at stand, forest, landscape, and ecosystem levels.

Like the FSC-BC final standard, the FSC-PC standard requires that inventory data be collected to allow for comparison against historical forest conditions; however, FSC-PC does not refer to an overarching concept such as RONV. Instead, it takes a mapping approach (Noss 1997), and a list of HCVFs categorized by location and forest type are provided as "one resource to guide forest managers as they determine the conservation value of a forest" (FSC-PC 2002, 63).[65] It requires consultation with stakeholders and scientists "to confirm that proposed HCVF locations and attributes have been accurately identified," and it stipulates that a "transparent and accessible public review of proposed HCVF attributes and areas" must be conducted on public land (Indicator 9.2.a).[66] Thus, in contrast to the BC final standard and similarly to the FSC-Maritimes standard, the FSC-PC standard uses peer-reviewed gap analysis studies to identify ecosystems requiring protection.

Clear-cut logging is explicitly permitted in the Pacific Coast standard as a restorative measure where "(1) native species require openings for regeneration or vigorous young-stand development, or (2) it restores the native species composition, or (3) it is needed to restore structural diversity in a landscape lacking openings" (Indicator 6.3.f.1).[67] In contrast, the BC final standard uses RONV as a "rule" to establish whether clear cutting, variable-retention logging, selective logging, or some other silvicultural approach should be taken. The FSC-PC standard also employs a variation on the IRA approach used in the BC final standard, identifying four categories of streams, with different protection regimes specified for each.[68]

Pesticides (including fungicides and herbicides) may be used in the Pacific Coast region, but only when research or empirical experience demonstrates that less environmentally hazardous practices are ineffective. Chemical applications must be narrowly targeted to minimize effects on non-target species, a restriction that presumably discourages aerial spraying. To reduce and/or eliminate chemical use, the FSC-PC standard requires that "managers employ silvicultural systems, integrated pest management and strategies for controlling pests and/or unwanted vegetation that result in the least adverse environmental impact" (Indicator 6.6.b).

Unlike other standards, including FSC-BC, the FSC-PC standard provides that, for portions of the FMU managed as plantations, FSC's Principles 1 to 9 do not necessarily apply.[69] This means that P10 is the major governing principle with respect to plantations in the Pacific Coast, a feature that will tend to facilitate plantation development in the region. However, under P10, FSC-PC establishes both upper limits on harvest opening size (Indicator 10.2.a)[70] and a sliding scale that relates the size of the FMU to the amount of area allowed for in a plantation management regime (Indicator 10.5.a).[71]

### The FSC-Boreal Standard

The FSC-Boreal standard contains a number of elements similar to those in the FSC-BC final standard. For example, FSC-Boreal requires that inventory data be used to generate an overarching conception of the forest's pre-industrial condition (PIC).[72] But where the BC final standard treats RONV as a "rule" for establishing ecological boundaries within which forest managers must operate, the FSC-Boreal standard employs PIC more as a "tool" to improve managers' understanding of forest dynamics without necessarily proscribing management options that depart from the PIC.[73] Although the PIC concept is intended to guide rather than constrain, managers are nonetheless specifically enjoined from adopting strategies that "attempt to mimic extreme events of low frequency" (Indicator 6.3.7).[74] Like the standards for our other comparator jurisdictions, the FSC-Boreal standard contains provisions for determining the presence of important ecosystems requiring protection via peer-reviewed gap-analysis studies.[75]

The FSC-Boreal standard provisions for riparian management are similar to those of the BC final standard. It permits partial harvesting within inner riparian reserves (IRRs) to "a minimum width of 20m from the treed edge of permanent water bodies."[76] These may be combined with additional reserves equivalent to at least an additional 45 metres, on average, as measured from the end of the IRR (Indicator 6.3.17).[77] This is similar to the RRZ and RMZ approach of the BC final standard.

Unlike the BC final standard, and in stark contrast to the FSC-M standard, the FSC-Boreal standard has relatively lenient provisions for pesticides, permitting their use where properly established standard operating procedures exist. Applicants are, however, encouraged to use integrated pest management programs wherever possible and to demonstrate continual reduction of chemical pesticide use with "the eventual goal ... [of] complete phase-out over time" (Indicator 6.6.3).

The FSC-Boreal standard's provisions on plantation conversions[78] are similar to those in the FSC-BC final standard.[79] However, interestingly, the FSC-Boreal standard explicitly addresses an apparent inconsistency in the generic FSC principles and criteria, where Criterion 6.10 permits some level of conversion of forestland to plantations while Criterion 10.9 appears to prohibit any conversion of natural forestland to plantations after November 1994. In the FSC-Boreal standard, Criterion 6.10 takes precedence over Criterion 10.9, thus permitting the certification of those plantations that were established after 1994 "where there are conservation benefits" (FSC-Boreal

2004, 125). The FSC-Boreal standard is the only one to directly address this issue, which has created some interpretative difficulties for certifying bodies in the past.

**Conclusion**

All the standards in our comparator jurisdictions adopt what can be described as a sustainable ecosystem-based approach to forest management, which aims to protect old growth, high conservation value forests, and biodiversity (see Table 9.1). Of the several versions of FSC regional standards considered here, British Columbia's is one of the more thoroughgoing; it obliges forest managers to do the following:

- conduct extensive inventories and collect substantial data;
- employ a broad-based definition of HCVF;
- conduct a detailed HCVF assessment;
- operate within the parameters of RONV;
- set aside substantial portions of the FMU in protected reserves;
- consult widely with experts, communities, stakeholders, and directly affected persons.

The FSC-BC final standard approach to identifying and protecting HCVFs goes beyond the mapping requirements adopted elsewhere. Managers are directed to employ a precautionary approach, and they bear the onus to demonstrate their adherence to this approach when other parties dissent on the adequacy of the HCVF plan. The size of forest reserves is determined by a complex arrangement that considers the extent to which any given ecosystem is protected outside the FMU. The final standard is also more explicit than its counterparts with respect to the level and character of the required consultation and involvement of qualified individuals, directly affected persons, indigenous peoples, and other relevant interests.

The generic FSC criteria related to clear-cut logging and silvicultural practices include requirements to maintain ecological functions and values (Criterion 6.3), control erosion, minimize forest damage during road construction and harvesting, and protect water resources (Criterion 6.5).[80] Of the standards considered here, FSC-BC's adopts one of the more innovative approaches to forest management by linking it to RONV and requiring that forests be maintained in or returned to their natural condition (see Table 9.2).[81] Most forests in British Columbia currently consist of uneven-aged stands, so this approach proscribes the application of clear-cut logging in many regions, especially on the Coast.[82] Provisions for clear-cut logging and silvicultural practices vary considerably across the comparator jurisdictions, but all appear more permissive than the BC final standard.

*Table 9.1*

___

**Comparison of provisions related to old-growth forest and biodiversity protection**

*FSC-BC*
- A "range of natural variation" (RONV) report is required. Environmental risk assessment is required. Measures are in place to protect red, blue, threatened, and endangered species.
- Network of protected reserves is established.
- Qualified specialists conduct high conservation value forest (HCVF) assessment with consultation from directly affected persons and relevant interests. Precautionary approach is used to manage HCVFs.

*FSC-Sweden*
- Montane forests to be treated specially; virgin-type montane forest to be excluded from harvesting.
- In non-montane areas, forests with old-growth characteristics to be managed for biodiversity. Key habitats (as defined by the National Board of Forestry) to be protected.
- At least 5% of the productive forest area to be managed for biodiversity values.

*FSC-Maritimes*
- Environmental impacts to be assessed prior to and following timber harvesting.
- Threatened and endangered species listed in federal and provincial legislation to be protected.
- Old-growth stands not to be harvested. Areas with unusually high native species or ecosystem diversity to be identified and protected. Large landowners to reserve a portion of their FMU to IUCN I or II levels. Forest conversion to plantations to occur in exceptional circumstances.
- Manager to actively support multi-stakeholder initiatives to establish systems of protected areas. HCVFs to be assessed using a consultative process and to be managed and monitored appropriately.

*FSC-Rocky Mountain*
- Assessment of current conditions to be completed.
- Sensitive, rare, threatened, or endangered species to be protected by designating conservation zones. Timber harvesting is permitted in conservation zones, provided the habitat is not degraded.
- Old-growth forests normally designated as HCVFs. Existing old-growth may be logged if replacement old-growth is recruited, resulting in no net loss of old growth.

*FSC-Pacific Coast*
- Manager identifies and describes ecological processes, common plants and animals and their habitats, rare plant communities, rare species/habitats, etc. Current ecological conditions are to be compared with historical conditions. Assessments to be used to develop options to maintain/restore ecological functions of forests.

___

▶

◀　*Table 9.1*

---

- Conservation zones to be established for existing rare species. Sample areas to be designated as ecological references to create or maintain representative system of protected areas, and/or to protect sensitive, rare, or unique features. HCVFs may be actively managed.
- Consultations to be held with stakeholders and scientists to confirm accuracy of HCVF locations. Transparent and accessible public HCVF review required for public forests. HCVFs managed for timber to maintain area and quality of conservation attributes.

### FSC-Boreal

- Assessment of environmental impacts required. Forests' pre-industrial condition (PIC) is characterized and benchmarks of current forest condition are established. List of species at risk is prepared and updated annually. Peer-reviewed scientific gap analysis is completed to determine need for protected areas.
- HCVFs are not converted to plantations; maximum of 5% of productive forest area to be converted to plantations. HCVFs are mapped according to the HCVF national framework. Qualified specialists, directly affected people, and indigenous peoples are involved in HCVF assessment.

---

*Table 9.2*

---

**Comparison of provisions related to clear-cut logging and silvicultural methods**

### FSC-BC

- Restoration forestry to be carried out in previously poorly managed natural forests and former plantations.
- Forests are to be managed within range of natural variation.
- Minimum requirements specified for retention of dominant and co-dominant green trees and snags and for average stand-level retention based on natural disturbance type.
- Connectivity corridors to be established.

### FSC-Sweden

- In managed forests, preference to be given to natural regeneration and to harvesting by means of selective felling.
- Trees with high biodiversity to be protected.
- Major landowners to implement landscape ecology planning.

### FSC-Maritimes

- Indigenous tree species appropriate to a site are to be present in frequencies similar to natural forest.
- Clear-cutting and even-aged management to be used only when appropriate to restore natural forest type.
- Management to strive to approximate typical natural forest characteristics for an ecosite, to achieve a functional level of connectivity, and to avoid excessive fragmentation.

---

▶

◄  *Table 9.2*

---

**FSC-Rocky Mountain**
- Silvicultural practices to generate stand conditions that emulate natural disturbance regimes.
- Even-aged management used only when ecologically appropriate. High-grade logging not to be practised. Forestry operations to minimize habitat fragmentation.

**FSC-Pacific Coast**
- Type 1 forests (at least 20 acres of never-logged contiguous forests), on non-tribal lands, are not to be harvested.
- Type 2 forests (old, unlogged stands smaller than 20 acres, of which at least 3 contiguous acres have been logged) and Type 3 forests (residual old-growth trees and/or other late successional or old-growth characteristics) to maintain late-successional and old-growth structures, functions, and components; legacy trees, old and large trees, snags and woody debris to be retained.
- In harvest openings larger than 6 acres, 10% to 30% of pre-harvest basal area to be retained with varying levels of green-tree retention, depending on opening size, legacy trees, adjacent riparian zones, etc.
- Even-aged silviculture may be used where native species require openings for regeneration, etc.; harvest blocks in even-aged stands to average 40 acres or less, with no individual block larger than 60 acres.

**FSC-Boreal**
- Forests pre-industrial condition (PIC) to be depicted over long-term planning horizon (40 to 100 years).
- Silvicultural practices (giving preference to natural over artificial methods) are to maintain stand structure over time, ensure effective and timely regeneration, and minimize impacts on wildlife habitat, etc.
- Under-represented forest ecosystems (compared to PIC) are to increase over time via restoration forestry.
- Management strategies not to mimic extreme events of low frequency; managers to consider disturbance mosaic in planning harvesting.
- Connectivity to be maintained; standard operating procedures to be developed that describe practices to avoid and minimize loss of productive land, soil rutting, nutrient loss, negative hydrological impact, etc.

---

At the international level, the FSC captures riparian management in a single Criterion, 6.5, which requires written guidelines to control erosion, minimize forest damage,[83] and protect water resources. Protection of riparian areas and habitat is a critical feature of most of the regional standards examined here, although the FSC-Sweden standard appears rather weak in this respect (see Table 9.3). Each of the comparator jurisdictions has made considerable efforts to ensure that their respective standards protect water bodies and courses from the effects of logging. However, the BC final standard

*Table 9.3*

## Comparison of provisions related to riparian management

### FSC-BC
- The manager implements measures to maintain ecological integrity of aquatic ecosystems.
- Machine-free zones established on all streams, lakes, wetlands, and marine shorelines.
- Integrated riparian assessment completed under new Criterion 6.5*bis* and implemented based on detailed framework outlined in Appendix B.

### FSC-Sweden
- No new ditches to be dug on previously unditched forestland.
- Roads across watercourses to be constructed and maintained to preserve functional integrity of river beds.
- Forestry along water bodies shall promote continuously forested and, if possible, layered transition zones.

### FSC-Maritimes
- Buffer zones of 30 metres to be maintained adjacent to all water bodies wider than 1 metre; zones of 15 metres for water bodies less than 1 metre wide; limits to be modified depending on slope, erosion hazard, wind-throw hazard, government watersheds, waterfowl production areas, wildlife travel corridors, and recreational waters.
- A no-logging zone to be established as one-third of the width of the buffer zone, with logging in remaining area maintaining or enhancing natural forest function.

### FSC-Rocky Mountain
- Manager to adequately protect all streams, lakes, wetlands, and associated riparian zones.
- Streamside management zones (SMZs) to be established and maintained adjacent to all water bodies. SMZs to be designed to control soil erosion and stream sedimentation, stabilize surface and ground water flows and water temperatures, and provide organic debris and habitat. Management in SMZs to take a conservative approach that puts aquatic and riparian protection above timber production.

### FSC-Pacific Coast
- Four categories of streams identified; each must be managed with specific inner and outer buffer-zone sizes.

### FSC-Boreal
- Forests surrounding or adjoining permanent water bodies must be protected by riparian reserves that exclude all forestry activity.
- Inner riparian reserves must be 20 metres from treed edge of permanent water bodies with partial harvesting permitted.
- Additional reserves must be required such that minimal total area within these additional reserves is equivalent to an additional 45 metres.
- Non-permanent water bodies (ephemeral and intermittent streams) to be considered.
- Standard operating procedures to be developed for riparian management.

is one of the more explicit, requiring managers to abide by the highly specific thresholds outlined in the new Criterion 6.5*bis*. The approach includes a peer-reviewed IRA, which has a detailed framework set out in Appendix B, "Requirements for Riparian Management."[84]

The generic FSC criteria relating to chemical use covers the use of chemical pesticides in management systems, disposing of chemicals, and using biological control agents and genetically modified organisms. The FSC-BC standard adopted a narrower conception of chemical pesticides and does not require their immediate cessation, a less restrictive approach than is found in the FSC-M standard, which requires the immediate discontinuation of "biocide" use (see Table 9.4). Nonetheless, FSC-BC's requirement that managers phase out the use of pesticides "no more than 5 years from the date of initial certification" of the FMU is more demanding than other comparator jurisdictions (Indicator 6.6.1).

*Table 9.4*

## Comparison of provisions related to use of chemical and biological agents

**FSC-BC**
- Where chemical pesticides are used, plans are in place to continually reduce and finally eliminate their use over a period of no more than 5 years; evidence of use of integrated pest management.
- Chemicals prohibited by FSC not to be used.
- Exotic biological control agents to be used only to control exotic species of plants, pathogens, insects, or other animals, and only when other methods ineffective.

**FSC-Sweden**
- Pesticides with specified hazardous designations are prohibited on forestlands.
- Plants are procured only from nurseries that aim to minimize the application and negative effects of chemicals.
- Biological control agents are applied using techniques that will avoid detriment to the environment and humans.

**FSC-Maritimes**
- Management is committed to not using biocides in its forestry practices.
- Biodegradable oil and other biodegradable products to be used when available.

**FSC-Rocky Mountain**
- Integrated pest management and vegetative control strategies to be preferred to chemical treatment; if chemicals used, a written prescription is to be prepared, fully describing risks and benefits.
- Exotic, non-invasive predators or biological control agents to be used only to control exotic species and subject to peer-reviewed scientific evidence.

▶

◄   *Table 9.4*

**FSC-Pacific Coast**
- Silvicultural systems with the least adverse environmental impact should be employed with the goal of reducing or eliminating chemical use. Chemical pesticides, fungicides, and herbicides to be used only when and where research or experience demonstrates that less environmentally hazardous practices are ineffective. Chemical application to be narrowly targeted to and minimize effects on non-target species. Chemicals to be used only when they pose no threat to domestic water, aquatic habitats, and habitats of rare species.
- Exotic biological control agents to be used only to control exotic species of plants, pathogens, insects, and other animals, and only when other methods are ineffective and when peer-reviewed scientific evidence is available of their non-invasiveness.

**FSC-Boreal**
- Applicant to develop and use integrated pest management.
- Demonstrated continual reduction in chemical use to eventual goal of complete phase-out over time.
- Chemicals prohibited by FSC not to be used.
- Standard operating procedures to be developed for chemical handling.
- Biological control agents to be used only when non-chemical pest control methods are deemed ineffective.

The FSC-BC approach to plantations differs substantially from that of the other relevant comparator jurisdictions (see Table 9.5). This reflects, in turn, difficulties interpreting the concept of plantation in the BC context and more general confusion within the FSC over the concept's definition. FSC-AC's Policy and Standards Unit launched a review of its approach to plantations in 2004, appointing a Plantations Review Policy Working Group in 2005 to canvass these and related issues.[85] Pending final revision of the FSC's plantation policy, the organization has prepared two advice notes to provide concrete guidance on the interpretation of C10.9 (FSC-AC 2002e) and on conversion of plantation to non-forest land (FSC-AC 2004b).

In summary, our assessment of the manner and extent to which ecological values are protected in other FSC jurisdictions illustrates the relative rigour of the FSC-BC final standard across five key ecological values: old-growth and biodiversity protection, clear-cut logging, riparian management, chemical use, and plantation establishment. The BC final standard is not the most explicit or strict on each of these issues (for example, provisions for chemical use are clearly more demanding in the FSC-M standard); overall, however, the standard developed for British Columbia establishes more explicit, quantified thresholds, requires more elaborate consultative procedures, and demands more data collection than the other comparator standards.[86]

*Table 9.5*

## Comparison of provisions related to plantation establishment

### FSC-BC

- Conversion to plantations not to exceed 5% of timber-harvesting land base (THLB); plantations not to exceed 10% of THLB.
- Plantations to be located according to scale of forest; plantations not to occur on HCVFs.
- Plantations to provide long-term benefits and promote landscape connectivity. They are not to increase pressures on natural forests.

### FSC-Sweden

- Not applicable.

### FSC-Maritimes

- Conversion of natural forest cover to plantations is prohibited.
- Plantation management to be based on long-term plan to restore the natural forest for the district and soil site.
- Species selection to be based on moving stand toward natural forest type.
- Plantations of exotic trees are limited to no more than 5% of contiguous portion of ecosite and are permitted only when clear evidence is provided of their non-invasiveness and compatibility with ecosystem, and when they are a step toward establishment of natural forest.
- Management to be clearly committed to using no biocides in its forestry practices.

### FSC-Rocky Mountain

- Management plan to identify existing plantations and those scheduled to be restored to more natural forest conditions.
- On areas converted to plantations, even-aged harvests lacking within-stand retention are to be limited to 40 acres or less in size unless a larger opening is scientifically justified.
- Regeneration in previously harvested areas to reach a mean height of at least 10 feet or achieve canopy closure before adjacent areas to be harvested.
- Use of exotic species in plantations is contingent on peer-reviewed scientific evidence of non-invasiveness.
- A graduating percentage of previously converted forestlands is to be restored to natural and semi-natural forest cover (i.e., in plantations larger than 10,000 acres, a minimum of 25% to be restored).

### FSC-Pacific Coast

- Plantations defined as "tree-dominated areas substantially lacking in natural forest attributes ... that usually require human intervention to be maintained." FSC Principles and Criteria 1 to 9 do not necessarily apply to portion of FMU being maintained under plantation management.
- Average harvest openings in plantations on soils capable of supporting natural forests to be 40 acres or less, with no individual opening less than 80 acres.
- Plantations on soils capable of supporting natural forests to have 4 dominant and/or co-dominant trees and 2 snags per acre retained.

▶

◄  *Table 9.5*

- Regeneration in previously harvested areas to reach mean height of at least 10 feet or achieve canopy closure before adjacent areas are harvested.
- FMUs in plantations on forest soils must maintain or restore natural forest vegetation according to sliding scale (e.g., FMUs larger than 10,000 acres must have 50% in natural forest, with 30% in long rotation and 20% in late-seral stage).

**FSC-Boreal**
- Forest conversion not to occur on HCVFs.
- Maximum of 5% of productive forest area to be converted to plantations.
- Proportion of the forest management area to be restored to natural forest cover.
- Measures to be taken to minimize outbreaks of pests, diseases, fire, and invasive species.

The FSC-BC final standard is, however, notably shorter and more flexible than its predecessors – the third draft (D3) and the preliminary standard – with respect to several requirements. Unlike other comparator standards – and in notable contrast to D3 – the final standard contains no major failure provisions, which gives certifiers and forest managers increased leeway to trade off modest performance in some areas against superior performance in others. In addition, the final standard is substantially shorter than the D3/preliminary standard with respect to indicators of environmental values. The D3/preliminary standard contained 105 indicators for Principles 6, 9, and 10; the final standard has reduced this number to 88.

The FSC-BC final standard undoubtedly sets a high bar for forest management practices in terms of protecting environmental values. It is harder to judge whether it is overly onerous when compared to the equivalent standards in other jurisdictions. Although many people engaged in industrial forestry perceived the FSC-BC final standard to be rigorous relative to neighbouring US jurisdictions, such as the Pacific Coast and Rocky Mountain regions, it should be borne in mind that these two standards apply almost exclusively to private, not public, lands. Similarly, although critics of the FSC-BC final standard often contrast it unfavourably in terms of "onerousness" with the original FSC-Sweden standard, a new Swedish standard is currently being developed to enhance compliance with applicable FSC requirements. Thus, the initial gap between FSC-BC and FSC-Sweden is likely to narrow substantially.

Our comparison also highlights the many similarities between the BC and Boreal standards, although the latter emerges as somewhat less demanding across a range of provisions. The degree to which this reflects natural differences in forest characteristics (highly biodiverse coastal temperate rainforests

versus moderately biodiverse interior boreal forests) or differences in negotiation arrangements is an especially interesting question and highlights
the importance of better understanding the broader political, regulatory,
and institutional context within which FSC standards are developed. It is
to these topics that we now turn in Part 3 of this book.

# Part 3
# Governance within and beyond the FSC System

In our Introduction, we defined and juxtaposed the companion concepts of "governance" and "government." We contended that while all governments engage in governance, by no means do all governance arrangements necessarily involve government. In this sense, we argued that "governance" should be thought of as an umbrella term to denote arrangements for *steering and coordinating the affairs of interdependent social actors based on institutionalized rule systems* that depart from the traditional strong centralized model of "government" that was characteristic of many Western liberal democracies in the generation that followed the Second World War (Benz, quoted in Treib, Bähr, and Falkner 2005, 5).

Rather than seeking to classify governance arrangements that employ the public-private dichotomy, we argued for a more nuanced taxonomy that facilitates an exploration of such arrangements by way of several relevant criteria. In Part 3 of *Setting the Standard*, drawing on the research we have done, we propose to embark on a multi-dimensional governance analysis of the FSC that focuses on three core analytic dimensions: politics, regulation, and institutional form.

In Chapter 10 we consider the role of politics as a means of understanding the dynamics of power within the FSC system. By analyzing six of the key political networks (environmental, large producers, social, indigenous peoples, certifiers, and donors) that contend for influence within the FSC regime, we reflect upon how and to what extent the FSC system succeeds in creating political space for these networks to understand, and negotiate for, their competing interests.

In Chapter 11 we turn our attention to the regulatory dimension of the FSC system by exploring the regulatory role performed by the FSC in setting standards for sustainable forest management. To this end, we distinguish between and assess the relative merits of competing forms of standards (performance, technology, and management), paying particular attention to the question of why performance-based regulation has emerged as a preferred form of policy instrument. Drawing on this analysis, we end the chapter with a detailed assessment of both the content, and the negotiations surrounding approval, of the FSC-BC final standard.

Finally, in Chapter 12, we undertake an institutional analysis of governance within the FSC system. Using a comparative constitutional approach, we examine the FSC's formal architecture and deliberative processes to consider the extent to which the regime emulates liberal democratic constitutional norms, principles, and mechanisms.

# 10
# A Political Network Analysis of FSC Governance

Our account of the negotiation of the FSC-BC final standard (see Chapter 4) highlights the crucial role that political networks play within the FSC. In many ways, we would argue that the final standard detailed and analyzed in Chapters 6 through 9 is a political artifact: a culmination and reflection of the strategies and resources that the political networks engaged in its negotiation brought to the bargaining table. In this chapter, we undertake a more systematic and in-depth exploration of the FSC's political networks with the aim of illuminating the nature and significance of this "political dimension" of the FSC system. We commence our exploration with some reflections on the nature and locus of politics as it is conventionally played out in Western liberal democracies. We then juxtapose this state-based "party politics" with what might be termed "network politics," a methodology for understanding political constellations and dynamics within the FSC system. To this end, we review the academic literature on networks, situating our conception of "homogenous" networks within the broader literature on public policy, transnational advocacy, and social network theory. We then chronicle what are, in our view, the FSC's six major political networks, concluding with some observations on the interaction among these various networks and, more generally, on the implications of this mode of politics for the future of the FSC governance system.

## The Nature and Locus of Politics

Politics is invariably a contest over values and interests. The greater the diversity of interests and values, the more intense the politics, with robust institutions required to mediate resulting conflicts. States are quintessentially political in this sense in that they are called upon to mediate conflict between diverse and often competing interests and values, whether through authoritarian or democratic means. Authoritarianism has been a highly prevalent form of state structure throughout history, but it is often said that we live today in a "democratic moment," a context in which most states in the world are either actual or proto-representative democracies (Benhabib 1996). In states such as these, citizens as members of the polity vote in elections for a government constitutionally limited in terms of power and duration.

"Politics," in popular discourse, is typically taken to refer to the actions of political parties and interest groups *within states* to influence voters and gain electoral advantage. This popular conception of politics is reflected in the work of pluralists such as Dahl (1961) who conceive of states as "polyarchies" in which a plethora of relatively equally matched interests engage in an overt struggle in the political marketplace to secure favourable policy outcomes. Neo-pluralists update this benign conception of politics by recognizing the privileged position that business occupies within most representative democracies. In Lindblom's (1982) formulation, the market acts as a "prison," constraining the freedom of political actors with respect to a wide range of policy options because of their need to satisfy the demands of business, to which responsibility has been delegated for the production of goods and services.[1]

Regardless of one's conception of politics, it is essential at the outset to identify the arena in which the contest occurs. The arena can, of course, vary from the local to the provincial, national, or international. Despite a wide diversity in population, resources, culture, and climate, these arenas all share one notable feature: they are territorially defined entities that coordinate their affairs internally through unitary or federal structures and externally through intergovernmental agreements. Within conventional political theory, the core territorial institution is the state, which is constituted as a social contract between individuals who come together to create a commonwealth to escape the vicissitudes of the "state of nature" (Hobbes and Locke) or the corrupting consequences of civilization (Rousseau). Leaving aside the peculiar nature of this social contract – it does not have an expiry date, binds all future progeny of the initial signers, and binds even those who consider themselves not to have signed it – the state that emerges consists of three core elements: a people or "polity," a territory, and sovereignty.[2]

The state became such a powerful social and political institution in the twentieth century that, until recently, it tended to obscure the potential for alternative organizational modes and arrangements. Even today, in an era of globalization, we continue to seek state-based solutions to global phenomena, with countries establishing international environmental "regimes" to prevent ozone-layer depletion, global warming, the trade in toxic waste, and nuclear proliferation (Keohane 1984; Young 1989). The proliferation of international regimes has been a key feature of the world system since the end of the Second World War. However, notwithstanding the large number of intergovernmental organizations that now populate the global system, their effectiveness and efficiency continues to be questioned. Although the

Montreal Protocol on Substances that Deplete the Ozone Layer is arguably testament to the potential effectiveness of the interstate approach to deliver successful outcomes, the specific features of the ozone-depletion issue – incontrovertible evidence of a cause/effect relationship, a small number of affected companies, and readily available substitutes – stand in stark contrast to the diffuse nature of other environmental threats. The dysfunctionality of intergovernmentalism is perhaps nowhere more evident than in the forestry sector, where agreements since the 1980s have done little if anything to halt deforestation or forest degradation anywhere in the world (Humphreys 2006).

It is not only states that have been beholden to inter-*national* solutions to global problems. In FSC's competitor schemes we discover a similar logic at work. The International Organization for Standardization, for example, is a federation of autonomous national standard-setting bodies whose representatives meet in coalitions of convenience to negotiate international standards in designated areas. Likewise, the Programme for the Endorsement of Forest Certification is an international organization whose certification standards are developed almost exclusively by nationally driven interests and processes. Two key weaknesses tend to characterize arrangements of this kind. First, dominant interests at the national level reflecting economic imperatives become magnified at the international level, marginalizing other actors representing the environment, communities, workers, and indigenous peoples. Second, interests lying outside the state are explicitly excluded from the negotiation process, enabling those within the state to strike bargains that generate negative externalities for those outside.

In recognition of the limits of conventional approaches for analyzing social phenomena – including policy making – in a globalizing world, analysts have turned to networks to better understand emerging social practices. There is now a vast and diverse literature on networks and networking, and a comprehensive review of this literature is not possible here. Instead, in this section we aim to situate our conception of networks within the broader economic, sociological, public policy, and international relations (IR) literatures; select a small number of criteria to facilitate our analysis; and engage in a preliminary mapping of six FSC networks that played a crucial role in the negotiation of the FSC-BC final standard.

Scholars from a range of disciplines employ the network concept; however, they use it in different ways. Thus, sociologists analyze *social* networks, economists focus on *business* networks, IR theorists investigate *transnational advocacy* and *transgovernmental* networks, and political scientists study *policy* networks. Although the network idea is common to each of these disciplines, network actors and purposes vary, as do the methods used by analysts to chart the networks' key features. Building on this tendency to "adjective-ize" the concept, and in order to distinguish FSC's networks from others identified in the literature, we characterize them as "homogenous networks": networks of actors who share dominant values and an interest in forest management.

Theorists within various disciplines develop network models to elucidate key aspects of interactions among targeted actors. The nature of each model, therefore, reflects certain significant features of these actors and their relational contexts. For example, economists interested in investigating a business network define it as a "group of agents that pursue repeated, enduring exchange relations with one another" (Poldolny and Page, quoted in Rauch 2001). Such repeated, enduring exchange relations are seen to occur among the diaspora of Chinese business executives, a network that is viewed in part as solving the problem of opportunism, with violators of agreements quickly blacklisted via word-of-mouth communication (Rauch 2001). An alternative conception of an economic network is Gereffi's (1994) concept of a global commodity chain. Gereffi defines global commodity chains as "sets of inter-organisational networks clustered around one commodity or product, linking households, enterprises, and states to one another within the world-economy" (2). Although "repeated, enduring exchange" is a feature of Gereffi's networks, his actors are more heterogeneous than Rauch's, incorporating a diversity of individuals and organizations that are linked together in the process of commodity production, distribution, and consumption.

The economics literature by no means ignores cultural and social variables, but it does tend to view networks instrumentally, as arrangements to achieve narrow economic objectives. This instrumentalist tendency is present also in transaction cost economics, which views networks as an alternative way to organize production relations, in place of integration within a single firm or separation via arm's-length contracts. When transaction costs associated with certain aspects of production through markets are high – such as costs associated with information, organization, time, lawsuits, intellectual property, and so forth – a firm will integrate those activities into its own operation. When they are low, conversely, the firm will prefer to operate at arm's length via market transactions. Networks, in this body of literature, constitute a third alternative to vertical and horizontal market relations. In Powell's (1990) view, they are "a separate, different mode of exchange, one with its own logic," which are especially useful when the value of a commodity – information, for example – is hard to assess.

The instrumentalism of the economic conception of business networks stands in stark contrast to sociology's more semiotic and relational approaches. Sociologists have had a long-standing interest in the interconnections between individuals and organizations, with efforts to map social networks dating back to the 1930s (Berry et al. 2004). Since then, a variety of sophisticated social-network approaches have been developed, culminating most recently in an approach known as actor-network theory, or ANT (Latour 2005). In an extended discussion of ANT, Dicken et al. (2001, 101-5) identify its four key concerns as practices, subjectification, spatio-temporal becomings, and technologies of being. By focusing on practices, ANT highlights how social networks "shape the conduct of human beings towards others and themselves in particular sites" (Thrift quoted in Dicken et al. 2001, 101). Moreover, such practices cannot be derived from a single logic of action, but, via subjectification, individuals are viewed as embodied, affective, and dialogical in a world that is "always in process" and mediated via different technologies of being.

Interestingly, networks within ANT are conceptualized as *hybrid collectifs* – combinations of human and non-human entities. As Dicken et al. (2001, 102) underscore, ANT assumes "that it is the non-human artefacts (for example computers, container port cranes, trains), tools (for example reports, maps) and rules (for example laws, policies) that enable social beings to develop and maintain modern social relations; relations that span out across space at all scales via networks ... We are all, to use Haraway's (1991) terminology, cyborgs; the global economy is driven by cyborgs." ANT's rich conception of networks is a useful counterbalance to the rational-actor model embedded in economic accounts of network structure and formation, but its complexity makes it unsuitable for our more limited purposes here.

If economic and sociological theory provide, respectively, a too impoverished and too complex perspective on social networks, might we find a more balanced approach in the public policy and international relations literature? The network concept, deriving from Katzenstein's (1987) and Atkinson and Coleman's (1989) seminal works on "policy networks," is now ubiquitous in public policy. Reworking Katzenstein's "macro" analysis on strong and weak states, Atkinson and Coleman employ three variables – state autonomy, state concentration, and mobilization of business interests – to identify six potential policy-network types at the "meso" level: state directed, pressure pluralist, clientele pluralist, parentela pluralist, corporatist, and concertation. Their approach has proved fruitful, giving rise to a wide range of applications in the field of public policy; in practice, however, it has been difficult to differentiate between some of these policy-network types. Moreover, the approach remains embedded in a broader public policy tradition that privileges the role of government in the policy process. Hence, policy networks are conceptualized as "power relationships between *government* and interest groups, in which resources are exchanged" (Borzel quoted in Thompson and Pforr 2005, 1; emphasis added), which limits the utility of the concept from our perspective.

Operating within an international relations framework, Raustiala (2002) and Slaughter (2004) elaborate a *transgovernmental* approach to networks that elucidates how governments are increasingly linking up relevant arms of their bureaucratic apparatuses to cope with globalization on the one hand and increased domestic regulatory demands on the other. According to Raustiala, the state is not in decline per se; rather, it is transforming itself by disaggregating its functions with respect to international cooperation: "Domestic officials are reaching out to their foreign counterparts regularly and directly through networks, rather than through state-to-state negotiation of the kind that dominated 20th century cooperation" (11). Transgovernmental networks are not new – Raustiala and Slaughter both cite the 1936 Convention on the Illicit Traffic in Dangerous Drugs as an early example – but they have been facilitated by technological innovation, the rise of the regulatory state, and globalization. Raustiala details the operation of three such networks: securities regulation via the International Organization of Securities Commissions, competition policy via the US Department of Justice and its proposed International Competition Network, and environmental regulation via the International Network for Environmental Compliance and Enforcement.

Keck and Sikkink (1998) also build on the policy-network approach to develop their conception of transnational advocacy networks. They argue that "networks are communicative structures" in which group actors operate to "influence discourse, procedures, and policy" (3). Networks are "political spaces" where "differently situated actors negotiate ... the social, cultural, and political meanings of their joint enterprise" (3), and they are "forms of organization characterised by voluntary, reciprocal, and horizontal patterns of communication and exchange" (8). In an important observation, Keck and Sikkink explain that they refer to these communicative structures as "networks" and not "coalitions" because they want to highlight "the structured and structuring dimension in the actions of these complex agents, who not only participate in new areas of politics but also shape them" (4).

In practical terms, Keck and Sikkink identify the actors participating in transnational advocacy networks as "(1) international and domestic non-governmental research and advocacy organizations; (2) local social movements; (3) foundations; (4) the media; (5) churches, trade unions, consumer organizations, and intellectuals; (6) parts of regional and international intergovernmental organizations; and (7) parts of the executive and/or parliamentary branches of governments" (1998, 9). Although these distinctions are important for their purposes, given their focus on understanding the process of advocacy within the international system and the ways in which actors operating within different social spheres link up to promote a particular policy solution, we view many of these "actors" as networks in their own right.

Keck and Sikkink's approach is not the only one to adopt the concept of networks to understand aspects of international politics. Notably, Dicken (2003) uses the approach to analyze the reshaping of the global system in the twenty-first century, and Peter Haas (1992) has used the network concept to identify groups of actors embedded in "epistemic communities" who are working to shape international norms in the environmental, development, and human rights fields. In Haas' formulation, actors operating within the global system come to share a similar account of the existence and causes of, and solutions to, a particular problem. By constituting themselves as epistemic communities and partaking in international negotiations, they play an important role in shaping the regime's ultimate normative structure – its rights, rules, and compliance mechanisms.

The economic, sociological, public policy, and IR literatures sketched out above indicate the rich and diverse lineage of the network concept. Where economics tends to view networks in narrowly instrumental terms, sociological approaches such as ANT adopt broad approaches that view networks as both constituting and constituted by social actors. The public policy approach tends to privilege certain actors – governments and firms – while Keck and Sikkink's focus on advocacy networks embraces a wide array of somewhat diverse actors.

These approaches illustrate the diversity in application and flexibility of the network concept. In our account of the FSC, however, we return to the basic definition of networks given at the outset of this chapter – networks as interconnections between individuals and organizations – and to Raustiala's focus on organizations of self-similar groups. Within the FSC, politics takes place through networks of actors who share a dominant interest within the value hierarchy. Actors differ in their ranking of the range of interests bound up in forest certification (i.e., profitability, sustainability, community, indigenous peoples, auditability, "worthiness"). Individual actors may disagree significantly on any final rank order, but they tend to unite around their dominant or highest-ranked interest. It is this dominant interest that creates the "glue" that holds our homogenous networks together, enabling actors to "self-recognize" each other even when they have never before met.

## Politics within the FSC System

The FSC eschews an international, state-based, territorial approach to standard setting in favour of an approach that is simultaneously global and interest based. In the FSC approach, the multiple "peoples" represented within the intergovernmental approach are replaced, in effect, with a single global, sectoral "polity" consisting of all those with an interest in sustainable forest management.[3] The interests embraced in this approach do not differ greatly from those present in most nationally based polities and include networks of forest owners and managers, workers, communities, indigenous peoples, environmentalists, and the range of downstream businesses that constitute the forest products chain, including sawmillers, chippers, wholesalers, retailers, and consumers. Uniquely, however, these interest-based networks are organized globally within FSC's three chambers based on the Brundtland Commission's three-legged stool of sustainability. The basic insight of this analogy is that if the three legs – economic, environmental, and social – of the sustainability "stool" are not of equal length, the resultant policy will be unbalanced, and sustainability will not be achieved.[4]

The FSC's global aspirations and its desire to balance economic, environmental, and social interests impel it toward a novel mode of politics. Political parties are inherently territorial entities and experience immense difficulties in "scaling-up" that are akin to those experienced by the European Parliament in its efforts to forge European-wide political identities out of the diversity of national parties. In place of the party-political contest of national politics, therefore, politics within the FSC occurs as a process of contestation that takes place within and between a few key political networks operating, for the most part, through the three formally established chambers.

Although at first sight it may appear that three of these networks are constitutionally mandated by the FSC-AC, in fact, on closer inspection, it is evident that no FSC chamber is exclusively composed of a single homogenous network. The economic chamber, for example, encompasses a large-industry network, a small woodlotters network, and a certifiers network, each of which extends well beyond the FSC. The environmental chamber, perhaps the most homogenous, likewise includes both mainstream and radical, small and large, preservationist and conservationist groups that differ fundamentally in terms of strategies and tactics, and many of whose members do not belong to FSC. The extreme heterogeneity of the social chamber, which consists of numerous networks of workers, indigenous peoples, community forestry groups, and human rights and Third World development organizations, explains in part its greater difficulty in coordinating action around shared policy goals (see discussion below).

Although there is a relationship between FSC's chambers and our homogenous networks, it is neither straightforward nor unproblematic. The large-industry network is housed, sensibly enough, within the economic chamber but extends well beyond it in terms of membership. Certifying bodies are also housed in the economic chamber, although they constitute a separate network in their own right. Likewise, in the social chamber the indigenous peoples network is clearly distinguishable from the nascent community forestry network also represented there, although there is also some overlap. The best fit between network and chamber is in the environmental chamber, although even here the broader network contains several sub-networks based on size, professionalism, budget, ideology, and strategy, and it extends well beyond the FSC.

In addition to these networks that are formally, if not necessarily comfortably, incorporated within FSC's three-chamber system, we map one other network that played a crucial role in FSC politics but sits outside the system.

This is the donors network, which not only funds FSC-AC but also provides substantial sums of money to support national and regional standards development. It remains a shadowy presence within the organization, preferring to stay in the background, but it has, nonetheless, played an important role in the FSC's establishment and development.

A variety of quantitative and qualitative methodologies exist to analyze networks. Quantitative methodologies include block modelling, Euclidean distance analysis, regression analysis, dynamic-network modelling, and event history analysis. These approaches work best when large volumes of high-quality numerical data are available, a situation that does not pertain here. Instead, we employ another well-known methodology for analyzing networks: the qualitative, case-study approach. This approach better suits our purpose, given that our goal is to identify and "map" the contours of each of the FSC's homogenous networks. In conducting this network mapping, we focus on basic network parameters: the extent of the network in terms of number and types of actors involved and the "strength" of the network in terms of the frequency and quality of the transactions taking place.

### Environmental Network

Environmental civil society organizations (ECSOs) have been instrumental in raising public and political consciousness about an extensive array of environmental issues around the world. In Canada, it was ECSO concerns about old growth, public land, and First Nations issues that contributed to the larger environmental battles in British Columbia. Through early market campaigns, the environmental network recognized the importance of forest certification as an instrument for improving forest management and as a complement to strong regulatory measures, public oversight, and preservation. The interests of the environmental network are summarized in a "Joint NGO Statement on Forest Certification" (2003) signed by thirty-seven organizations.[5] The statement endorsed the FSC, noting that it "is the only certification system that is currently worthy of our support" and that "comes close to effectively addressing the range of social and environmental elements critical to the responsible use of the world's forests."[6]

Environmental network members involved in FSC-Canada include a who's who of Canadian ECSOs (see Table 10.1). Greenpeace (Tamara Stark), the David Suzuki Foundation (Cheri Burda), and the Sierra Club of BC (Lisa Matthaus) played especially key roles in British Columbia.

Of the several networks described here, the ECSO network is one of the most robust, being the most formally organized and interactive, which gives rise to a greater density of transactions. An example of its strong internal linkages is Good Wood Watch, a web-based initiative coordinated by West Coast Environmental Law (WCEL).[7] The original aim of this initiative evolved in tandem with the FSC-BC process, from simply providing information

*Table 10.1*

**Canadian members of FSC-AC: Environmental chamber**

| Group | Region | |
|---|---|---|
| | BC | Canada |
| Conservation organizations | David Suzuki Foundation Ecotrust Canada Greenpeace Canada Sierra Club of Canada (BC Chapter) Wildsight (formerly East Kootenay Ecological Society) | Alberta Wilderness Association Canadian Parks and Wilderness Society (Edmonton Chapter) Ecology Action Centre Ecotrust Canada Nature Trust of New Brunswick Sierra Club of Canada UQCN (Nature Québec) Wildlands League World Wildlife Fund Canada Yukon Conservation Society |
| Academic institutions | | Faculty of Forestry, University of Toronto |
| Individuals | Graeme Auld Karen Tam Wu | Jamal Kazi |

*Note:* The ECSO network extends well beyond those who are formally members of the environmental chamber.
*Source:* Compiled from FSC Membership List (FSC Doc. 5.2.2), December 2007.

"about ecoforestry and eco-certified forest products" to helping "concerned citizens navigate forest certification in BC and learn more about the FSC, ecoforestry and eco-certified products" (Clogg 1999; GWW 2005). Led by ecoforestry practitioners as well as producers and vendors of certified forest products, the initiative was sponsored by six BC-based ECSOs: Greenpeace Canada, Sierra Club of BC, David Suzuki Foundation, and Wildsight,[8] all members of the environmental chamber; WCEL, from the social chamber; and Friends of Clayoquot Sound.[9]

The coherence of the environmental network is evident in comments made by our interviewees. One observed that it "has become more organized to be sure they have enough buy-in from all the different constituencies" (personal interview 2004), an observation of special relevance in the BC case, where the environmental community put in place strong mechanisms for reporting back. Another interviewee noted that "if you want to go to work on the Great Bear agreement, it's fine if ForestEthics, Sierra Club, and Greenpeace feel 'this' – but if you can't get some level of buy-in from Valhalla and Raincoast, ... it's going to hurt" (personal interview 2004).[10] In this context, strong reporting relationships became crucial. The same interviewee noted that "for environmental groups there were people – for example, that [names of two forest campaigners] and I were clearly responsible to – that we clearly represented, people who wanted to have their input to a certain degree, who wanted to set a negotiating bar."

FSC-BC's close-knit environmental network operated effectively within the province, ensuring representation by FSC-BC members and non-members alike, but ECSO representatives elsewhere, whether inside or outside the FSC, did not always share its perspective. Although the BC ECSO network was largely united in the belief that promoting ecoforestry should be the overarching goal of the FSC and were suspicious of some certifying bodies' (CBs) commitment to this goal, other environmentalists within the broader FSC adopted a more conventional approach to forest management and were less skeptical of the CB's motives. This was especially true of those ECSOs that represented the South who regarded CBs as playing a key role in the struggle against rampant unplanned and corrupt forest management practices.

These tensions within the environmental network between ecoforestry and sustainable forestry advocates and between North and South were evident at the two informal meetings held between BC, Canada, and international FSC members at the 2002 General Assembly to discuss the "extraordinary solution" for British Columbia (described in Chapter 4). Indeed, in the course of seeking that meeting, there was considerable discussion within FSC-Canada and FSC-AC on how to position the idea within a broader FSC context so that it did not appear to be merely a special arrangement for British Columbia. Initially, southern members of the FSC-AC board were reluctant to endorse preliminary accreditation for precisely this reason. The fact that preliminary accreditation was eventually adopted and implemented testifies, however, to the strength of the environmental network within the FSC system and to BC ECSOs within the network. While they were unable to push the third draft (D3) all the way through to accreditation, they were able to avoid having it formally returned to Canada and BC with a request that further work be done.

## Large-Producers Network

The FSC aims to transform forest company behaviour, and it has been closely monitored by the commercial forest sector since its inception. Varying industry perspectives on the FSC experiment can be conceptualized by locating companies within one of three concentric circles. At the core is a small group of firms that are FSC members and that either are or are committed to becoming FSC certified. In Canada, these include Tembec, Alberta-Pacific Forest Industries (Al-Pac), and Stora Enso. Outside the core in the inner circle is a group of companies that are potential FSC members. They have adopted a wait-and-see attitude to FSC certification, monitoring its development and rationally assessing its costs and benefits. Some may have undergone a "scoping" exercise to determine their preparedness for FSC certification should it become a market requirement. Further out again, in an outer circle, is a large group of firms that are either indifferent to or skeptical of the FSC. Indeed, some in this outer circle regard the concept of forest certification as unnecessary, infeasible, ineffective, and costly.

As of 2007, there were thirty-one members in the FSC-Canada economic chamber. Just over a third of these members were producers, divided equally between large- and small-scale operations.[11] There are few processors and only a single retailer in FSC-Canada's economic chamber, reflecting the "extroverted" nature of the country's forest industry, with much processing and retailing occurring outside the country. FSC-Canada's economic chamber also contains a relatively large number of consultants and service providers.[12] Members of the large-industry network believe that some individuals and organizations – such as the Ecoforestry Institute and Will Horter (executive director of the Dogwood Initiative) – do not belong within the economic chamber, arguing that, despite their residual economic interest in the forest, their reliance on membership fees, donations, and grants for revenues means they should belong to either the environmental or social chamber.

The tensions between large- and small-business networks created difficulties when it came to negotiating common policy positions within the economic chamber; the fact that the BC small-business representative supported D3 while the large-business representative did not is a case in point. In the BC case, however, it was the large-producers network that played the lead role throughout negotiations toward the final standard. Lignum and Western Forest Products (WFP) took an early interest in FSC certification on the advice of BC-based consultant Patrick Armstrong (Moresby Consulting). When the FSC-BC steering committee was formally established in 1999, Bill Bourgeois (Lignum) became the large-industry representative, with Sandy Lavigne (WFP) serving as his alternate. Following its January 2001 commitment to have all its operations across Canada FSC certified, Tembec Inc. also became a key player, with two senior Tembec employees, Chris McDonnell and Troy Hromadnik, elected FSC-Canada working group chair and FSC-BC steering committee co-chair, respectively.

Members of the large-industry network (notably Tembec, Lignum, and WFP) uniformly reported that their rationale for engaging in the BC standard-setting process was to facilitate their participation at all levels of the FSC to ensure their interests were represented. These companies sought a Swedish-style BC standard that would be relatively discretionary and not too costly to implement, that would allow them to meet emerging internal corporate social responsibility mandates, and that would increase their competitiveness in global markets. To achieve these objectives, large producers established an informal caucus around individuals such as Bourgeois and Lavigne, who in turn relied on their general knowledge of the industry to guide them on what might be feasible for and beneficial to their own and similarly situated operations.[13] Information on the state of FSC-BC negotiations was communicated through existing industry channels and on an informal, ad hoc basis.

Bourgeois had broad discretion to determine the network's negotiating position and exercised this discretion based on his perceptions of the structure and operation of the global forest products market. A critical modus operandi of the network was to commission consultants to prepare reports on the anticipated impact of proposed measures. For example, Bourgeois was instrumental in hiring Zielke and Bancroft of Symmetree Consulting Group Ltd. to prepare a forest sector submission on D2 for the FSC-BC Technical Advisory Team (TAT). This submission was prepared on behalf of fifteen companies and three associations.[14] Both this submission and a later report by Zielke and Bancroft (2002), which modelled the potential timber supply impacts of D3, detailed deficiencies in the emerging drafts and suggested requisite changes.[15] Large producers eventually had a substantial impact on the FSC-BC final standard as it moved through FSC-Canada to FSC-IC, and their concerns about the negotiating process and outcomes were recapitulated in the FSC-IC's Accreditation Business Unit (ABU) report.

In summary, large producers were effective in representing their interests, despite their informal structure. They were able to influence outcomes because of their structural position as producers in the FSC system and their capacity to commission and publish detailed and often persuasive studies outlining key concerns. As D3 moved from FSC-BC to FSC-Canada and FSC-AC, large producers lobbied those in formal positions of power within the FSC system, especially the executive director. Our interviewees noted that there was a concerted campaign by large producers to block accreditation of D3 via emails, telephone calls, and conversations targeting FSC-AC's executive director. Industry also worked behind the scenes to influence opinion within the organization, making its views known to CBs and FSC-AC officials and board members. Its arguments were more persuasive at the national and international levels than the regional level, mainly because FSC-Canada and FSC-AC were keen to avoid accrediting another under-used Maritime-style standard and placed more trust in the capacity of CBs to assess the adequacy of forest management practices.

## Social Network

Social issues have proved to be one of the FSC's biggest challenges, in part because they bring together a diversity of networks – including unions, communities, academics, and indigenous peoples' organizations – within a single chamber. The FSC's vision of sustainable forestry includes providing social benefits in the broad sense, minimizing negative social impacts, enabling both local people and society at large to enjoy long-term benefits, and providing strong incentives to local people to sustain the forest resources and adhere to long-term management plans. When it comes to dealing with wider social impacts, however, it can be difficult to determine where the responsibility of a forest manager seeking certification ends. Where institutional systems for social security and welfare do not exist, to what extent should managers (and CBs) be expected to address these issues? At the 2002 FSC General Assembly, FSC-AC launched a social strategy that recognized the concerns of FSC's "social constituencies," formulated core social values, and sought to translate these concerns into specific outputs, objectives, and activities closely connected with the core business of the organization (FSC-IC 2002c).[16] FSC-AC has also launched the Small and Low-Intensity Managed Forests Initiative to assist small, non-industrial forest management operations and has developed policy guidelines for incorporating International Labour Organization (ILO) conventions into the FSC principles and criteria, both under the umbrella of this social mission.[17]

A large number of FSC members around the world share a common vision of the FSC as an organization that is fundamentally about setting high environmental and social standards. However, it has proved difficult to attract social networks to participate in the organization's social chamber. This chamber remains relatively small,[18] yet it also has the most heterogeneous membership, including workers, communities, and academics.[19] In Canada, for example, social chamber members include an environmentally active legal organization (WCEL), a First Nation (Shuswap Nation), and an academic (Constance McDermott) (see Table 10.2).

The diversity of networks housed in the social chamber has necessitated the use of formal mechanisms to canvass their interests with respect to regional standards. Interviewees suggested that "the social chamber outreach was the most formal, in the sense that we actually had a social chamber forum in which we invited social chamber members, we have a listserv, things like that" (personal interview 2004). The social chamber helped facilitate a series of outreach activities, including letters to chamber members, a chamber forum, workshops for communities and IWA locals, and an independent review document of comments on D2, which was published and distributed by WCEL (Clogg 1999).[20] Members of the social chamber also worked to facilitate broader feedback on approval of D3: "Many, many BC members sent letters supporting the BC standard and urging them to move it forward, to try and demonstrate broad support in BC" (personal interview 2004).

The social network that formed around FSC-BC played a key role initiating standard development in British Columbia. Members of the Kootenay Centre for Forestry Alternatives were involved in the early organization of the FSC in British Columbia as part of the ad hoc group of volunteers that developed D1 before the establishment of an official FSC-BC steering committee. Concerned that "there did not seem to be as much active engagement as we'd

*Table 10.2*

**Canadian members of FSC-AC: Social chamber**

| FSC Level | Group | |
|---|---|---|
| | BC chapter | Canada working group |
| Organizations | First Nations Summit Society Kootenay Centre for Forestry Alternatives West Coast Environmental Law | Falls Brook Centre National Aboriginal Forestry Association (NAFA) Pulp, Paper and Woodworkers of Canada (PPWC) |
| First Nations | Shuswap Nation Tribal Council | Algonquins of Barriere Lake |
| Individuals | Cam Brewer Dr. Constance McDermott | |

*Source:* Compiled from FSC Membership List (FSC Doc. 5.2.2), December 2007.

have liked," network members placed "a lot of extra efforts ... on reaching non-members, and getting that feedback." This was partly due to over-representation of contractors, staff, and volunteers of the FSC among the network members: "The vast majority of social chamber members in BC were either involved in the steering committee, the standards team, or were staff for FSC," leaving "only a handful of members" who were not otherwise engaged and making recruitment a priority (personal interview 2004). Jessica Clogg (WCEL)[21] and Greg Utzig (Kootenai Nature Investigations)[22] are prime examples of individuals within the social network who were instrumental in ensuring coordination with other FSC networks and providing technical contributions on sustainable ecosystem forest management.

**Indigenous Peoples' Network**
FSC-Canada has been a leader within the FSC on indigenous peoples' issues, acting as a centre of excellence to address the difficult issues under Principle 3, which deals with indigenous peoples' rights (FSC-CAN 2001). At the international level, the social chamber has explicit responsibility for indigenous issues. In recognition of the importance of Canada's First Nations, however, FSC-Canada went beyond the standard three-chamber arrangement and systemically entrenched the involvement of indigenous peoples in a fourth chamber, a unique arrangement in the FSC system (NAFA 2001).[23]

Although many First Nation peoples are employed in Canada's resource industries, demographically they experience disproportionate levels of poverty, illness, and unemployment. Historically, they have also been under-represented in both government policy bodies and corporate decision making (Hessing and Howlett 1999). There has been significant movement toward recognizing and respecting Aboriginal rights and title in recent years (see Chapter 8), but many challenges lie ahead, particularly in the context of Aboriginal community economic development. The indigenous network in British Columbia and Canada saw the FSC-BC standard-development process as a promising opportunity to define and endorse systems of consultation and protocols for forest resources management from an Aboriginal perspective.[24]

Although the work of the indigenous peoples chamber has centred on issues of consultation and Aboriginal rights, the role of First Nations forest managers is also growing. As of September 2005, First Nations-managed forestlands accounted for an impressive 50 percent (87,980 hectares) of the FSC-certified area in British Columbia.[25] Although the volume harvested on these two licences (Iisaak Forest Resources' public land on the west coast of Vancouver Island,[26] and Inlailawatash Holdings' communal land near North Vancouver[27]) represents only 16 percent (31,215 cubic metres) of the current total FSC-certified provincial allowable annual cut, there is huge potential for growth in this aspect of the indigenous network (Abusow 2005).

FSC-Canada recognized early on that there was a need to strengthen and maintain a strong network of key individuals and groups in the Aboriginal community to support the standard-development process. Aboriginal peoples face myriad difficulties related to their geographic isolation in northern regions and the poor real and virtual communications between groups (FSC-CAN 2001), so FSC-Canada worked with partners such as the National Aboriginal Forestry Association (NAFA) to disseminate information about standards development and build a communications program that would facilitate input from First Nations and address indigenous expectations for engagement at national, regional, and community levels.[28]

In March 2003, FSC-Canada and NAFA jointly appointed an independent indigenous advisory council (IAC) in response to recommendations in an FSC-Canada strategy document (Collier et al. 2001).[29] In their initial meeting, the IAC members and advisors approved a legal memo that emphasized the importance of ensuring the Crown upheld its duties to treat First Nations honourably.[30] It also underscored the limitations of existing provincial consultation guidelines and forest management legislation in relation to Aboriginal and treaty rights (FSC-CAN IAC 2003). Although these recommendations were directly tied to the development of the Boreal standard, the IAC also made recommendations for advising FSC-AC, forest companies, and CBs on the interpretation of treaty and Aboriginal rights and for collaborating with other organizations on indigenous land-use issues.[31]

Close scrutiny of the indigenous network reveals an impressive level of commitment on the part of a relatively small number of individuals, not only in British Columbia but also in Canada as a whole. Many of these members move through several roles, centres of decision making, and levels of the larger FSC system, accumulating and disseminating their expertise. In recent years, this small pool of highly qualified individuals has been under immense pressure to represent, interpret, and deliver upon a complex range of issues. Throughout the FSC-BC standard-development process, the Shuswap Nation Tribal Council was particularly well represented, with a number of its members – most notably Dave Monture, David Nahwegahbow, George Watts, and Russell Diabo – holding various formal positions in the FSC system. As FSC-AC board chair, David Nahwegahbow (social chamber) was in an especially influential position as D3 moved from FSC-Canada to FSC-AC.[32]

### Certifiers Network

Certifying bodies are a critical component of the FSC system by virtue of their ongoing role in evaluating the degree to which a forest operation conforms to FSC standards. Where no national or regional standards exist, CBs are permitted to assess operations against the FSC's generic principles and criteria, using their own company-based indicators and verifiers. Their general expertise in forest management, coupled with a right to interpret the FSC's principles and criteria in specific circumstances, enables CBs to wield considerable influence within the FSC system. However, the influence of this network is limited by several constraining factors. CBs compete with each other for business, so the degree to which they cooperate and share information is limited. In addition, the largest CB, SmartWood, is a not-for-profit organization, which drives a wedge between it and other companies that operate on a for-profit basis.[33] Moreover, CBs are geographically dispersed, often operating on different continents, which limits the potential both for conflict and cooperation. The CBs caucus at the FSC General Assembly and other meetings to pursue their interests, but they have not, to date, formalized this arrangement to permit ongoing discussions throughout the year.

As of early 2006, FSC-AC had licensed thirteen CBs to conduct forest management audits on its behalf; all thirteen received "worldwide" accreditation status, enabling them to certify operations globally.[34] However, in practice, the business of forest management certification is allocated between four main companies, with SmartWood currently having the lion's share of total area certified (see Table 10.3). SmartWood is particularly dominant in the North American market, accounting for 76 percent of total certification by area, significantly ahead of its rival, Scientific Certification Systems (SCS), at 20 percent.

CBs have a vested interest in growing the certification business, in standard setting, and in FSC-AC's procedures to accredit them.[35] From their point of view, the higher the bar is set in a standard, the greater the costs of certification and the smaller the pool of interested clients. At the same time, they recognize that weak standards are not in their interest either, because they

*Table 10.3*

**Proportion of total certified area by region and certification body (2005)**

| Region | SmartWood (%) | QUALIFOR (%) | SCS (%) | Woodmark, Soil Association (%) | Other (%) |
|---|---|---|---|---|---|
| North America | 76 | – | 20 | 1 | 3 |
| Europe | 18 | 51 | – | 21 | 10 |
| Africa | – | 73 | – | 27 | – |
| Latin America | 61 | 14 | 23 | 1 | 1 |
| Asia-Pacific | 37 | 43 | 13 | 7 | – |

*Source:* www.certified-forests.org, October 2005.

can undermine the rationale for certification. FSC-accredited CBs have a commercial and contractual arrangement with FSC-AC and pay special attention to the arrangements it adopts to accredit, monitor, and license them. Our interviewees confirmed that certifiers do not lobby FSC-AC directly over specific standards. However, they do exercise discursive influence with industry and FSC-AC due to their technical expertise and in-the-field exposure to various national and regional standards. As such, they are frequently called on to field-test draft standards to assess their clarity, auditability, feasibility, and efficiency.

Both SmartWood and SCS had serious reservations about FSC-BC's D2 and D3 across all four of these criteria, particularly their feasibility and efficiency. There has also been speculation that they resented the complexity and detail of these drafts and the implication that certifier discretion needed to be fettered.[36] Although CBs may not have actively lobbied FSC-AC regarding D2 and D3, their reservations concerning the various drafts of the BC standard undoubtedly influenced industry and FSC-AC views as D3 moved through the FSC hierarchy. SmartWood was in a position to be especially influential because it participated in field trials of D2 and later certified Tembec to the BC preliminary standard, a process that inexorably led to the standard's revision.[37] These revisions, as noted in Chapter 4, included the development of a stand-alone standard for small operators, deletion of all major failure provisions, simplification of language, and a reduction in the number of indicators in, and the length of, the FSC-BC final standard.

**Donors Network**

FSC-AC's business model is heavily dependent on donations from philanthropic foundations (see Table 10.4) (see also Bartley 2007). Although FSC-AC has sought to generate income from internal sources via membership and accreditation fees and logo royalties, these remain only a small proportion

*Table 10.4*

**Breakdown of FSC-AC revenues, 1999-2004 (US$)**

| Year | Total revenue | Donations | Accreditation program | Membership | Other |
|------|-----------|-----------|-----------|-----------|-----------|
| 1999 | 763,606 | 446,204 | 186,827 | 49,000 | 81,575 |
| 2000 | 1,808,472 | 1,534,662 | 173,132 | 50,131 | 50,547 |
| 2001 | 3,426,705 | 3,118,596 | 240,332 | 62,365 | 5,412 |
| 2002 | 2,042,547 | 1,273,545 | 502,732 | 54,768 | 211,502 |
| 2003 | 3,039,441 | 2,122,377 | 711,631 | 68,195 | 137,238 |
| 2004 | 3,729,625 | 2,470,939 | 991,207 | 167,360 | 100,119 |

*Source:* FSC-IC Financial Report 2004.

of its overall revenue. The donors network consists of a large number of philanthropic foundations, mostly based in the United States (see Table 10.5). To date, the biggest direct grant to FSC-AC was US$5 million from the Ford Foundation in 2001. FSC-AC has also received many other large grants over the years, including US$750,000 from the Rockefeller Brothers Fund (2000) and US$425,000 from the MacArthur Foundation (1999).

In addition to direct grants to FSC-AC, FSC-US,[38] and FSC-Canada, the donors network has contributed substantial sums to related organizations, including the Rainforest Alliance (which houses SmartWood) and the Certified Forest Products Council (CFPC), whose major activity is developing the US market for certified timber products.[39] The Pew Charitable Trusts granted US$2.2 million to support the FSC-Canada Boreal standard-development

*Table 10.5*

**Major foundations making grants to the FSC system, 1993-2005**
(alphabetical by country)

| Region | Country | Foundation |
|--------|---------|------------|
| North America | Canada | Richard Ivey Foundation |
| | | Walter and Gordon Duncan Foundation |
| | US | Flora Family Foundation |
| | | Ford Foundation |
| | | Kohlberg Foundation |
| | | MacArthur Foundation |
| | | The Pew Charitable Trusts |
| | | Rockefeller Brothers Fund |
| | | Summit Foundation |
| | | Surdna Foundation |
| | | Wallace Global Fund |
| Europe | UK | Esmée Fairbairn Foundation |

process, and the Tides Foundation received substantial funds from other donors, amounting to almost US$400,000, which it deployed to support the FSC-BC standard-development process.[40] Finally, to these contributions should be added some relatively large grants (totalling US$2.8 million) to the FSC Global Fund Inc., a Washington-based independent fund established by US foundations in 2001 to support the FSC.

Donor support played a crucial role in establishing and running FSC-Canada. As was outlined in Chapter 2 (see Table 2.2), 90 percent of FSC-Canada's revenues from all sources derived from donations, mainly from foundations. Donors are in the business of supporting worthy humanitarian and environmental causes with financial resources. What causes they select depends on their individual foundation charters and general trends within the donor community. In the 1990s, donors to sustainable development and environmental causes shifted their emphasis from national to global approaches and from community- and government- to market-based strategies. As noted, therefore, to a considerable extent FSC-AC was "founded" by donors to test the viability of this new approach. Without such financial and strategic support, third-party forest certification would never have evolved beyond a fringe idea.

Fran Korten, chief executive of the Positive Futures Network, and Michael Conroy, senior program officer of the Ford Foundation, were central figures in a donors network that included Michael Northrop (program officer, Rockefeller Brothers Fund), Michael Jenkins (MacArthur Foundation), and Melissa Dann (program officer, Wallace Global Fund). Today, thirty-five to forty donor organizations meet biannually to discuss key trends and consider coordinated funding to worthy causes.[41] This donor group was deliberately instrumental in their actions and, according to Northrop, catalyzed approximately fifteen of the current FSC-related institutions in the United States: "We didn't have the institutions that we needed. We started talking about the need to create a Forest Stewardship Council office in the United States with the help of Michael Jenkins at the MacArthur Foundation ... in 1994. As we got deeper into the issue we thought that there needed to be similar institutions working on corporate demand. It took us almost three years, ultimately – between early 1994 and late 1996 – to help birth an institution" (GrantCraft 2005).[42]

Despite evidence in recent FSC-AC financial statements that the organization is weaning itself from dependence on donors, it continues to rely heavily on this support. This is problematic for two reasons. First, there appears to be little formal accountability to the donors network on the one hand and FSC members on the other. FSC-AC financial accounts do not set out in detail the precise nature of donor support, nor are members apprised of the views of the donor community with respect to forest certification and whether, for example, proposed shifts in policy direction have their origin in concerns expressed by donors. Second, donor priorities can change; if they do, FSC's effectiveness could be significantly affected should it become necessary to secure alternative funding sources.

## Network Politics and the FSC-BC Final Standard

Politics within "pluralist" democratic states typically reach a crescendo at election time, when political parties compete with each other for the votes of the electorate. Once elections have occurred, however, the party that gains power stands ready to dominate the policy process. Although there is a tendency for such parties to adopt "centrist" political platforms, inevitably they tilt toward the interests of business or labour. Likewise, the policies that are implemented are coloured by a party's orientation, enabling some interests to be overruled and others indulged. Electoral and policy-making processes thus tend to be adversarial, winner-take-all affairs.

Network politics within the FSC provide a striking contrast to conventional party politics. Instead of establishing an electorate from a territorially defined polity, which then vote this or that political faction into power, the FSC encourages the establishment of a global forestry polity by enlisting networks to join three chambers relevant to their dominant interests.[43] All of the networks examined here, with one exception, are accommodated within FSC's formal institutional arrangements. In this regard, the donors network stands outside as an interesting anomaly – neither explicitly recognized nor overtly vocal.

Throughout the BC standard-development process, from D1 to the final accreditation, networks operated in a strategic environment and championed their interests through overt and more opaque channels. Although loosely monitored by FSC-Canada and FSC-AC, the negotiations that culminated in the D1 standard largely unfolded according to an internal, provincial logic. Recognition that the standard emerging from the exclusive approach used to negotiate D1 could not be legitimately defended resulted in a tacit agreement within and across networks to restructure the process to give more weight to the perspectives of large-producer interests, mainstream environmental groups, and indigenous peoples.

Following the sidelining of D1, the development of D2, and the controversial approval of D3, actors operating within two networks – the large producers and certifiers – expressed concern about "prescriptiveness," a pejorative term employed to capture the twin concepts of "lack of managerial discretion" on the one hand and "height of the bar" on the other (see Chapter 11). As large producers directly used their economic and formal negotiating power within FSC-BC's steering committee to increase managerial discretion and lower the bar, CBs influenced the broader FSC network by discursively explaining their concerns, validating, in the process, some of the objections of large producers.

If regional standards setting depended completely on local network politics, D3 would have been approved as the FSC-BC final standard, given prevailing FSC rules that permitted voting if consensus could not be reached. However, some powerful interests saw D3 as an attempt to trump rather than accommodate progressive large-producer interests. As a result, large-producer and certifier networks refocused their efforts on influencing senior FSC decision-makers, with the former actively lobbying FSC-Canada and FSC-AC members to prevent accreditation of D3. Unfortunately for them, FSC-Canada's prevailing decision-making rules were not consensus based, but rather majority based. As a result, although only five FSC-Canada board members voted in favour of D3 on the day, with no support from the economic chamber (one against and one abstention), the motion to endorse the FSC-BC standard was nonetheless carried in conformity with FSC-Canada's by-laws and forwarded to FSC-IC.

From the perspective of large-producer and certifier networks, a full-fledged crisis was now emerging in the FSC system. Donors were also beginning to ask questions within their network about the feasibility and functionality of the FSC model, which was supposed to accommodate progressive large-producer interests, not frustrate them. Moreover, splits were developing within the environmental network between FSC-BC members – who were a powerful vocal minority within the FSC system – and members elsewhere. Southern members and those who held interests important to the South (including indigenous peoples, community forestry, and proponents of Third World development), were perplexed by the failure to achieve consensus through the FSC-BC process and were reluctant to tailor policy to the BC case, especially given how well that process was funded in comparison to the budgets of other FSC regions. Other social and environmental representatives regarded the process by which FSC-Canada had approved D3 as fundamentally flawed because consensus had not been obtained and the objections of the economic chamber were overridden. In their view, it was unwise to approve a standard that did not have a significant level of support from large industry.

In the charged atmosphere that developed over the summer of 2002, network politics within the FSC played out in an attempt to influence the outcome by lobbying FSC-AC – especially its executive director and board – with respect to the pending vote on D3 accreditation. By then, the issue had become highly polarized. A categorical decision to accept or reject D3 risked splitting the organization into "winners" and "losers," jeopardizing FSC's foundational premise that competing interests committed to a spirit of give-and-take can reach agreement on what constitutes well-managed forests through equitable, rational deliberation.

With time running out at FSC-AC, networks within the FSC targeted the upcoming Oaxaca General Assembly as an important forum in which to make their case. As this meeting approached, however, the strategic environment shifted dramatically. FSC-Canada and FSC-BC were making progress toward the "extraordinary solution" of preliminary accreditation, and this was taking the heat out of the debate as several proponents turned their attention to the standard's practical implementation. Meanwhile, BC environmentalists were coming to appreciate that it was unlikely that FSC-AC would accredit a standard that did not enjoy significant large-industry support. The General Assembly meeting was thus something of a political watershed. BC environmentalists put forward resolutions that would have entrenched a more explicit ecosystem-based approach to forest management, but these went down to defeat. In contrast, resolutions that affirmed the obligation of national initiatives to approve regional and national standards using a consensus-based approach prevailed (FSC-IC 2002b).

## Conclusion

Contestation over values and interests takes a very different form in the FSC than it does in the more familiar party-political elections of modern democratic states. The latter enables a territorially delimited polity to vote on who will rule within its borders, but national political processes suffer two major deficiencies in an era of globalization. Most obviously, elected governments give expression to interests defined nationally; the interests of non-citizens can only be included via intergovernmental negotiations. This typically results in lowest-common-denominator outcomes and also thwarts the "global public interest" due to the one-state one-vote and GDP-weighted voting procedures that tend to prevail in intergovernmental forums. Moreover, the adversarial nature of modern democratic politics results in winner-take-all electoral competitions in which the victorious party, for a variety of reasons, is neither accountable to nor interested in robustly representing environmental, indigenous, community, and various other values and interests.

The FSC solution to these two deficiencies is both innovative and pragmatic. By encouraging pre-existing homogenous networks to organize into chambers and requiring them to engage in intra- and inter-network deliberations, FSC creates a political space where no single interest may trump another. Policy thus emerges from the discursive efforts of each network to articulate a bottom line, influence other networks within its own chamber, and have the ensuing accommodation reflected in the final policy outcome. This deliberative policy process does not merely reflect interests and values, it also constructs and reconstructs them.

Nonetheless, it could be argued that the FSC's process also results in lowest-common-denominator policy outcomes. However, there is a critical difference between national and intergovernmental policy processes on the one hand and the FSC's on the other. In the former, a state's interest is refracted through a national policy lens that tends to privilege business interests. Intergovernmental bargaining merely magnifies this tendency, as states motivated by the desire for domestic competitive advantage seek to protect their forest industries from "unfair" foreign competition. This typically results in lowest-common-denominator outcomes that tend to reflect business interests and values, marginalizing those of other actors. In contrast, within the FSC, the lowest-common-denominator solution to policy issues must reflect a bargain that is minimally acceptable to the major networks housed within its six sub-chambers representing economic, environmental, and social interests from both the North and the South. In practice, however, the bar is higher than that, and policy emerges as a compromise between the needs of industry, the demands of environmentalists, and the rights and aspirations of people as expressed by unions, communities, and indigenous peoples groups. As such, the FSC is much more than a policy instrument or a market mechanism; rather, it could be argued, it is a new political space.

# 11
# A Regulatory Analysis of FSC Governance

As an emerging experiment in governance, one of the most challenging new roles undertaken by the FSC is that of "regulator." Earlier, we defined "regulation" as an institutionalized rule system that sets and enforces standards aimed at influencing the behaviour of interdependent social actors in order to promote mutually agreed-upon values, principles, or objectives (see Chapter 1). It is thus through regulation that the overarching shared values that bring "the governed" together in a common enterprise are translated into directives, procedures, and requirements that, in turn, become its operating legal code. In Chapter 12, we will consider how the FSC regulates *internal* relations among its constituent institutions and interests as measured in democratic constitutional terms. In this chapter, we assess how the FSC regulates, through the standard-setting process, the *external* activities of FSC-certified firms in the realms of civil society and the marketplace.

In making this assessment, it is critical to recognize the broader context within which the FSC has undertaken this regulatory role, particularly the vigorous and ever-evolving debate (within both scholarly and public policy circles) about the nature and future of regulation in the environmental and resource-management contexts. Throughout much of the 1980s and early 1990s, this debate tended to focus on the relative merits of market-based versus more traditional regulatory approaches (Latin 1982, 1985; Ackerman and Stewart 1985; Anderson and Leal 1991; Hahn and Stavins 1991; Tollefson 1998). Increasingly, however, attention has shifted to a consideration of the relative merits of specific forms of regulation (Ayres and Braithwaite 1992; Pildes and Sunstein 1995; Chertow and Esty 1997; Gunningham and Grabosky 1998; Gunningham and Sinclair 2002; Coglianese and Lazar 2003; Fiorino 2006). In this reconfigured debate, performance-based regulation has quickly come to assume a dominant role. Both in hard- and soft-law settings, proponents of performance-based regulation have contended that it should be presumptively preferred to management- or technology-based

alternatives due to its superior capacity to reward firms for meeting designated standards in a cost-efficient fashion. Performance-based approaches
(also known as results- or outcome-based regulation) are also championed
as being inherently less prescriptive and, accordingly, less costly to comply
with than their management or technology-based counterparts.

The stakes of what, on first impression, appears to be a rather abstract
and rarefied intellectual contest should not be underestimated. In the United
States, for almost a generation now, the future of US environmental and
public safety laws has hung in the balance as powerful interests have as-
sailed the existing regulatory framework to promote a wholesale reconfiguration that would see traditional prescriptive, command-and-control-based
regimes replaced by a largely voluntary, corporate-defined performance
approach.[1] Indeed, a strong appetite for regulatory reform aimed at significantly expanding the role of performance-based regulation appears to exist
in most OECD countries. For example, the Canadian government has recently committed to a far-reaching "smart regulation" strategy that strongly favours performance-based regulation over more traditional prescriptive
regulatory forms.[2] Likewise, an ideological commitment to the merits of
performance-based regulation provided the conceptual foundation for the
BC government's decision to repeal the "prescriptive" Forest Practices Code
and replace it with a results-based regime. The potential and limits of
performance-based standards have also played a central role in the burgeoning debate over the future of voluntary codes, certification, and other
alternatives to state regulation.

The negotiations within the FSC that culminated in approval of the BC
standard in many ways reflect, and can be elucidated by, this broader and, at
this juncture, still fledgling debate over regulatory form. Like many standard
setting entities, both governmental and non-governmental, the FSC has
expressed a commitment to and a preference for employing performance-
based regulation. But as the negotiations and controversy surrounding the
BC standard emphasize, there are varying opinions about how this commitment should be interpreted and implemented given the diverse values FSC
standards exist to promote.

In this chapter, we argue that determining whether and when to use a
particular form of regulation requires a clear understanding of the distinctions between performance-, management-, and technology-based standards.
These distinctions, in turn, make choosing the optimal form of standard in a
given scenario a highly context-dependent exercise. Indeed, in many contexts
the heterogeneity of the regulated community and/or the problems being
addressed may require a hybrid approach in which several forms of standard
are employed simultaneously or sequentially. In light of the foregoing, we
would argue that it is important to understand both the underlying reasons

for the rush to embrace performance-based standards and the need to be skeptical of unequivocal assertions about the superiority of this approach to standard setting.

In the next section of the chapter, we consider the standard-setting process with a view to providing a coherent account of similarities and differences between performance-, management-, and technology-based standards. We also examine the considerations that should inform how and when they are employed. In the following section, we reflect on the reasons why performance-based regulation has secured ascendancy and how this has affected FSC policy and practice with respect to standard setting. In the penultimate section, we link this discussion back to the four key areas of the FSC-BC final standard discussed earlier in this book (tenure and use rights, community and workers' rights, indigenous peoples' rights, and environmental values). For each of these areas, we examine in detail the form of standard that emerged from the FSC-BC process. In the final section, we offer some observations about the implications of the foregoing discussion for standard setting generally, as well as some thoughts on standard setting in the broader FSC network and within FSC-BC.

## Crafting Standards: An Overview

### Setting Standards to Protect Environmental Values

Before one can assess the nature and relative merits of differing forms of a standard, it is first instructive to consider how standards are developed in general terms. To do this, we will consider how environmental protection standards, particularly those in the pollution-control context, are developed. Standard setting is typically conceived of as a four-stage chronological process that consists of (1) setting an objective, (2) developing criteria, (3) establishing an ambient quality standard, and (4) defining an individualized operational standard.

All standard-setting processes, whether undertaken by governments or private entities, should, in theory, proceed from a clear definition of the overarching outcome(s) that the regime exists to achieve. Defining these objectives is a quintessentially political choice, a judgment about values (Franson, Franson, and Lucas 1982). Often, the objective will be stated in broad, categorical terms (i.e., securing safe drinking water; protecting biodiversity). However, this is not always true. A case in point is the statement of objectives set out in the regulations enacted by the BC government to implement its new results-based forest practices legislation. For each of the different forest values identified in the legislation (soils, timber, wildlife, biodiversity, and so on), the discrete objective is stated as being to conserve the value in question "without unduly reducing the supply of timber from BC forests" (*Forest and Range Practices Act*, s. 510).

Where the standard takes the form of hard law, the articulation of these threshold objectives will either be set out in legislation or, more often, by means of a legally enforceable regulation (as is the case with recently passed regulations under British Columbia's new results-based *Forest and Range Practices Act*). Where the standard is being developed by a non-governmental entity, these outcomes will usually form part of an organizational mission statement or code of practice. In the FSC context, this objective-setting function is performed by its ten governing principles.

Having identified the objective that a standard exists to protect, the next task is to identify the criteria by which to assess whether that objective is being achieved or compromised. This is often a daunting scientific undertaking. Whether the objective is safeguarding potable water or protecting ungulate winter habitat, the range of potential threats, and the nature of the interaction amongst these threats, can be highly complex. Inevitably, there are knowledge gaps in the "dose-response" relationship (Coglianese and Lazar 2003). As a result, a pragmatic approach is frequently adopted and criteria are formulated on the basis of a limited number of parameters (i.e., threats, risks, etc.) of greatest and/or most obvious concern. In the hard-law context, criteria definition will most likely be found in executive orders or regulation schedules. In the FSC context, this role is ostensibly performed by the criteria that accompany its ten principles.

The third generic step in the process is the formulation of ambient quality standards. This process has both a technical and a political dimension. It involves taking the scientific data generated by the criteria-identification process and evaluating how, in light of this data, the objective(s) identified can best be achieved, taking account of various political considerations, including the costs, benefits, and risk acceptability. Identification of the ambient standard is often framed in terms of the carrying capacity of the receiving environment. Ambient standards can take various forms. Sometimes they are expressed as an upper limit on the mass or concentration of harmful substances in the receiving environment; alternatively, they may prescribe minimum acceptable limits with respect to the presence of desired substances. When governments engage in this type of analysis (many Canadian governments do not), ambient standards are typically established informally in the form of policy. Although in theory voluntary private codes should likewise incorporate ambient quality targets, in many cases they do not. Perhaps the most notable way the FSC-BC standard grapples with this challenge is through its use of the concept of range of natural variability (RONV), discussed in Chapter 9.

The final stage in a standard-setting process involves moving from a statement of the desired ambient outcome to the crafting of a standard that defines the permissible behaviour of the individual party being regulated. In the context of point-source air and water pollution, this is typically referred to as a "discharge standard." Because standard setting in the resource management context remains in its infancy, there is no equivalent terminology, so we propose to use the term "resource management standard." Once again, this process has both scientific and political elements, requiring a complex assessment of both the incremental impact of the activity being regulated and the costs/benefits and compliance issues associated with different standard prescriptions. Governments frequently express standards of this kind in pollution permits adapted to suit the particular circumstances of the applicant and the airshed or watershed within which they are proposing to operate.

In Canada, the prevailing practice with respect to air- and water-pollution permits has been to express discharge standards in performance-based terms, often as an upper limit for the mass or concentration of harmful substances that the permit holder is authorized to release into the environment over a designated period of time. In contrast, the United States has adopted a distinctly different approach to permitting. There, in recognition of the difficulties associated with individually calibrating discharge standards in a manner that does not compromise ambient standards, the dominant approach has been to rely on technology-based permits. Under this approach, the permissible level of discharge is defined with reference to the level of environmental protection the permit holder could achieve using the best available technology (BAT).

Most forest-certification standards do not contain provisions that directly address the question of on-the-ground performance by stipulating what we would refer to as a resource management standard. Normally, environmental impacts are addressed by a generic requirement that the manager comply with all applicable forestry and environmental laws; sometimes this is bolstered by various planning and management obligations. In this regard, the FSC model is unique in that it aspires – through its national and regional standard-setting processes – to translate the organization's principles and criteria into measurable and verifiable managerial obligations expressed in terms of indicators.

One final point deserves mention before we consider the relative merits of different forms of standard. In the typical standard-setting process set out above (most notably in the context of air- and water-pollution control), the obligation of the permit holder is defined in terms of a discharge standard that is derived from an assessment of the receiving environment's relevant carrying capacity (the ambient quality standard). In the context of air and water pollution, the reason for adopting this approach arises from the necessity of factoring into the equation discharges from other point and non-point sources. It should be borne in mind, however, that where the regulator's concerns are limited to the environmental and other impacts of a single operator (as is often the case in the forest management context), they may choose to employ a standard that describes either what actions

*Table 11.1*

**Nexus between stages of production and forms of standard**

| Stage of production | Planning | Acting | Outputs |
|---|---|---|---|
| Form of standard | Management-based | Technology-based | Performance-based |

*Source:* Adapted from Coglianese and Lazar 2003.

the operator must take or avoid (a resource management standard) or what environmental conditions must be maintained during and after resource development has occurred (an ambient quality standard). As we will see later in this chapter, both of these approaches to standards drafting can be found in the FSC-BC final standard indicators (where the requirements of responsible forest management at the level of the forest management unit are defined).

**Performance-, Technology-, and Management-Based Standards**

Regulators use three main forms of standard to influence behaviour. Before discussing their relative merits, it is worth offering some ideal-type definitions. *Performance-based standards* are intended to be a specific and measurable indicator of a desired objective to be achieved, leaving broad discretion as to how to achieve this objective to the regulated entity. In contrast, in their pure form, *technology-based standards* are silent about desired outcomes; instead, they specify a particular technology, input, or mode of operation that is thought to optimally promote the overall objective of the standard. Finally, *management-based standards* (also known as process-based standards) eschew specification of both outcomes and means, opting instead to promote achievement of identified policy objectives by obliging operators to undertake specified management-planning activities.

As noted earlier, in regulating air and water pollution, Canadian governments have tended to employ performance-based standards; their American counterparts have elected to use technology-based ones. Over the last two decades, considerable research has been carried out to examine the experience of using both forms of standard in these and other jurisdictions. In contrast, both the experience with and the literature considering the efficacy of management-based standards has been much sparser. A helpful way to conceive of the relationship between these three forms of standard has recently been offered by Coglianese and Lazar. They argue that, in developing standards, regulators can choose to intervene at various stages of a firm's production process: the planning stage, the acting stage, or the output stage. The goal of such intervention is to correct market failure by seeking to promote the production of social-good outputs (i.e., environmental protection) without unduly fettering the production of the private-good outputs (i.e., timber production). As Table 11.1 illustrates, management-based standards represent an intervention in a firm's planning processes; technology-based standards can be seen as an intervention in the operational or "acting" stage; and performance-based standards constitute an intervention at the output stage, specifying what social outputs must be attained.

*Prescriptiveness and Standard Form.*
A key feature of the debate over the relative merits of these three forms of standard is the often-asserted proposition that performance- and management-based approaches are superior to technology-based standards because they are less prescriptive. While this "prescriptiveness" critique is seldom fully elaborated, it typically refers to the extent to which a particular form of regulation constrains a firm's actions and decisions by imposing additional (and potentially unnecessary) costs. The schemata proposed by Coglianese and Lazar (2003) helps to demonstrate the fallacy of this critique. When they are seen as reflecting different "moments" for intervention in the production process, all three forms of standard are potentially prescriptive in the relevant sense. How prescriptive they are depends, in turn, more on the content as opposed to the form of the standard in question.

Although technology-based standards are clearly prescriptive in terms of production methods, they are notably non-prescriptive in terms of management planning and, more significantly, outcomes or outputs. Similarly, the prescriptiveness of management-based standards can vary widely from a simple exhortation that operators implement a generic environmental management system, to a standard that tightly specifies the nature of planning to be undertaken and requires government signoff before production activity can proceed.

Performance-based standards also vary considerably in their prescriptiveness, depending on whether the standard in question is loosely or tightly specified. The specificity of a standard in this regard is often closely related to whether the standard is *qualitative* or *quantitative* in nature. To illustrate, assume that the standard's objective is maintaining slope stability and preventing soil erosion. A qualitative performance standard could be framed so that it turns this desired outcome into a mandatory obligation by implicitly adding the words "thou shalt" (i.e., *thou shalt* maintain slope stability and prevent soil erosion). Conversely, the same outcome could be sought through the use of a quantitative performance standard framed as follows: "No tree felling shall take place on slopes that exceed thirty five degrees."[3] This latter formulation – clearly a performance-based standard – is, on its face, highly prescriptive in that it overrides a manager's discretion as to where and when to log.

Arguably, by foreclosing the potential for a manager to seek to achieve the expressed objective (maintaining slope stability and preventing soil erosion), the quantitative standard is more prescriptive than management- or technology-based standards that could be crafted to secure the same goal. Under a management-based standard, where the slope in question exceeds thirty-five degrees, a requirement that the manager conduct slope stability tests or third-party assessments prior to logging might be triggered. Under such an approach, if the tests confirmed that logging could occur without compromising the slope stability or creating soil erosion, the manager would be at liberty to proceed. It is also possible to posit a technology-based standard that preserves a relatively broad scope of managerial autonomy. For example, such a standard could stipulate that logging can take place on slopes of over thirty-five degrees, but only if a specified mode of low-impact harvesting (i.e., helicopter logging) was employed.

### Private Autonomy versus Certainty

The corollary of the prescriptiveness critique discussed above is the assertion that performance standards, relative to other forms of standard, are superior in terms of preserving private autonomy. The preceding hypothetical situation shows that this generalization is misleading. Performance standards can sometimes preserve private autonomy; however, the extent to which they do so will depend on the congruency between the objective that the standard exists to advance and the form of standard itself. The greater the congruency, the more discretion the regulated party will enjoy in determining how to meet the standard. Thus a performance standard that transforms the desired objective (stable slopes) into a generic mandatory obligation ("thou shalt protect slope stability") preserves much greater private autonomy than one that specifies a narrower or subsidiary goal (prohibiting or restricting logging on slopes of greater than a specified steepness). As a general proposition, therefore, qualitative performance standards rate much higher than quantitative ones in terms of private autonomy. Indeed, as we have argued, quantitative performance standards are often highly prescriptive.

It is important to recognize that preserving private autonomy comes at a price in terms of certainty that the promulgated standard will achieve the policy objective it exists to serve. In other words, all other things being equal, it can be argued that the more generic and imprecise the formulation of a performance standard, the higher the risk the objective of the standard will be compromised. Whether, and in what circumstances, the risks inherent in performance-based standards are worth the potential benefits in terms of promoting other values (such as reducing industry regulatory costs, promoting innovative practices, etc.) is a question that must be assessed on a case-specific basis. Relevant to this determination will be the nature and importance of the objective the standard exists to promote and the pervasiveness, severity, (ir)reversibility, and probability of the risk that the regulated activity will compromise this objective. Also relevant will be the circumstances, and particularly the compliance history, of the party being regulated. Different strategies may be advisable depending on whether the regulator's goal is to create incentives for industry leaders or, alternatively, to police the activities of their laggard counterparts (Gunningham and Sinclair 2002).

Another set of issues relate to the "ownership of uncertainty" (Coglianese and Lazar 2003). Assuming we can assess who benefits and who loses from the uncertainty that some types of performance standard introduce, should the beneficiaries be called upon to bear a higher level of accountability in the event that an objective is compromised? In other words, should greater private accountability be the quid pro quo for greater private autonomy? Or should parties regulated under an open-ended performance standard be able to defend their failure to meet the standard by claiming "due diligence," as their counterparts regulated under more prescriptive regimes are able to do?

Protecting private autonomy by enacting a standard that confers broad managerial discretion to choose how compliance is to be achieved can also have a cost in terms of transparency and public participation. The move toward an approach under which managers are mandated to develop in-house strategies to meet the requirements of a pure performance standard inevitably leads to heavy reliance on complex, predictive models of performance. Even assuming members of the public are given access to this modelling information (which is by no means assured), their ability to effectively participate in regulatory decision making will decline as the complexity of the analysis increases. This professionalization of standard setting also has the potential to undermine the capacity for meaningful regulatory oversight by government or private standard-setting bodies, by forcing regulators to rely more heavily on third-party experts (e.g., academics, consultants, certifiers) or, alternatively, accepting on faith analyses provided by private managers.

### Measuring, Evaluating, and Verifying Performance

It is broadly accepted that performance standards work best "when actual performance can be measured, evaluated and verified" (Coglianese and Lazar 2003). As we have noted, in standards where there is a close congruency between the objective to be achieved and the form of standard, uncertainty arises, with the result that measurement, evaluation, and verification of performance can be undermined.

To minimize this uncertainty (and enhance the potential for accurately measuring, evaluating, and verifying performance), one obvious option is to employ quantitative performance standards. This prospect raises its own questions. One set of questions concern what is being measured and to what extent it serves as a reliable indicator that the objective the standard exists to promote is being achieved. In theory, the most reliable quantitative indicator would be one that offers a direct measure of the condition of an ambient environmental value (i.e., ambient water or soil conditions; the continued viability of specified listed species in the area being logged). Because such ambient values are, by definition, affected by a variety of other natural and non-natural factors, however, quantitative performance standards are typically crafted so as to measure the incremental impact of the activity being regulated. They thus normally take the form of discharge or resource management standards (i.e., they prescribe acceptable discharge limits or restrict logging on slopes of more than thirty-five degrees).

Even carefully crafted discharge or resource management standards do not, of course, eliminate uncertainty with respect to measuring whether the objective is being met. For instance, there will always be some uncertainty in what we referred to earlier as the dose-response relationship – in other words, the nature of the relationship between the behaviour the standard targets and the impact of that behaviour on the ambient environment. As well, because of the diversity of environments within which the activity being regulated occurs (of which forestry is a prime example), a generalized standard that may be consistent with the achievement of the objective in one context may not be appropriate in others. Indeed a one-size-fits-all quantitative performance standard can be under-protective of the desired objective in some contexts and over-protective in others. Thus, under particularly sensitive soil conditions, logging on slopes of less than thirty-five degrees may well have significant environmental consequences. Conversely, where soil conditions are less sensitive, or if a low-impact logging method is used, it is plausible that harvesting could occur on slopes steeper than the standard prescribes without the objective of the standard being placed in jeopardy.

There is also, in this context, the looming question of who is responsible for measuring, evaluating, and verifying performance. The logistical and resource challenges associated with these various tasks were a primary reason why, in its first-generation environmental laws, the US government opted to employ a technology-based approach. For similar reasons, many have been critical of the BC government's decision to replace the Forest Practices Code with results-based legislation, particularly given dramatic and ongoing budget cutbacks at the Ministry of Forests and Range. The dilemma, of course, is that any potential gains in terms of the cost of regulatory oversight that might be achieved by moving toward a performance-based model are likely to be nullified if responsibility for measuring, evaluating, and verifying performance remains with the regulator. Thus, while costs to business under a performance-based regime are likely to go down, costs to government (particularly those associated with compliance and enforcement) may well go up. This is particularly so given the diversity of regulatory compliance arrangements that performance-based standards contemplate.[4]

Governments are, of course, not alone in expressing concerns on this front. To the extent that open-ended performance-based standards (as opposed to more easily auditable management- or technology-based standards) play a key role in voluntary codes, the task of certifiers is made more difficult. More importantly, however, unless there is broad confidence that the standards articulated are subject to rigorous and regular measurement, evaluation, and verification, their perceived legitimacy (and ultimately their viability) will be put in jeopardy.

## Choosing the Optimal Form of Standard

A consensus appears to be emerging that choosing the best form of standard requires careful consideration of a variety of context-specific factors, including the logistics associated with measuring, evaluating, and verifying achievement of the objectives of the standard, as well as the nature of the "regulated community."

It is claimed, for example, that where it is a relatively straightforward matter to assess whether a standard's objective is being achieved (by means of measurable indicators or other well-correlated proxies, for example), and if it can be done at a relatively low cost, a preference should be given to performance-based standards. Alternatively, where it may be difficult or costly to monitor firm-level performance, and where the regulated sector is relatively homogenous, the best regulatory option may be one that is technology based. This is especially likely to be true when the relationship between action and output is well understood and can be easily evaluated. However, where the regulated community is composed of "heterogeneous enterprises facing heterogeneous conditions," there is emerging agreement that a strong theoretical justification exists for management-based standards, all other things being equal (Gunningham and Sinclair 2002). Adopting this option, it is said, allows operators to deploy their informational resources and act on their private incentives insofar as they are not inconsistent with achieving the standard's broader public objectives (Coglianese and Lazar 2003).

To date, reflections on these themes have primarily been offered in the context of the choices confronting government regulators, but we would argue that they provide a highly instructive vehicle for evaluating the FSC-BC final standard provisions dealing with tenure and use rights, community and workers' rights, indigenous peoples' rights, and environmental values. Before embarking on this assessment, however, we first explore the broader context within which the FSC's approach to, and recent policy pronouncements on, standard setting have emerged.

### The Ascent of Performance-Based Regulation

Although there appears to be a small but growing cohort of opinion advocating the merits of employing a context-driven approach to making decisions about what form of regulatory standards best advance the public interest, a strong preference for performance-based regulation persists in much of the standards debate in academe and beyond. Frequently, ideology plays a key, if often unspoken, role within this debate. Yet, although many proponents of performance-based regulation tend to be harsh critics of what they would style "command and control" statist regulation, performance-based regulation has managed to secure a remarkably diverse constituency, including many supporters within civil society. In this section we reflect on some of the main factors that explain the genesis and crystallization of this constituency. We conclude with some observations on how this has influenced the FSC's approach to and policy on standard setting.

Over the last twenty years, political life in Western liberal democracies has been increasingly preoccupied with renegotiating the role of the state. Trade liberalization has brought with it growing global competition that has, in turn, translated into pressures on domestic governments to rein in social spending and reduce the regulatory burden shouldered by business. Aided by the fall of the former Soviet Union, neo-liberal, market-based policy diagnoses and prescriptions have succeeded in shifting the goalposts of political debate, achieving the status of contemporary common sense. In the regulatory realm, support for traditional command-and-control regimes has dissipated dramatically, replaced by calls for performance- or results-based regulatory models that harness the power of the market. The drive toward regulatory reconfiguration is also being propelled by international trade agreements that exhort governments to eschew prescriptive regulation in favour of more ostensibly neutral performance- and market-based regulatory alternatives.

Environmental regulation, whether government-sponsored or under the auspices of voluntary codes such as those promoted by the FSC, has been a key venue for the debate over the need to move toward more performance-based approaches. In our view, two factors stand out in terms of understanding the origins and nature of this debate: continuing reformist pressures targeting the US federal environmental regulatory regime and concerns about the susceptibility of prescriptive, state-based environmental standards to international, trade-based challenges.

### The Regulatory Reform Debate in US Environmental Law and the Birth of "Smart Regulation"

The performance-based critique of command-and-control-style regulation can, in many ways, be seen as the result of an ongoing, sometimes overtly ideological, battle over US environmental policy. Paradoxically, federal American environmental laws became a target for sustained critique not because they had failed, but in large measure because they had succeeded. Most of these laws – most notably the *Clean Air Act*, the *Clean Water Act*, and the *Resource Conservation and Recovery Act* – were enacted within a single decade, between 1970 and 1980. Unlike other federal jurisdictions, where national governments, for various constitutional and political reasons, have tended to eschew taking a leadership role in tackling environmental issues, in the United States these laws reflected a broad consensus that the federal government not only had the legal authority (founded on the powerful Commerce Clause) to take leadership on environmental issues, but that it was also politically obliged (and, arguably, best situated) to do so.

The American approach was anomalous in terms of the scope of the federal role; it was also unique in terms of its reliance on a particular mode of regulation. Virtually all the federal environmental laws enacted in this period employed best available technology (BAT) standards. Under these standards, regulated facilities were obliged to install the most stringent pollution-control technology available, up to the point that the costs of doing so would cause them to shut down (Hirsch 2001). For the most part, BAT standards were sector specific and were typically applied uniformly on a national basis. These American laws were also unique insofar as they provided legal means (known as "citizen suits") by which citizens could challenge permit holders and governments for failing to comply with or effectively enforce legal obligations prescribed by these permitting arrangements (Thompson 2000).

The decision to adopt the BAT approach was justified on various grounds: among other benefits, they minimized government monitoring and enforcement costs, enhanced public scrutiny and participation in compliance efforts, reduced the potential for "regulatory capture" and other forms of political manipulation, and diminished interstate competition that might arise from the potential for pollution havens (Latin 1985).

By most accounts, over the ensuing quarter-century these laws, and the programs that were funded to implement them at the federal and state levels, brought about significant improvements in environmental quality (Latin 1985; Chertow and Esty 1997). These improvements were particularly notable in relation to discharges from single point sources (Percival 1996). Before long, however, these first-generation environmental laws came under sustained criticism.

Although the critics usually conceded that these first-generation laws had been successful in "picking the low-hanging fruit," they argued that the BAT model was in many ways highly inefficient. They claimed, for example, that one-size-fits-all approaches (such as BAT) ignore basic cost-benefit analysis by failing to take into account the social benefits of reducing pollution in a given setting (relative to the costs of securing those benefits) or the differential ability of firms within a given sector to meet the BAT standard. Critics also railed at the costs to business of the culture of legal adversarialism that, they claimed, the laws promoted. Moreover, they contended that, although a case could be made that BAT was an effective means of regulating industry laggards, it provided no incentive for more responsible firms to make research and development investments in new pollution-control technologies. For these reformers, the policy prescription was clear. Having combated the first-generation challenges of curbing the most serious and pressing major sources of pollution, the time had come to enact a second generation of laws that were capable of leveraging further pollution reductions in circumstances where the marginal cost of achieving progress was typically high and that could also address discharges from small, non-point sources, including those in the consumer, services, and agricultural sectors (Ackerman and Stewart 1985).

The reformers' critique did not go unchallenged (Latin 1982, 1985; Thompson 1996; Steinzor 2001). Defenders of the prevailing legal regime pointed to its record of achievement in enhancing air and water quality, to the additional regulatory costs associated with requiring federal agencies to undertake additional cost-benefit and risk analyses, to the dangers of leaving to the market (through pollution-trading regimes and other less state-centric instruments favoured by reformers) determinations with respect to risk exposure at the local level, and, perhaps above all, to the state's obligation to actively engage in safeguarding environmental values as public goods (Thompson 1996, 2000; Steinzor 2001).

Through much of the late 1980s and early 1990s, this debate was cast as a classic showdown between proponents of free markets and defenders of a strong, interventionist state. By the mid-1990s, however, with the arrival of the Clinton administration, the tenor of the debate changed. At this point, it was becoming clear that the ambitious legislative aspirations of the reformers were unlikely to be realized in the short term, if at all. At the same time, however, their critique was having an impact in a more interstitial fashion. Starting in the early 1990s, the US Environmental Protection Agency drew on the reformers' arguments to initiate a variety of programs aimed at combatting the perceived rigidities and inflexibilities associated with first-generation laws (Hirsch 2001). Among them were programs that facilitated pollution credit trading and provided new incentives, including regulatory relief, for superior environmental performers.

By the mid-1990s, a debate that had focused on US environmental law and been dominated by American legal scholars and economists began to expand outward. There was a growing sense that the regulatory questions being mooted in relation to the future of US environmental law were of much wider significance. The most important of the new contributions to the debate is an approach to regulatory reform that its Australian originators termed "smart regulation" (Gunningham and Grabosky 1998; Gunningham and Sinclair 2002).

The thesis promoted by Gunningham and his colleagues is deceptively simple: a pluralistic, flexible, context-specific approach should be taken as instrument choice in the realm of environmental policy and beyond. Drawing on case studies of regulatory innovation in North America, Europe, and Australia, they contended that framing the choice of instrument in ideological terms – state versus market – was unhelpful. Moreover, it was important not to underestimate the successes of, or the continuing need for, command-and-control-style regulation. However, they argued that, in many cases, better environmental outcomes could be leveraged by deploying a mix of policy instruments tailored to the particular goals and circumstances of the regulatory context. Governments could also achieve better results (given static or declining budgets) by harnessing resources from outside the public sector and engaging directly with regulated parties, communities, and civil society organizations. Thus, they asserted, governments should "steer not row" and should, where possible, "govern from a distance." To this end, governments should be ready to engage in continuing governance experiments aimed at identifying and testing new ways to share governance responsibilities by empowering business and civil society through information, networks, and partnerships.

The debate over "smart regulation" has produced a remarkable outpouring of scholarly work. Many governments have adopted the term to provide intellectual justification for a wide range of public-sector regulatory reform initiatives. In the United States, for example, federal agencies are now directed by executive order (originally issued by President Clinton and later affirmed by President Bush) to employ performance-based regulation wherever feasible. Many federal agencies, including the Environmental Protection Agency, the Nuclear Regulatory Commission, and the federal Highway Administration, to name a few, are now piloting performance-based regulatory programs. The influence of the smart-regulation critique has also resonated in potentially sweeping law and policy reforms taking place in many other Western liberal democracies, including Canada, the United Kingdom, and Australia.

### Trade Law and Standard Setting

The drive toward performance-based regulatory approaches has also been propelled by concerns about the implications of international trade rules governing product labelling and certification. Here the overarching issue relates to the applicability of the principle of non-discrimination between "like products," one of the basic tenets of international trade law.

Under most international trade regimes, member countries are obliged not to discriminate between like products either directly (i.e., by levying differential tariffs or taxes) or indirectly (i.e., by subjecting like products to labelling requirements). Labelling requirements that provide consumers with information relevant to the use or consumption characteristics of a product, arising from how it was produced (i.e., its physical characteristics or safety), are considered legitimate. In trade-law parlance, such requirements are referred to as process or production method (PPM) standards. However, labelling requirements that identify the social or environmental impacts or characteristics of a product, information that is unrelated to its end use or consumption, are generally deemed to be an illegal non-tariff barrier and, as such, in violation of international trade law.

The presumptive illegality of non-product-related PPM standards, a presumption fortified by early World Trade Organization (WTO) jurisprudence, including the Tuna-Dolphin II litigation, has come to be regarded by many governments as a justification to eschew any direct involvement in environmental labelling (Charnovitz 2002). This is especially so in relation to forest products, where state-sponsored labelling schemes are likely to be challenged as impermissible, non-product-related PPM standards (Hock 2001) because, from the perspective of end-use suitability, forest products from sustainably managed forests are unlikely to differ from those harvested in a less-sustainable manner.

The perception that state-initiated forest product labelling schemes would be judged non-product-related PPM standards led many sustainable forestry reformers to adopt a strategy that focused on the development of non-state certification schemes. This strategy, which led directly to the establishment of the FSC, was based on the assumption that private, voluntary, eco-labelling schemes were exempt from challenge under international trade law because the applicable WTO provisions only apply to such schemes if they can be said to have "a governmental character."

However, there are now concerns that WTO rules could be brought to bear against voluntary labelling schemes controlled by non-state entities, such as the FSC. This growing concern is in part a function of developing countries' continuing opposition to non-product-related PPM standards, an opposition that was strongly reaffirmed at the WTO meetings in Cancun in 2003. There is also a growing appreciation that the long-term prospects for the FSC (and analogous schemes) depend on the organization's ability to influence government procurement policies. Because government procurement policies must comply with WTO rules, non-state certifiers and standard-setting bodies such as the FSC are being forced to re-evaluate whether their certification and labelling protocols are WTO-compliant.

As a result, many international voluntary standard-setting and accreditation organizations that focus on social and environmental issues (including the FSC) have recently established an umbrella organization known as the International Social and Environmental Accreditation and Labelling Alliance (ISEAL). The purpose of ISEAL is to support "credible standards and conformity assessment by developing capacity building tools to strengthen members' activities and by promoting credible voluntary social and environmental certification as a legitimate policy instrument in global trade and development" (ISEAL 2006a, 1). A key priority of ISEAL has been to develop the Code of Good Practice for Setting Social and Environmental Standards. The code is intended to provide a benchmark to assist its member organizations "to improve how they develop social and environmental standards" (ISEAL 2006a, 1). In developing this code, ISEAL has examined the question of whether, and to what extent, its members should take steps to ensure their standards comply with WTO non-product-related PPM requirements.

The strategy adopted by ISEAL has largely been precautionary, one aimed at pre-empting trade-law-based challenges. To this end, ISEAL is committed to ensuring that the "social, environmental and/or economic objectives of a standard promoted by its members are "clearly and explicitly specified" and that such standards are "no more trade restrictive than necessary" (ISEAL 2006a, 5). In implementing this commitment, ISEAL members commit to ensuring that their standards "shall be expressed in terms of a combination of process, management and performance criteria, rather than design or descriptive characteristics. Standards shall only include criteria that contribute to the achievement of the stated objectives. Standards shall not favour a particular technology or patented item" (ISEAL 2006a, 5).

Notably, although the ISEAL code does not preclude the use of standards that employ non-performance-based criteria (by explicitly contemplating that standards may combine process, management, and performance criteria), the terms "process," "management," and "performance" are left undefined. By excluding standards that are design or descriptive based, or that favour a particular technology, it would appear that only standards that purport to describe or require a particular mode of producing a product (or developing a resource) are prohibited under the ISEAL policy. Thus, both performance-based and management-based standards are deemed permissible.

### FSC Policy on Standard Structure and Content

Like ISEAL, FSC-AC has consistently professed a commitment to employing a performance-based approach to standard setting. As we shall see in the next section, this commitment is underscored in unequivocal terms in the Accreditation Business Unit (ABU) report on the FSC-BC third draft (D3) standard. In 2004, no doubt due to the ongoing controversy over that standard, the FSC-IC adopted a policy – "Structure and content of forest stewardship standards" – that, among other things, addresses this important topic (herein the "FSC standards policy" or the "policy") (FSC-AC 2004c).

Much of this policy focuses on the content and form of FSC standard indicators that serve to define management-level obligations for FSC operators. It begins by observing, "There are many different ways in which a forest management unit's overall level of compliance with a set of such indicators may determine whether a forest management unit merits certification" (FSC-AC 2004c, 3). According to the FSC, however, the lack of a uniform approach to drafting standards across the FSC network has created serious difficulties for those responsible for developing standards; for companies that want to ensure they are being assessed on a fair basis against peers and competitors; and for stakeholders who want to understand the rationale for certification decisions. Consequently, the FSC standards policy aims "to improve consistency and transparency in certification decisions between different certification bodies and in different parts of the world, and thereby to enhance the credibility of the FSC certification scheme as a whole" (3).

In at least one respect, the policy goes farther than those recommended by ISEAL. According to the policy, "each indicator shall specify outcomes or levels (i.e., thresholds) of performance that are measurable during an evaluation." This does not mean that a standard cannot require a manager to undertake a "particular procedure," but where such a requirement is imposed, it shall be "accompanied by the further requirement for evidence of implementation and an indication of the expected outcome in the forest management unit" (FSC-AC 2004c, 7). Consistent with ISEAL, the policy expressly forbids standards that are defined in terms of "design or descriptive characteristics" or that "favour a particular technology or patented item." Moreover, it strongly urges drafters to eschew "subjective" language such as "ordinarily," "substantial," "wherever possible," and "best available" (FSC-AC 2004c, 7).

Table 11.2 is a schematic depiction of how the FSC standards policy maps onto the typology introduced in the "Crafting standards" section of this

*Table 11.2*

**FSC-AC policy on standards regarding various stages of forest management**

| Stages of forest management | Planning | Operations | Outputs |
|---|---|---|---|
| Potential forms of standard | Management-based | Technology-based | Performance-based |
| FSC policy | *Permitted* (though must be accompanied by evidence of implementation and anticipated outcomes in management unit) | *Prohibited* (design- and technology-based requirements concerning forest operations are precluded) | *Preferred* (with indicators to identify measurable outcomes or performance levels) |

*Source:* FSC-AC, "Structure and content of forest stewardship standards" (FSC-STD-20-002), 2004c.

chapter. The depiction underscores that the goal of these policies is to promote standards that identify measurable outputs accruing from responsible forest management. Despite this first-order preference for performance-based indicators, the policy recognizes the need for standards to provide direction to managers at the planning stage. Consequently, management-based indicators that address planning are permitted, but only where they are tied to implementation requirements and anticipated outcomes in the management unit. What the policy appears to forbid are standards that impose requirements with respect to the means by which forest operations are actually conducted: what we earlier referred to as technology-based standards.

### Evaluating the FSC-BC Final Standard

In this section, we will draw on the preceding analysis to consider in detail the structure and content of the FSC-BC final standard in terms of its consistency with the direction provided by the FSC standards policy discussed above. To do so requires an initial scrutiny of the Accreditation Business Unit's critique of the FSC-BC D3 standard and a closer look at the newly adopted FSC standards policy.

### The ABU Critique of the FSC-BC D3 Standard

As noted in Chapter 4, during the process that culminated in the granting of preliminary approval to the FSC-BC D3 standard, the standard met with considerable criticism from business interests – criticisms that were largely echoed by the FSC-ACs ABU report (FSC-AC ABU 2003). These criticisms focused on two main themes: the burdensomeness of the BC standard and its overly prescriptive and procedural nature.[5]

As a key condition for granting preliminary approval to the FSC-BC regional standard, the ABU directed FSC-BC to revise the text of its standard "to ensure that indicators and verifiers have a first order preference for not prescribing specific and explicit ways of achieving compliance with standards criteria and that, as a rule, indicators are focused on output and specific performance level rather than process and procedures." Although it acknowledged that the draft BC standard would likely "result in a high level of performance in the field," the ABU criticized the standard's "highly prescriptive" and procedural nature, contending that this was inconsistent with the intent that FSC standards be performance-based. It also concluded that procedural approaches of the type found in the BC standard "result in significant limitations on flexibility and adaptability of forest management in the field" and place auditors and certification bodies in the position of having to make value judgments. In the ABU's words: "It is FSC's intent to avoid such procedural approaches, allow for flexibility in forest management while guaranteeing high quality outcome and limiting value judgment by auditors and certification bodies as much as possible."

The ABU identified a large number of provisions in the BC standard that it felt required revision because of their prescriptive or procedural nature. Throughout the annex to their report, the ABU use the terms "prescriptive" and "procedural" synonymously. This usage appears to be a function of the fact that the authors of the ABU report consider mandated processes or procedures to be "prescriptive" insofar as they prescribe a particular method of complying with the relevant principle or criteria. The provisions of concern due to their prescriptive/procedural nature are primarily clustered in the standards accompanying Principle 3 (Indigenous peoples' rights), Principle 4 (community relations and workers' rights), Principle 5 (benefits from the forest), and Principle 6 (environmental impacts).

As we argued in the "Crafting standards" section of this chapter, it is a mistake to subscribe to the notion that process standards (in our terminology, "management standards") are inherently more prescriptive than other forms of standards (performance- or technology-based). All three forms of standard entail prescription by requiring firms to perform actions that they might not choose to undertake if left to their own devices. As we have argued, the key conceptual distinction between these three forms of standard is the stage of the production process (planning, operations, or output) that is targeted by the standard employed. Determining whether, and to what extent, a given standard is prescriptive (i.e., constrains a firm's actions or decisions in the production process) requires a context-specific analysis of how that standard affects action and decision making at the firm level.

Similarly, it is wrong to suppose that the prescriptiveness of a standard will necessarily be correlated to the burdensomeness of that standard at the firm level (in terms of production costs). A highly prescriptive standard (whether management, technology, or performance based) will not necessarily entail greater firm compliance costs than a relatively non-prescriptive standard. The principal variable that determines a firm's compliance cost is the magnitude of the new investment necessary to achieve compliance.[6] Although it is likely that complying with a highly prescriptive standard will have cost implications at the firm level, these may be significantly less than the cost of complying with a relatively non-prescriptive standard that establishes a high level of performance that is not achievable using existing production methods.

In short, we would contend that rather than focusing on the prescriptiveness of a standard, standard setters should be mindful of a variety of other context-dependent considerations, including the nature of the objectives the standard exists to promote, the challenges associated with monitoring and evaluating performance, and the nature of the regulated community. In the FSC context, given the diversity of the objectives that FSC standards seek to advance, the dynamic nature of the forest environment, and the diverse range of prospective companies seeking certification, we would argue that the optimal approach to standard setting is one that deploys both management- and performance-based standards.

### Revisiting the FSC Policy on the Structure and Content of Standards

Notwithstanding the ABU's preoccupation with prescriptiveness and its apparent disapproval of management-based standards, it would appear that the FSC-IC's standards policy (developed and published after the ABU report on the BC standard) in large measure concurs with the approach to standard setting we elaborate above. Nowhere does the policy refer to the analytically unhelpful concept of prescriptiveness. Moreover, although the policy appears to give preference to performance-based standards, it explicitly recognizes that management-based standards may be employed when they are accompanied by (1) a "further requirement for evidence of implementation" and (2) an "indication of the expected outcome in the management unit" (FSC-AC 2004c, 7). It is worth reflecting on these two caveats with respect to the use of management-based standards. In neither case does the policy provide illustrations to aid in interpretation. In our view, requiring evidence of implementation of management planning suggests the FSC is concerned about ensuring effective monitoring of planning activities. Hence, it would appear that this caveat is intended to require firms to establish record-keeping and reporting procedures that will facilitate assessment of the effectiveness of the management planning function.

It is more difficult to divine the intent underlying the policy's related requirement that management-based standards provide an "indication of the expected outcome in the management unit." By definition, management-based standards do not prescribe outcomes beyond preparing and implementing specified management planning activities: this is what distinguishes management-based standards from their performance-based counterparts. Equating the term "outcome" with the objective of the standard does not provide assistance either. This is because, as discussed in the "Crafting standards" section of this chapter, standards are not intended to define objectives. Under the FSC regime, objectives are set out at the level of principles and criteria. For example, the objective of Criterion 5.6 is to ensure that the rate of harvest does not "exceed levels that can be permanently sustained." Under the FSC-BC standard, managers are required (under Indicator 5.6.2) to carry out a timber harvest analysis that considers prescribed factors (i.e., growth and yield projections) that are ostensibly designed to advance this objective. Thus, it is not at all clear how management-based standards such as these could comply with the policy's direction that they indicate an "expected outcome in the management unit."

Finally, it is worth noting that the policy purports to preclude the use of indicators that are "defined in terms of design ... characteristics." This restriction appears to be based on ISEAL policy, which in turn is driven by trade-law-compliance concerns (discussed earlier in this chapter). Although we view it as unlikely that this prohibition significantly constrains standard setting in the forestry setting (given that management- and performance-based approaches are generally better suited to this context), it could nonetheless be argued that operational requirements framed with reference to a definable model or conceptual design (such as the RONV or integrated pest management provisions of the FSC-BC final standard) could be characterized as running afoul of this prohibition.

## Analyzing the FSC-BC Final Standard

### Tenure, Use Rights, and Benefits from the Forest

The provisions of the BC standard that we have grouped under this heading address two principal objectives: ensuring that forest managers respect the legal and customary rights of local communities and maintaining a rate of harvest that can permanently be sustained (Criteria 2.2 and 5.6).

As Table 11.3 illustrates, the form of standard that has been developed in support of each of these objectives combines management- and performance-based elements. In relation to local communities, a management-based approach is used to ensure managers inventory local tenure and use rights. To ensure that these rights are respected, the standard then employs performance measures, including an obligation on the manager to secure affected rights holders' consent to proposed forest development and a requirement that, when disputes arise, the manager will abide by recommendations that emerge from a prescribed dispute resolution process. A similar approach is adopted in relation to managing harvest rates. Here the standard sets out in considerable detail the elements that must be contained in a manager's timber harvest analysis (Indicators 5.6.1 and 5.6.2). This is coupled with a performance-based obligation on managers to demonstrate that projected harvesting activity will not exceed the long-term harvest rate (Indicator 5.6.4). With respect to the harvesting of non-timber forest products, a single performance-based measure is employed (Indicator 5.6.3).

### Community and Workers' Rights

Under the rubric of community and workers' rights, the principal objectives of the standard are two-fold: to ensure that communities secure employment, training, and related benefits from forest development; and to ensure that forest planning and operations incorporate the interests of directly affected people (Criteria 4.1 and 4.4).

The form of standard developed by FSC-BC to achieve these objectives is strongly performance-based, focusing on outputs as opposed to mandating planning inputs (see Table 11.4). The only management-based element in this part of the standard is a requirement that managers develop and

*Table 11.3*

**FSC-BC indicators addressing tenure, use rights, and forest benefits:**
**Principles 2 and 5**

*Criterion 2.2*   Local communities with legal or customary tenure or use rights
shall maintain control ... over forest operations unless they delegate control with free and informed consent to other agencies.

*Criterion 5.6*   Rate of harvest of forest products shall not exceed levels that
can be permanently sustained.

| FSC-BC indicators | Form of standard |
|---|---|
| In consultation with local people, the manager identifies any legal or customary tenure or use rights in the management unit held by people who reside within or adjacent to it (2.2.1). | Management-based |
| The manager obtains free and informed consent from local rights holders to any portion of the management plan that affects their rights and resources (2.2.2 [a]). | Performance-based |
| If local rights holders dispute that current or proposed management protects their rights and resources, the manager implements recommendations developed through a Criterion 2.3 dispute resolution process (2.2.2 [b]). | Performance-based |
| The rate of timber harvest for the management unit is based on an analysis incorporating a prescriptive list of factors (5.6.1). | Management-based |
| The rate of timber harvest is determined in a manner that adequately reflects reliability and uncertainty associated with inventory data, management assumptions, growth and yield projections, and analysis methodologies (5.6.2). | Management-based |
| The manager ensures that the rate of harvest reflects the best available inventory and productivity data, provides for sustainable production, and is adjusted when monitoring indicates over-harvesting (5.6.3). | Performance-based |
| The manager demonstrates that the average of the present and projected timber harvests over the next decade, and the averages of projected timber harvests over all subsequent decades, do not exceed the projected long-term harvest rate, while meeting the FSC-BC standards over the long term (5.6.4). | Performance-based |

implement a plan for public participation that accommodates directly affected persons (Indicator 4.4.1). All remaining requirements set out in the
standard are performance-based, including obligations with respect to the
employment of local forest workers (Indicators 4.1.1 and 4.1.2), training
(4.1.5), assistance to displaced employees (Indicator 4.1.7), use of local goods

*Table 11.4*

## FSC-BC indicators addressing community and workers' rights: Principle 4

*Criterion 4.1*   Communities within or adjacent to the forest management area should be given the opportunity for employment, training, and other services.

*Criterion 4.4*   Management planning and operations shall incorporate the results of evaluations of social impact. Consultations shall be maintained with people and groups directly affected by management operations.

| FSC-BC indicators | Form of standard |
|---|---|
| Local forest workers are employed on the management unit and paid wages and benefits consistent with regional BC industry average (4.1.1; 4.1.2). | Performance-based |
| Managers provide training opportunities and/or collaborate with other local partners to ensure local people receive enhanced employment qualifications, forest workers receive training to comply with FSC-BC standards and applicable law, and employees receive skills upgrading to facilitate advancement within manager's operation (4.1.5). | Performance-based |
| Managers help displaced employees make the transition to new work (4.1.7). | Performance-based |
| Managers use local goods and services (4.1.8). | Performance-based |
| The manager develops and implements a plan for ongoing public participation that accommodates directly affected persons (4.4.1). | Management-based |
| Directly affected persons are provided with information used in making management decisions in a manner that allows them to understand potential impacts on their rights or interests (4.4.2). | Performance-based |
| Steps sufficient to protect the rights and interests of directly affected persons are developed and agreed to through the public participation process and implemented by the manager (4.4.3). | Performance-based |
| Where disputes arise, approved alternative dispute resolution process established under Criterion 4.5 is used (4.4.4). | Performance-based |

and services (Indicator 4.1.8), and managerial relations with directly affected persons (Indicators 4.4.2 to 4.4.4). As was discussed in Chapter 7, in these areas the final standard departs from its D3 predecessor, which adopted a more management-based approach.

*Indigenous Peoples' Rights*

The principal objectives of the standard in terms of indigenous peoples' rights are three-fold: to ensure that indigenous peoples control forest management of their lands and territories; to ensure their resources, tenure rights, and special sites are protected; and to ensure they receive appropriate compensation for use of their traditional knowledge (Criteria 3.1 to 3.4).

To achieve these objectives, the drafters of the FSC-BC final standard have adopted a strongly performance-based approach that requires managers to seek and secure the consent or satisfaction of relevant First Nations to proposed forest management activities (see Table 11.5). Adoption of this approach has yielded a standard that from a manager's perspective is clearly a highly demanding one, particularly insofar as the designated outcome (consent/satisfaction) is ultimately not one over which a manager can exercise control.

This performance-based approach is also evident in other aspects of the standard. For example, whether a manager has recognized and respected indigenous rights is measured by an affirmation to this effect by relevant

*Table 11.5*

**FSC-BC indicators addressing indigenous peoples' rights: Principle 3**

| | |
|---|---|
| *Criterion 3.1* | Indigenous peoples shall control forest management on their lands and territories unless they delegate control with free and informed consent. |
| *Criteria 3.2 and 3.3* | Protection of resources, tenure rights, and special sites. |
| *Criterion 3.4* | Compensation for traditional knowledge. |

| FSC-BC indicators | Form of standard |
|---|---|
| Manager recognizes and respects First Nations' rights over their traditional territories (3.1.1). | Performance-based |
| If First Nation elects, protocol agreement concluded (3.1.2). | Management/ performance hybrid |
| Manager obtains free and informed consent of all affected First Nations to management plan (3.1.4). | Management/ performance hybrid |
| Forest activities are carried out to maintain resource and tenure rights unless First Nations sign off (3.2.1). | Performance-based |
| Forest activities are carried out to protect special sites unless First Nations sign off (3.3.1). | Performance-based |
| Traditional knowledge is only used where mutually agreed, and IP receive fair compensation for such use (3.4.1). | Performance-based |

First Nations (Indicator 3.1.1). Similarly, the standard directs managers to ensure that their activities do not adversely affect indigenous resource and tenure rights or special sites unless relevant First Nations are satisfied with measures to offset this impact or otherwise provide compensation (Indicators 3.2.1 and 3.3.1).

Elsewhere in this section of the standard, the drafters have employed a hybrid approach that combines management- and performance-based elements. One example is the provision relating to protocol agreements. At the request of a First Nation, a manager is required to negotiate such an agreement, which must address a variety of prescribed issues (Indicator 3.1.2). This provision can be thought of as a hybrid management-/performance-based instrument, combining a strong planning component that is tested by a performance standard in the form of the First Nation's consent to the final form of the agreement. A similar approach is adopted for the requirement that a manager secure free and informed consent to the management plan (Indicator 3.1.4). Here the glossary definitions of "management plan," "joint management agreement," and "consulting with First Nations" spell out a variety of planning obligations for managers. Ultimately, however, the adequacy of a manager's efforts to implement these planning obligations is tested by a performance-based measure – namely, First Nations' consent to the resulting management plan.

*Environmental Values*

With respect to environmental values, FSC standards call on managers to embrace a variety of objectives, which include protecting old-growth areas and ecological diversity; promoting responsible logging and silviculture methods; maintaining and restoring riparian ecosystems and functions; and promoting the development and implementation of non-chemical methods of pest management.

As we discussed in Chapter 9, it was in this area that the process of crafting the FSC-BC standard was particularly contentious and difficult. This is, of course, hardly surprising. In this area, more than any other, there were diverse views regarding not only the form of the standard but also, more fundamentally, its content. The multifaceted and interrelated nature of the objectives at stake, the complex and dynamic nature of the forest environment within which these objectives were being pursued, and the heterogeneous nature of the regulated community are all factors that work in favour of a standard that combines performance- and management-based elements. In large measure, this is precisely the form of standard that ultimately emerged: one in which management- and performance-based requirements were closely linked.

Instances where the standard deploys a hybrid management-/performance-based approach abound (see Table 11.6, pp. 270-71). They include:

- completing an RONV (or equivalent) analysis (Indicator 6.1.7); management of forestry consistent with this analysis (various indicators under Criterion 6.3)
- mapping species habitat (Indicator 6.2.1); demonstrating measures to protect species (Indicator 6.2.2)
- developing a network of protected reserves (Indicators 6.4.1); managing the network to maintain/restore ecological integrity (Indicators 6.4.2 and 6.4.5)
- conducting an HCVF assessment and developing a program to monitor HCVFs (Indicators 9.1.1 to 9.1.3 and 9.4.1); adaptive management of HCVFs in accordance with precautionary principle (Indicators 9.3.2, 9.4.2, and 9.4.3)
- establishing a landscape connectivity strategy; implementing strategy (Indicator 6.3.11)
- establishing a strategy for managing aquatic ecosystems to maintain ecological integrity; implementing strategy (Indicator 6.5.5)
- developing an assessment regime for maintaining/restoring riparian ecosystems; implementing the regime in accordance with a retention budget (indicators under Criterion 6.5)
- developing plans to phase-out chemical pesticides within five years of certification (Indicator 6.6.1).

Where this hybrid approach is not employed, the standard is expressed in a more purely performance-based fashion. Qualitative performance-based provisions include requirements to carry out restoration forestry (Indicators 6.3.1 and 6.3.2), implement integrated pest management (Indicator 6.6.1), and avoid the use of specified chemicals and exotic biological control agents (Indicators 6.6.1 and 6.8.1). Quantitative performance requirements are employed in an even more limited fashion, appearing in only three places in the standard: namely, tree retention minimums (Indicator 6.3.9), machine-free set asides in aquatic areas (Indicator 6.5.6), and riparian reserves (Appendix B, 94).

## Conclusion

At the outset of this chapter, we argued that the policy debate over the future of environmental and natural resource regulation has increasingly focused on the relative merits of competing forms of standard. This emerging appreciation of the salience of standard form should be welcomed insofar as it reflects a recognition that it is misleading and unhelpful to frame the choice confronted by policy makers (and, more broadly, society at large) as being between market-based instruments and command-and-control regulation. Yet because the analysis of the form of standards remains very much in its infancy, conceptual confusion and obfuscation present serious challenges.

*Table 11.6*

## FSC-BC indicators addressing environmental protection: Principles 6 and 9

| *Criteria regarding* | Old-growth and biodiversity protection. |
| *Criteria regarding* | Logging and silvicultural methods. |
| *Criteria regarding* | Riparian management. |
| *Criteria regarding* | Chemical and biological agent use. |

| FSC-BC indicators | Form of standard |
| --- | --- |
| Range of natural variation-based (or RONV-compatible) report completed; environmental risk assessment completed (6.1.7). Forests are managed within RONV (or RONV compatible) (various indicators under 6.3). | Management/ performance hybrid |
| Habitat of red, blue, threatened, and endangered species mapped (6.2.1). Manager demonstrates measures to protect species (6.2.2). Appropriate training programs established (6.2.4). | Management/ performance hybrid |
| Network of protected reserves is established based on external disturbance (6.4.1). Management of reserve network maintains/restores ecological integrity (6.4.2) and generally limited to low-impact activities (6.4.5). | Management/ performance hybrid |
| Qualified specialists conduct HCVF assessment (9.1.2; 9.1.3) in consultation with relevant interests (9.1.1). Manager establishes program to monitor HCVFs (9.4.1). HCVFs are managed adaptively (9.4.2; 9.4.3), consistent with "precautionary approach" (9.3.2). | Management/ performance hybrid |
| Restoration forestry is carried out in previously poorly managed natural forests and former plantations (6.3.1); regeneration surveys confirm restoration progress (6.3.2). | Performance-based |
| Minimum requirements specified for retention of dominant and co-dominant green trees and snags and for average stand-level retention based on natural disturbance type (6.3.9). | Performance-based |
| Landscape connectivity goals established, and connectivity management strategies implemented (6.3.11). | Management/ performance hybrid |
| Implementation of identified measures to maintain ecological integrity of aquatic ecosystems (6.5.5). | Management/ performance hybrid |
| Machine-free zones established on all streams, lakes, wetlands, and marine shorelines (7-metre minimum and other requirements) (6.5.6). | Performance-based |

▶

◀  *Table 11.6*

| | |
|---|---|
| Maintain and/or restore riparian functions by completing assessment and implementing maintenance regime that meets/exceeds retention budget as specified in Appendix B to FSC-BC fiscal standard (6.5). | Management/ performance hybrid |
| Plans to be developed to phase out chemical pesticide use within 5 years of certification (6.6.1). | Management/ performance hybrid |
| Evidence of use of integrated pest management during phase-out (6.6.1). | Performance- based |
| Chemicals prohibited by FSC not to be used (6.6.4). | Performance- based |
| Limits on use of exotic biological control agents (6.8.1). | Performance- based |

This is as true for government decision makers as it is for non-governmental regulators charged with developing voluntary codes akin to the FSC's national and regional standards.

In this chapter we grappled with some of the reasons why, in this emerging academic and policy debate, performance-based approaches have played such a dominant role to date. We have also tried to offer some conceptual clarity to the debate, drawing on illustrations from the forest-certification context. In this conclusion we will offer two general observations on instrument form before making some specific comments on the FSC-BC experience.

Our first general observation concerns the need to work toward a common understanding and language to describe the endeavour of standard setting. In this regard, we would argue that employing a typology that distinguishes between management-, technology-, and performance-based standards has strong analytic utility. We would also argue that conceiving of these forms of standard as representing interventions at relatively discrete stages of the production process helps to make clear that all forms of standard have elements of prescription, putting to rest the popular notion that there is a dichotomy between prescriptive standards and performance-based standards. Similarly, we would argue that it is important to appreciate that there is no necessary or obvious correlation between the prescriptiveness of a standard and the business costs associated with complying with that standard.

Our second general observation is that standard setters should strive to be agnostic as to the relative merits of varying forms of standard, mindful that determining the optimal form of standard will always be a context-specific endeavour. This means keeping an open mind about what form of standard is best suited to a particular application. The overarching concern at all times should be to craft a standard that will optimize fulfillment of its desired objective(s). In tackling this challenge, as we discussed in "Crafting standards," standard setters need to be alive to a diverse range of considerations. These include deciding what form of standard will ensure that managers have an incentive to gather and deploy reliable data relevant to optimizing realization of the objective the standard aspires to promote. The form of standard adopted should also, as far as is practicable, be conducive to independent measurement, evaluation, and verification. And the standard should be in a form that is well suited to the nature of the external environment in which it will operate and to the nature of the parties that will be subject to its requirements. As experience accumulates with differing forms of standards (and their hybrid cousins), we would expect that the merits of approaching the standard-setting task in this flexible, context-driven fashion will become increasingly apparent.

The publication of its 2004 "Policy on the structure and content of standards" indicates that the FSC-AC is developing a more coherent approach to the choices that surround the form of regional and national standards. Prior to this policy, as our critique of the initial ABU report on the FSC-BC standard demonstrates, there was considerable confusion and a lack of clear direction within the FSC in this critical area. This said, we would argue that the policy requires considerable further development and elaboration. To this end, we would argue that the policy should provide clear working definitions of the various forms of FSC-compliant standards, drawing on the emerging standard-setting literature and the standard-setting experience in both the regulatory context and the voluntary sector. It should also clarify that both management- and performance-based approaches are indispensable, and equally valid, tools to promote responsible forest management. Greater clarity is required in terms of what the FSC considers to be an impermissible design- or technology-based standard. We agree that technology-based standards of the type employed in the air- and water-pollution context are not likely to have significant application in the forestry setting; nonetheless, it could be argued that a requirement to manage forests in accordance with RONV or to implement an integrated pest management system represent design-based standards precluded by current FSC policy. Given this uncertainty, FSC-AC policy should be clarified to ensure that systems or models such as these can be deployed in appropriate situations.

As it undertakes the task of reforming its policy on standard structure and content, we believe the FSC can learn from the FSC-BC final standard. Our review of this standard suggests that the drafters were keenly aware of the need to approach their task in a context-sensitive fashion, mindful of the strengths and weaknesses of both management- and performance-based approaches. In many areas, albeit through an arduous iterative process, the standard that emerged is highly performance-based. In other areas, particularly in relation to environmental values, the emerging standard integrates both approaches. As well, by developing a stand-alone standard for small operators, FSC-BC addressed concerns raised earlier in the standard development process about the accessibility and usability of its standard. Although the ultimate success of the FSC-BC final standard must be gauged – at least to some extent – by uptake in the field, it is fair to say that it represents a highly instructive and not insignificant achievement in the emerging art of crafting standards.

# 12
# An Institutional Analysis of FSC Governance

On paper and in practice, no other forest-certification scheme rivals the FSC in terms of the sophistication and complexity of institutional governing arrangements. To achieve the ambitious aspirations articulated in its principles and criteria, it has combined a complex global democratic architecture with a deep deliberative process to promote dialogue, equality, and transparency. By deploying an array of constitutional checks and balances, the FSC responds to the identified needs and priorities of its constituents. Given the diverse interests housed within this organizational structure, including the ever-present North-South tension, the magnitude of the challenge that this entails – to mediate disputes, oversee standards harmonization, and continually update its regulations and procedures to take account of emerging knowledge and stakeholder concerns – can scarcely be overstated.[1]

In this chapter, we undertake the task of understanding and characterizing the institutional dimension of the FSC by examining its formal architecture and deliberative processes. We do this by considering the extent to which the FSC model elaborates established liberal democratic constitutional norms and mechanisms and noting some striking similarities. Viewed through the lens of comparative constitutionalism, we contend that the FSC emulates much of the architecture and many of the processes of modern democracies across a range of criteria – from the duties it imposes on members (e.g., to affirm and uphold its principles and criteria and by-laws) and the various roles and responsibilities assigned to its "executive," "legislative," and "judicial" branches, to its internal regulatory, administrative, and dispute resolution mechanisms. But FSC does not just aim to emulate the institutional form of liberal democracy within a transnational context. Its broader objective is to transcend several of democracy's well-known deficiencies, which derive from the limited regulatory scope of the nation-state, the privileged position of business in policy making, and the majoritarian approach to deliberation and decision making.

To demonstrate how FSC emulates some of the key features of modern liberal democracies, the first part of this chapter is structured around the four basic elements found in most constitutions: a preamble; a charting of organizational competencies; an amending clause; and a bill of rights. This organizing framework enables us to compare and contrast the institutional features of modern democracies with those of the FSC, highlighting similarities and differences. However, we are conscious that the FSC remains a work in progress and that its institutional forms continue to evolve. The danger in our approach is that it provides a too-static analysis of the FSC, reifying elements of its institutional arrangements that are in fact in the process of further development. We return to this theme of static versus dynamic analysis in the chapter's conclusion.

According to Landes (1987, 60-61), a preamble sets out "the goals and principles of the polity" in motivational and rhetorical language. The organizational chart, in contrast, provides a "power map of the polity," delineating "whether the political institutions are to be federal or unitary, presidential or parliamentary in nature," and which tasks are to be performed by which level of government (60-61). The purpose of the amending formula is to allow for constitutional adjustment to changing circumstances over time, recognizing that such provisions are "often made difficult to use, on the assumption that the 'supreme law of the land' should not be routinely changed or manipulated" (60-61). Finally, constitutions often include a bill of rights that legally enshrines the fundamental rights and freedoms guaranteed to citizens. Using this framework, it becomes abundantly clear just how "state-like" the FSC's structure is, especially in its efforts to establish a set of governance checks and balances within its constitutional provisions to prevent "tyranny" in the form of single-interest, single-state, or Northern domination.

The FSC's "constitution" is inscribed in two foundational documents, its statutes and by-laws, which collectively establish its vision, structure, and process for amendment.[2] Although it currently lacks a constitutional bill of rights, in this respect the FSC is no different from many other states. Canada did not have a constitutionally enshrined bill of rights until enactment of the *Charter of Rights and Freedoms* in 1982. The Australian Constitution still lacks a bill of rights, a source of concern to many there in the increasingly illiberal era dominated by the "war on terror." And even the original American Constitution lacked a bill of rights, although this was soon changed in 1791 with the adoption of ten amendments to guarantee basic civil and political liberties such as free speech, freedom of assembly, and freedom from arbitrary arrest and detention.

## Preamble

The purpose of a constitution is to found a state consisting of a "people" occupying a given territory over which sovereignty is exercised. Perhaps the best-known, if rhetorical, example of a preamble is that from the US Constitution, which states, "We the People of the United States, in Order to form a more perfect Union, establish Justice, insure domestic Tranquility, provide for the common defence, promote the general Welfare, and secure the Blessings of Liberty to ourselves and our Posterity, do ordain and establish this Constitution for the United States of America." In contrast, the original Canadian Constitution is more prosaic, observing simply that several provinces have "expressed their desire to be federally united into one dominion."[3] Notably, however, neither preamble indicates how "the people" came to be formed; the drafters instead presumed the prior existence of Americans on the one hand and Canadians on the other.

In an analogous manner, FSC-IC presumes the existence of a global forestry polity in its by-laws and statutes. Noting that the purpose of the association is to "promote an adequate management of forests," Statute 4.4 specifies the contours of this polity as consisting of "developers of forest management policies, forest managers, legislators, and ... any other person interested in forest management." Unlike nation-states, however, the FSC is a voluntary rather than a compulsory polity: individuals, associations, and companies apply to join. Those who are accepted as members have voting rights and may stand for election to the FSC-AC board of directors. As discussed in Chapters 2 and 10, members of FSC-AC's global forestry polity are grouped into three constituencies or chambers, which are, in turn, split between North and South, constituting a unique system for aggregating and representing sectoral and global interests.

Those who join the FSC "polity" are obliged to endorse a vision of global forestry as "environmentally appropriate, socially beneficial, and economically viable management of the world's forests" (Statute 4.1). Notably, this vision does not specify forest type (boreal, temperate, tropical), scale of operation (small, medium, large), kind of enterprise (family farm, woodlot, cooperative, private firm, public corporation), or location (North or South). The absence of any qualifiers implies that no forest management operation is excluded, a priori, from achieving FSC-AC's vision and that FSC certification should be available to all regardless of size, kind, location, or forest type.

### Organizational Chart

In addition to a preamble that references a polity and sets out a statement of its purpose and values, an important element of a constitution is a charting of the basic structure of government in terms of institutional competencies. Democratic constitutions prescribe whether the polity is to be governed by a unitary state or federally, by a single legislative chamber (unicameral) or two (bicameral), by a prime minister or a president, and by an appointed or elected judiciary. The constitution may also imply or specify the process required for legitimate policy making. The general purpose of democratic constitutional provisions is to prevent tyranny and ensure that those in power are formally accountable to the people. Mechanisms to achieve this goal include provisions relating to elections, the division of powers between different state institutions, and the establishment of a system of checks and balances to prevent any one arm of government from unduly dominating the others.

In both parliamentary and presidential models, constitutions vest considerable power in the executive branch of government – the prime minister and cabinet in Canada; the president and White House staff in the United States. Executive power is deliberately offset, however, by countervailing forces in other parts of the system. In federations, these reside in sub-national levels of government that have exclusive jurisdiction in some legislative areas and co-responsibility in others. Neither level of government can abolish the other, so disputes must be negotiated or adjudicated through a high or supreme court. Unitary systems, of necessity, lack this federal division of power, but they, along with federations, make use of another mechanism – the separation of powers and the establishment of systems of checks and balances – most notably by distributing power to the legislative, executive, and judicial branches of government. In formal constitutional terms, the role of the executive is to propose policies and implement legislation; the role of the legislature is to consider legislative proposals and pass, amend, or block them; and the role of the judiciary is to interpret legislation and rule on its legality.

The above sketch will be familiar to many readers. In order to examine FSC-AC's arrangements, we need to focus likewise on the division and separation of powers set out in its constitutional documents. In the following discussion, we first consider the division of powers in the FSC, observing that although it was established as a "unitary" system, it does have several "federal" features – notably a system for devolving responsibility to the regional and national levels. Then we consider the separation of powers within the FSC between its legislative, executive, and judicial arms, examining how these interact in the making of FSC policy.

## Division of Powers

Governments are said to be "unitary" to the extent that power resides centrally and "federal" to the extent that power is shared geographically (i.e., to states, provinces, or territories). The FSC-AC's founding documents envisage a "unitary" governance structure, but it is nonetheless one that is under considerable systemic decentralizing or "federalizing" pressures (Meidinger 2006a). Two key issues in this regard are the extent to which FSC-AC ought to defer to national or regional standards recommendations and how the FSC-AC should deploy its powers to supervise the harmonization of neighbouring initiatives.

FSC by-laws state that the FSC-AC "shall encourage and support national and regional initiatives which are in line with the FSC mission," a goal it has actively pursued over the years (para. 71). Constitutionally, national and regional initiatives can assume several different forms: an FSC-AC-appointed contact person, a national working group, a national advisory board, or a national or regional office.[4] The by-laws give only the broadest outline of the responsibilities contemplated under each of these respective arrange-ments.[5] A contact person is appointed by FSC-AC; national working groups are self-constituted, their primary purpose being to facilitate consultation. An advisory board, on the other hand, is elected by national initiatives, is similar in composition to FSC-AC, and has a more specific range of tasks. These include promoting the FSC and its mission; providing information, support, and services; maintaining ongoing consultation on certification; facilitating and overseeing the process of developing national (and sub-national) standards; and reviewing and making recommendations to FSC-AC on national certifying bodies (CBs) seeking FSC accreditation (By-law 71c). A national office may be established wherever an advisory board exists and CBs are active. A national office is the bureaucratic arm of the national board and relates to it in the same way as the FSC-AC secretariat (staff) relates to the FSC-AC board (By-law 71d).

The basic division of powers between FSC-AC and initiatives, whether regional or national, is set out as a positive list of rights in By-law 71d that gives FSC-AC final approval of standards and accreditation of CBs, control over the use of the FSC's name and logo worldwide, power to determine the category under which an initiative is established, and, crucially, the right to withdraw recognition. The only obligation placed on FSC-AC is to consult with national and regional bodies before undertaking activities in a coun-try or region (By-law 71d). These rights combined with a single obligation clearly grant FSC-AC authority over initiatives at lower levels, including the potentially disciplinary step of withdrawing recognition from any initiative that does not comply with FSC-AC imposed requirements.

Despite this, there are numerous rights and obligations that are not clearly set out in FSC-AC's constitution, creating unresolved tensions between dif-ferent organizational levels. One of the most important of these is the power over "taxation" – in this case, the right to raise and distribute funds to run the FSC's operations. Because the by-laws are silent on the issue, the presump-tion is that national initiatives have the power to raise and spend their own money, and this they have certainly done. However, international and na-tional bodies have competed for donor funding, especially in North America, and greater clarity is required to spell out how regional and national initiatives are to be funded and what FSC-AC's responsibilities are in this regard.[6]

In the early days of FSC-BC, as we noted in Chapter 4, there was consider-able confusion about the status of the FSC-BC regional initiative in relation to the FSC-Canada working group and FSC-AC. Many provincial members viewed FSC-BC as quasi-autonomous with a de facto right to report directly to FSC-AC rather than FSC-Canada. This separatist ethos stemmed from the almost simultaneous emergence of both organizations after 1996, and it was greatly strengthened by FSC-BC's fundraising capacity, which outpaced FSC-Canada's in the early stages. FSC-Canada's lack of funds, small staff, and remote, infrequent board meetings in Ontario all contributed to a perception within FSC-BC that it should make its own decisions and report directly to FSC-AC.

From 2001 onwards, however, FSC-Canada – chastened by the experience of the FSC-Maritimes dispute and under pressure from FSC-AC to exercise a Canada-wide mandate – worked diligently to integrate regional initiatives into a more explicit national organizational hierarchy. By the summer of 2002, when approval of Draft 3 of the BC standard was pending at FSC-AC, BC's separatist ethos was waning and regional members were accepting, if reluctantly, the respective roles and legitimacy of FSC-Canada and FSC-AC. Indeed, there is clear evidence that the FSC-BC steering committee began winding down from at least the middle of 2003, although it remained formally constituted until at least March 2006 and, arguably, still exists today.[7] However, the separate FSC-BC website has been replaced by the FSC-Canada site, and the regular meeting minutes and announcements are no longer available.

**Separation of Powers: Executive, Legislature, Judiciary**
In Western liberal democracies, power is typically allocated among three branches of government: legislature, executive, and judiciary. A pure separation of powers would see citizens elect each body, but in practice this rarely occurs.[8] Power is similarly fused in the FSC system, which lacks the equivalent of a legislature to debate policy. In the absence of a legislative assembly, the FSC's executive and legislative powers are largely merged in the elected, nine-member FSC-AC board, which develops, debates, and interprets policy and monitors implementation in consultation with the FSC-AC secretariat. Nor is the FSC board entirely functionally separate from FSC's "judiciary": the six-member Dispute Resolution and Accreditation Appeals Committee (DRAAC). The board appoints the DRAAC, which is constituted according to specific membership criteria for ensuring chamber, North/South, and regional balance.[9] In the following sections, we will review this relative fusion of powers in the FSC, highlighting areas where a greater separation of authority may be desirable.

*Executive Government*
In parliamentary government, the executive consists of the prime minister, cabinet, and an extended bureaucracy organized into ministerial portfolios.

Bureaucrats, ideally, provide "frank and fearless" advice to ministers about the consequences of different policy options, leaving it to the prime minister and cabinet to make policy choices, which the bureaucrats then assume responsibility for implementing. In practice, of course, such a division of labour between policy formulation and implementation is difficult to maintain. A similar split between "political" and "administrative" branches is attempted at a formal level in FSC-AC's executive. Policy making is officially the responsibility of FSC-AC's elected, nine-member board, representing diverse constituencies and regions; implementation is the responsibility of the FSC-AC secretariat, consisting of the executive director and staff, which may also propose policy initiatives to the board for discussion and approval. As in the governmental context, however, this formal distinction between political and administrative rule frequently breaks down in practice, with bureaucrats taking on a role in policy making and board members intruding into the realm of implementation.

Within the executive, the dual role of parliamentary members – who represent both a specific constituency and, as ministers in cabinet, the government as a whole – is mirrored by the role of members of FSC-AC's board, who are not only tasked with representing the FSC's broader interests as an evolving international institution, but also aim to secure their own network's interests. Although the board's consensus-based decision-making process prevents one set of interests from trumping others, it can also lead to deadlock, and there is no deadlock-breaking power akin to that of prime ministerial authority under the Westminster model.

The analogy to parliamentary government does break down across other dimensions. Unlike members of governments, for example, FSC-AC board members are not full-time policy makers paid to consider the peoples' business. Most, in fact, conduct their FSC work on a part-time and voluntary basis, splitting themselves between service to the FSC and regular paid employment. The voluntary nature of the board makes it difficult for the executive to meet on a timely basis and cope with the heavy workload. Because the FSC is a global organization, some board members find themselves travelling considerable distances to attend international meetings, which entails even more work time. In addition, and unlike many government executives, the FSC-AC board conducts its business across different interests, cultures, languages, and levels of development. These differences make it difficult for the board to cohere as a single unit, in contrast to a cabinet, where prime ministerial authority, party affiliation, frequent meetings, cabinet secrecy, and fear of the opposition create a strong sense of solidarity.

Many of FSC-AC's management difficulties were aired in a 2001 "Change Management Team" report, which identified inexperience and lack of training as impediments facing new board members and also noted a general tendency for board members to over-involve themselves in technical matters – such as reviewing individual national standards – at the expense of strategic considerations. The Change Management Team recommended that roles and responsibilities of the FSC-AC's board be redefined with the aid of NGO management experts (FSC-AC 2001b).[10]

In addition to these managerial challenges, there is also a tension within FSC-AC between those responsible for policy making and those charged with policy execution.[11] As is detailed in Chapter 4, such friction was evident between the FSC secretariat and the FSC-AC board following FSC-Canada's approval and forwarding of the BC D3 standard to FSC-AC. The bureaucratic arm of FSC-AC, its Accreditation Business Unit (ABU), reviewed all draft standards to ensure they conformed to a set of pre-established technical requirements. The ABU concluded, on the basis of an internal standard checklist, that D3 was too long, too detailed, unclear, and prescriptive (see Chapter 11). Draft 3 and the draft ABU report were circulated to FSC-AC's standards committee, which had been established only six months earlier to reduce board members' workload and involvement in the technical issues of standards review and approval. Some members of the standards committee were unhappy with the ABU report, viewing it as unfair and unduly influenced by industry lobbying. Instead of accepting the ABU's view, and recognizing the controversial nature of the issues presented, the Standards Committee chose to refer the report to the full FSC-AC board for further consideration.[12]

Tensions between political and bureaucratic arms of an executive can also arise in connection with the power to hire and fire. Traditional public administration was based on permanent, life-long civil servant contracts, which muted political power over policy implementation since there was no ultimate power to dismiss bureaucrats that underperformed. Over the past two decades, governments have sought to secure greater bureaucratic responsiveness to political decision makers by placing senior civil servants on contracts and specifying more precisely the level of expected performance. The board has the power to hire and fire FSC-AC's senior bureaucrat (its executive director or ED), a power it has exercised from time to time. To date, three individuals have held this office: Tim Synnott (1994-2000), Maharaj Muthoo (2000), and Heiko Liedeker (2000 to 2007). Of the three, Muthoo's term was by far the shortest, and he lost the confidence of the board after only six months in office due to his non-consultative style. This was illustrated by the secretariat's sudden announcement of an in-principle agreement with the Ontario Ministry of Natural Resources (OMNR) to certify all the province's Crown forestlands with little if any consultation with FSC-Canada members.[13] As Muthoo's deputy at the time of the OMNR debacle, Liedeker became ED on an interim basis and was later appointed to the position following a search and interview process. Soon after, Liedeker presided over the organization's relocation to Bonn, Germany,[14] and he is generally regarded to have performed well, aided by the development of an enhanced standardization policy and standards approval protocols and by the board's ongoing efforts to clarify the nature of its role and maintain the line between policy development and administration. However, as this book was going to press, a short press release from FSC-AC announced Liedeker's "unexpected" resignation, noting that, at the request of the board, he had agreed to "continue to manage the FSC until a successor is successfully recruited." As yet, there is no commentary on his resignation beyond a note that the FSC is much better positioned today than it was in 2001 when Leideker took over (FSC-AC 2007b).

*Legislative Arrangements*
A key role of any political system is to pass legislation through parliament that establishes the law of the land in a specific issue area.[15] In parliamentary systems, legislation is introduced and subjected to three readings, although the first and last readings tend to be formalities. Under the practice of responsible government, derived from British historical experience, the executive proposes legislation that is then, in theory, extensively scrutinized and debated by the entire parliament, where necessary amendments are passed prior to approval. Notions of responsible government are seriously compromised by the rise of the party system, and today the prime minister and cabinet can almost always secure the passage of legislation with as many or few amendments as they decide to allow. The situation is different when there is an active upper house of parliament, as in the Australian system, where senators perform an effective scrutinizing role. Legislation that easily passes through the lower house because the government "has the numbers" can encounter serious criticism and undergo significant amendment if the upper house is more independent and represents a wider diversity of interests and values.

Law making assumes two guises in the FSC system. One is the development of soft-law standards at the regional or national levels, which is discussed in detail in Chapter 11. The other, of more present interest, is the broader legislative function assumed by the organization to regulate its internal constitutional affairs, where prevailing arrangements are relatively amorphous and unsophisticated. Functionally, this role is split between the General Assembly and the board. It is the responsibility of the board to consider resolutions passed at the General Assembly, and these are generally given high priority.[16] Between General Assembly meetings, the board makes policy decisions on its own or with the assistance of specially appointed committees, such as the recently formed Plantation Policy Working Group (PPWG), that consult FSC members and the broader global forestry polity on options.

Why did the FSC not adopt an arrangement equivalent to a legislature? There is no easy answer to this, but the decision likely reflects a desire to establish an organization that would not require large-scale operational costs. Operational cost was certainly a concern of the interim board in the early 1990s when, in response to advice from the board's consultant (James Cameron) concerning the costs and dangers of potential openness, it revised the draft charter to alter the structure of the FSC from an association to a foundation.[17] A foundation was deemed to be a less-costly institutional arrangement than a membership body requiring General Assembly meetings every three years. There is no doubt that a legislative body mediating between the General Assembly and the board would be costly, as it would be required to meet much more frequently than a General Assembly and would be much larger than the current board. Whether these pragmatic considerations should outweigh those linked to concerns over democratic accountability is a matter to which we return in the conclusion.

Legislation sets out the broad parameters of how a policy is to be interpreted and administered, but it cannot anticipate all eventualities. Consequently, the bureaucracy is normally given the task of developing the necessary associated regulations and policy guidance documents on an as-needed basis. In the FSC system, the elected FSC-AC board is responsible for determining a course of action, and the secretariat is charged with interpretation and implementation. The adoption of the "Policy for preliminary accreditation of national/regional forest stewardship standards" (FSC-AC 2003c) is a good example of how this system operates in practice (see Chapter 4). The impasse that developed over D3 of the BC standard in 2002 led actors at all levels of the FSC to search for an "extraordinary solution," a process that culminated in two meetings at the 2002 General Assembly that involved members of FSC-AC, FSC-Canada, and FSC-BC. The notion of preliminary accreditation was broadly accepted at the second meeting, with the FSC-AC board formally approving the principle at its subsequent meeting. At this point, responsibility for developing the policy was turned over to the head of the Policy and Standards Unit, Matthew Wenban-Smith, who sought advice from a consultant, James Sullivan, on its utility and feasibility. In early 2003, Wenban-Smith submitted a detailed proposal on the policy and how it would work to the FSC-AC board for approval and implementation.[18]

Should the FSC General Assembly have been consulted on the new policy for preliminary accreditation? The question is interesting because it helps clarify the roles of the FSC-AC secretariat, board, and the General Assembly in regard to constitutional change, policy making, regulation, and implementation. Preliminary accreditation was not a constitutional issue, because it did not alter FSC-AC's fundamental documents, its statutes or by-laws. Hence, there was no formal role for the General Assembly. Preliminary accreditation did, however, necessitate a change in the FSC's arrangements for implementing certification, so board authority was needed to formulate the new policy.[19] Once that authority was obtained, it was up to the secretariat to implement the new arrangements, adjusting existing regulations and generating new ones to achieve the policy objectives.

This analysis captures core elements of the different roles performed by FSC-AC's General Assembly, board, and secretariat, but it is evident these roles are not completely separated. In this, the FSC mirrors parliamentary practice insofar as the bureaucracy is often heavily involved in policy making, and, likewise, the Cabinet is often engaged in questions of implementation. In the FSC-BC case, members of the bureaucracy were involved in the conception and championing of the idea of preliminary accreditation. Jim McCarthy, FSC-Canada's ED, building on extensive experience with the Canadian Standards Association (CSA), was especially influential, proposing the idea at the FSC-BC level, explaining it to officials at FSC-AC, defending it at two General Assembly meetings in 2002, and generally working to have the concept accepted as a way out of the impasse.[20] There were conversations between McCarthy and Liedeker to discuss ways to overcome anticipated opposition from the South to exceptional arrangements for the North, and they agreed to present preliminary certification as a potential solution to standard-development impasses elsewhere.

### Judicial Arrangements
In a national political context, disputes inevitably arise that require adjudication. Democracies manage such conflicts via an independent judicial system that is designed to mediate, arbitrate, and pass judgment on the cases brought before it. In the liberal democratic model, the judiciary is expected to operate at arm's length from the political apparatus, be free from conflicts of interest, and be impartial in the administration of justice. Although it is not entirely free from corrupting influences – notably the capacity of the rich to "buy" justice by hiring the best lawyers and by continuing with litigation until others with fewer resources are bankrupt – the system works to deliver an element of justice at least some of the time. The FSC adopted its own dispute resolution system to handle conflict within the FSC system, but there is a general consensus that it is not working well and needs substantial reform. In this section, we review the structure and operation of FSC's "judicial" branch of governance – its Interim Dispute Resolution Protocol (IDRP) and Dispute Resolution Committee (DRC).

The FSC-AC has grappled with several high-profile conflicts over the years, including disputes over the certification of Leroy Gabon (1997), the endorsement of the Maritimes standard (2000), the Allagash certification (2001), the denial of accreditation to KPMG (2004) (Weber 2005b), and, of course, the challenges we have chronicled associated with accreditation of the FSC-BC D3 standard. Law professor Gregory Weber has taken a particular interest in FSC-AC's dispute resolution arrangements; he drafted the 1998 IDRP and conducted an assessment of the operation of its dispute resolution arrangements over the past decade (Weber 2005b). This assessment highlighted several deficiencies in current arrangements, including delays, procedural complexity, and the lack of safeguards against conflicts of interest. In addition, it is broadly acknowledged that FSC-AC is hobbled, in this and other areas, by a lack of capacity in terms of staff, resources, and training.

An overarching problem, however, is an insufficient separation of powers between the FSC's judicial and executive branches. Under the FSC-IC constitution and the IDRP, in the event of a dispute (for example, over a certification decision made by a certifier), the complainant is required to file an initial appeal with the certifier. If the dispute cannot be resolved at this stage, the next level of appeal is to an "independent committee" composed of two or three independent certification professionals not involved in the original assessment (FSC-AC 1998b, para. 4.1). Should these "informal" arrangements fail to resolve the dispute, the complainant may then appeal directly to the FSC-AC, which, in the normal course of events, forwards it to the FSC chair to mediate. The final level of appeal is to the FSC secretariat for screening and submission to the FSC-AC board for adjudication. The board may refer the matter to the DRC and, if that fails to resolve the matter, to the General Assembly, which has final authority. People are starting to recognize that this process relies far too heavily on those who are already burdened by FSC work, some of whom may have a vested interest in the outcome. Not only is the system open to charges of conflict of interest, its complexity creates significant delays and expense.[21]

There is general consensus that the FSC needs to improve its dispute resolution system. At a 2005 workshop convened in Manaus to consider this issue, several alternatives were canvassed, ranging from minor to far-reaching alterations to existing arrangements. Weber outlined four possible models at the meeting, all of which appear to fuse judicial and executive power to some degree, potentially jeopardizing the independence (real and perceived) of the process. For example, all of the models described by Weber contemplated a substantial role for organs of the FSC: the ABU, the board, and/or a board-appointed DRC. Of the several alternative models considered at the workshop, only the one proposed by Errol Meidinger envisaged the establishment of a new institution – an FSC disputes office.[22] Although this is not the place to propose a detailed alternative to FSC's dispute resolution arrangements, it is important to note that the general guiding principle governing such arrangements should be to establish a system that is as independent of the executive – the FSC secretariat and board – as practicable.

### Amending Formula

States adopt a variety of arrangements to amend constitutions. In some cases, constitutions can only be amended via citizen referendum; in others, national and sub-national legislatures make the decision according to various rules. In Ireland, for example, constitutional change must first be approved by both houses of Parliament and then be approved by a simple majority of the electorate in a referendum. Constitutional change in Australia can only be initiated with the support of both houses of Parliament, and the proposed alteration must be put to a referendum and secure a "double majority" before it is approved. Not only must a majority of the entire electorate vote in favour of the change, but electors must also be spatially distributed so that there is a majority of the electors in a majority of states in favour. Unsurprisingly, it has proved very difficult to change the state's constitution. Constitutional change is somewhat easier in Canada because referenda are not required. Instead, both houses of Parliament must pass the proposed amendment, which according to the "7/50" rule must then be approved by the legislatures of at least two-thirds of the provinces and territories, representing at least 50 percent of the population.

Within FSC-AC, the General Assembly has authority to amend FSC-AC's by-laws and statutes and make other major decisions. In this regard, FSC By-law 18 stipulates that the General Assembly "will normally restrict its decisions to revising the Statutes and Principles and Criteria, admitting and destituting members, electing the Board and being the final authority in dispute resolutions." Although the General Assembly has ultimate authority, the FSC-AC board plays a gate-keeping role and may block a resolution from proceeding to the General Assembly for a vote, although it does not appear it has ever done so.[23] Similar to other General Assembly resolutions, amendments to FSC-AC's constitution require the normal super-majority vote of 66.6 percent in favour to pass, setting a suitably high bar for constitutional change and effectively requiring endorsement by all three chambers.

There is clear evidence that the FSC-BC case prompted parties to seek to alter FSC-AC's constitution via resolutions to the 2002 FSC General Assembly.[24] One motion, for example, sought to affirm that consensus decision making should apply at both the international and national levels in order to ensure that no significant minority could be overruled (FSC-AC 2002b).[25] Had this principle been in effect at the time the FSC-BC standard was submitted for accreditation, FSC-Canada could not have approved D3 and forwarded it to FSC-AC. Instead, the matter would have had to be resolved differently – either by referring D3 back to FSC-BC with a request that more negotiations be entered into and consensus reached, or by establishing a process at the FSC-Canada level to achieve a similar result.

## Bill of Rights
As noted at the outset, the FSC's constitution does not contain a formal bill of rights, and in this respect it resembles many other national constitutions, past and present. Indeed, some might argue that such a feature is

unnecessary in a voluntary, membership-based, civil society organization. Nonetheless, in light of the preceding discussion, we would argue that, as part of a broader effort to clarify roles, rights, and powers within the FSC, promulgation of a bill of rights that defines members' rights in relation to other institutions and actors within the FSC system would have undeniable organizational benefits.

## Conclusion

Although it is still a young institution, the FSC has developed relatively sophisticated institutional governance arrangements that, as is revealed by our comparative constitutional analysis, often parallel those that have evolved at the state level in many liberal democracies. Flowing from a strong commitment to democratic participation, its institutional architecture contains a variety of important checks and balances. Diverse values and interests are grouped into a small number of formal constituencies and mediated by an elected board of directors that is accountable to its membership every three years at a General Assembly. These democratic structures are bolstered by an organization-wide commitment to consensus-based decision making. This serves to check the authoritarian and majoritarian tendencies of modern electoral politics, where political parties can secure power through the ballot box by appealing to and promoting the interests of specific sectors. In a globalizing world, moreover, the FSC has one major advantage over existing governmental arrangements: its inclusive approach to interest representation regardless of geography. This is in contrast to states, which by their very nature partition the world in two, giving voting rights only to those deemed citizens and excluding all others.

The FSC is formally "unitary" in design, but its global reach and the authority it confers on national and regional initiatives suggest that it may be subject to "federalizing" pressures over time. These may well be exacerbated by uncertainty about what might be termed the "division of powers" between the FSC-AC and its member initiatives, particularly, as we have noted, around fundraising and the thorny question of harmonization.

Our comparative constitutional analysis likewise suggests that challenges lie ahead for the organization in terms of "separation of powers." Currently, the FSC's legislative and executive powers are fused, to a considerable extent, within the board of directors. Although there is no doubt that this fusion is in large measure attributable to the daunting difficulties associated with constituting a full-time representative legislative branch, it also flows from deliberate constitutional restrictions imposed on the role of the General Assembly. Under the FSC model, the board is charged with determining policy while the secretariat is responsible for policy interpretation and implementation. Apart from electing the board, the jurisdiction of the General Assembly over the day-to-day affairs of the organization is limited to major decisions such as revising the statutes and the principles and criteria, dealing with key membership-related decisions, and overseeing appeals when significant disputes arise. Concerns have been expressed about the extent to which the board relies on the secretariat, and it has also been argued that there is an insufficient separation of powers between the FSC's executive and judicial branches. It appears that the FSC is taking steps to address this latter criticism. Finally, the FSC currently lacks a bill of rights, but we would argue that there are compelling organizational benefits that would accrue from amending the FSC's constitution to better define and enshrine members' rights.

In offering this overview and critique of the FSC's institutional arrangements, we are mindful of the dynamic nature of the institutional processes and relationships within the organization. Throughout its history, the FSC has demonstrated a capacity for institutional adaptation and innovation. When it was first established, it chose to defy the advice of consultants who argued that a charitable foundation was a preferable organizational vehicle to the vastly more ambitious organizational form it ultimately adopted. Its capacity for significant institutional changes is also underscored by constitutional reforms adopted in 1996 that split the environmental and social chamber in two, laying the foundation for its current tripartite chamber structure, and by FSC-Canada's establishment of a fourth indigenous chamber.

Flowing from our comparative constitutional analysis and a recognition of the FSC's institutional dynamism, we conclude this chapter by sketching out institutional reforms in three areas – legislative arrangements, judicial arrangements, and members rights – that would enhance accountability and due process throughout the organization.

Our first proposal addresses the uncertainty and potential conflict arising from the fusion of executive and legislative functions within the FSC. As we have chronicled, there is arguably a democratic deficit between its "polity" (which consists of individual and organizational members of the FSC) and its executive (which is composed of the nine-member board of directors and the secretariat). To address this deficit, we argue that the FSC should consider creating an independent legislature that would ensure wider debate of executive decisions and achieve a clearer definition and separation of institutional powers. One way to implement this proposal would be for members to elect a larger body, what might be termed an "FSC council," composed of thirty individuals, with five from each of the six sub-chambers to ensure balance. This council would then elect its own nine-member executive board to manage the FSC's daily affairs. The council could meet as often as necessary, but at least twice a year, and the board would manage the affairs of the FSC between council meetings. The board would bring important matters of policy before the council for deliberation and decision, and special (even virtual) meetings of council could be called to deal with emergency matters.

The board would be accountable to council, and the council would in turn be accountable to the larger membership at triennial General Assembly meetings.

Our second reform proposal seeks to address some serious problems in the FSC's dispute resolution arrangements. Current procedures, in our view, are too slow, overly complex, and lacking in independence. To address these problems, the FSC should consider adopting a more thoroughgoing separation of the judiciary from the executive and legislative arms by establishing an independent, standing, FSC disputes office. The disputes office would have a small budget and staff with a mandate to review appeal applications, place them on an appropriate track for resolution, oversee the resolution process to ensure timeliness, and schedule arbitral hearings when necessary for especially problematic cases. As is the case in modern democracies, the executive could nominate a panel of arbitrators for confirmation by the council. These arbitrators would be legally trained and could serve for a single, limited term. As with most modern judiciaries, the tenure of arbitrators would be protected during their term in office.

Finally, we believe that, as members of what we contend is an emerging "global forestry polity" (see Chapter 13), individuals and organizations that belong to the FSC deserve the protection of a bill of rights. Inspired by the challenge of adapting the bill of rights concept to the civil society context, a variety of leading international NGOs, including Amnesty International, Transparency International, Greenpeace International, and Oxfam International, recently came together to develop what they call the "International non governmental organisations accountability charter."[26] This charter is intended to promote transparency and accountability in the NGO sector by committing signatories to a broad range of actions, including accuracy in information, ethical fundraising, safeguards for whistleblowers, and production of detailed annual reports and audited accounts. The FSC is not currently a signatory to the charter, but we would argue that its endorsement and implementation would be a natural next step for an organization that prides itself on its commitment to democratic practice.

# Part 4
# Conclusions

In this final section of the book, we offer two concluding chapters. The first integrates and builds upon the theoretical analyses contained in Part 3; the second distills the lessons that emerge from our case-study and comparative research both in terms of standard setting and in terms of the future of the FSC as an experiment in global governance.

Chapter 13 offers some broader theoretical reflections on the implications and significance of our foregoing analysis for the study of governance. In this chapter, we elaborate the discussion of the FSC's political, regulatory, and institutional dimensions by means of a comparative analysis of the FSC and analogous certification systems in the forest context and beyond. We then advance and develop the argument that the FSC system merits recognition, both along these dimensions and in other key respects, as a sui generis form of governance, an experiment in what we term "global democratic corporatism."

In Chapter 14, we reprise our research into the FSC-BC case with a view to identifying the barriers that almost thwarted development of the final standard; the reasons why a consensus standard eventually emerged; the nature and significance of this final standard in comparative terms; and, finally, the lessons that can be drawn from FSC-BC for the FSC and analogous civil-society-led forays into the realm of global governance.

# 13
# Theorizing Regulation and Governance within and beyond the FSC Model

The topics of governance and regulation have, in recent years, generated a staggeringly broad and diverse literature from a variety of academic disciplines. The conceptual breadth and diversity of this literature is daunting, to the point that it has, arguably, hobbled constructive engagement and debate. Much of the scholarly output to date has been devoted to chronicling and cataloguing ostensibly "new" modes of governance and regulation, often at a relatively high level of theoretical abstraction. Detailed, case-based research aimed at illuminating, testing, and refining emerging theory has been relatively rare (but see Ansell 2000; Blatter 2001; von Bernstorff 2003; Cashore, Auld, and Newsom 2004). We have sought to remedy this by embarking on an in-depth, comparative study of one of the icons of this new governance literature.

Despite its iconic status, scholars have struggled to capture the essence and significance of the FSC phenomenon. It has been variously characterized as an exemplar of "non-state, market-driven" governance (Cashore, Auld, and Newsom 2004); a "transnational, rule-oriented, system" (Meidinger 2006a); and an archetype of "soft law" and "legalization" (Stokke 2004). To date, however, we would argue that these various efforts have failed to fully capture the uniqueness of what might be termed the "FSC experiment." A key reason why a more robust and fulsome understanding of this experiment has yet to emerge has been a tendency by many analysts to characterize the FSC and other competing forest-certification regimes collectively as instances of "private governance."

Conceiving of the FSC in this way, we would argue, tends to obscure the extent to which the FSC deserves recognition as a sui generis governance form. As such, we contend that, across a range of criteria, the FSC stands in sharp contrast not only to traditional public institutions (state forestry departments and intergovernmental agencies) but also to other "new" governance initiatives motivated by a similar regulatory vision (International Organization for Standardization or ISO, Sustainable Forestry Initiative or

SFI, Canadian Standards Association or CSA, and the Programme for the Endorsement of Forest Certification or PEFC) and even sister stewardship schemes, most notably the Maritime Stewardship Council (MSC).

This chapter seeks to synthesize and elaborate upon the theoretical insights generated by our study of the FSC-BC final standard. In the preceding three chapters we assessed the nature and significance of the FSC regime through what we argue are three distinct dimensions of new governance: political networks (Chapter 10), regulation (Chapter 11), and institutional design (Chapter 12). In this chapter we revisit these three dimensions of governance to reprise and elaborate the rationale for the approach we have adopted. We also intend to situate the FSC model along these various dimensions relative to both its forest-certification competitors and the MSC. In the balance of the chapter, we advance and defend the argument that the FSC embodies a new form of governance, which we would term "global democratic corporatism."

## Private versus Public Governance

At the outset of this book, one of our priorities was to establish some fixed conceptual definitions to lend a coherent frame to the analysis that followed. To this end, we deployed "governance" as an umbrella term to denote arrangements for "*steering and coordinating the affairs of interdependent social actors based on institutionalised rule systems*" (Benz, quoted in Treib, Bähr, and Falkner 2005, 5; italics added). This broad definition was counterposed to a more conventional understanding of "government," conceptualized in terms of an ideal-type strong, central state – reliant on command-and-control-style regulation – that was typically associated with the governing style prevalent in most liberal democracies in the post-Second World War era. In the context of this notion of "governance," we defined "regulation" as *an institutionalized rule system that sets and enforces standards aimed at influencing the behaviour of interdependent social actors in order to promote mutually agreed-upon values, principles, or objectives.*

The expansiveness of our conception of governance – while consistent with most definitions found in the broader governance literature – poses the very real theoretical risk of conceptual overbreadth. This risk flows from the potential that it will subsume without differentiation a range of governance phenomena that, on closer inspection, are much more diverse than a unitary conception would tend to suggest. For this reason, many scholars have posited that governance should itself be regarded as a continuum concept, framed in terms of the public-private distinction (Abbott and Snidal 2001; Jordan, Wurzel, and Zito 2005).

There are some compelling reasons to conceive of governance in this fashion. An obvious one is the prevalence of this nomenclature in popular discourse about governance. More importantly, at either pole of this public governance-private governance continuum there is a strong intuitive appeal to such designations. Thus, to use the popular analogy, there is a world of difference between arrangements where governments steer and industry rows and those in which industry is steering *and* rowing. Moreover, asserting the existence of a conceptual distinction between public governance initiatives and private ones can pre-empt the unhelpful rhetorical tendency to characterize all new governance initiatives as forms of government privatization.

To date, as we have noted, the dominant tendency has been to characterize the FSC as a privatized form of governance or regulation and, therefore, to portray it as analogous to competitor forest-certification regimes such as those of the Canadian Standards Association (CSA) and the SFI. Illustrations of this approach abound (see Lipschutz 2000; Pattberg 2005; Rhone, Clarke, and Webb 2005). The precise conceptual basis for this characterization is rarely explicit, even though it typically appears to turn on the fact that membership in such governance arrangements is voluntary and largely, if not exclusively, composed of "private" actors as opposed to "public" bodies or agencies. In fairness, we do not wish to suggest that previous commentators have been unaware of the problematic nature of the much-maligned private-public distinction. For instance, Meidinger, while formally classifying and analyzing the FSC alongside its competitor regimes as a form of private governance, emphasizes how the FSC model departs from its competitors in terms of its values-based aspirations and its organizational structure. As a result, he characterizes the FSC as a form of private governance somewhat advisedly (Meidinger 2000, 226).[1] Moreover, in his more recent writings, he has argued that the FSC and its competitor certification regimes comprise collectively a "multi-centred private-public governance regime" (Meidinger 2006b).

Cashore, Auld, and Newsom (2004) are similarly alive to the difficulties associated with deriving a conceptual framework for conceiving of the FSC governance experiment. While they embrace the language of private governance (5), their ultimate aim is to contrast the traditional concept of public authority under the Westphalian system (under which states possess exclusive sovereign rule-making authority) with an emerging new form of "non-state, market-driven" authority. As a pioneering attempt to frame the essence and significance of the FSC, this formulation has much to commend it. However, we would contend that the appellation "non-state" is too generic to provide a viable conceptual framing of the FSC as a mode of governance. Likewise, the centrality of the "market" in this definition is problematic because it fails to give sufficient descriptive prominence to the unique modalities through which the FSC seeks to achieve its mission and to the breadth and distinctiveness of the values it aspires to advance.[2]

Consequently, we would contend that the public-private distinction fails to capture the subtle and complex interpenetration of "public" and "private" actors within the governance arrangements that are coming to dominate the realm of environmental management; it conceals the reality that such arrangements characteristically occupy a social space replete with inter-dependence and hybridity (Wood 2002-3). Thus, although many emerging regulatory systems are championed and largely driven by private interests, they typically interact, often at several levels, with public authorities in the domains of law, regulation, and public policy. Observing this, Freeman notes (2000, 551) that although "public" and "private" retain value in terms of highlighting differences between governance forms, one must be vigilant to avoid assuming that there is ever "such a thing as a purely private or purely public realm."[3]

In a similar vein, Wood has cautioned that characterizing emerging governance initiatives in terms of orthodox dichotomies, particularly where corporate-dominated environmental management systems (EMSs) are concerned, can adversely narrow the analytic field of view, "deactivat[ing] the substantial political stakes of corporate environmental management by treating them as 'technical' matters to be resolved by neutral professional expertise and simultaneously as 'private' matters of consumer or commercial preference to be resolved by the market" (Wood 2002-3, 132). What might be termed the "ecological" conception of social reality emerging from Freeman's and Wood's work suggests that a botanical approach to instances of governance and regulation, which identifies several dimensions of difference, will prove more instructive than one premised on a single dichotomous continuum.

Moreover, just as uncritical designation of industry-dominated EMS regimes as "private" runs the risk of disguising the very public values at stake in such regimes, we would argue that designating governance initiatives such as the FSC as "private" risks understating the public values and benefits such regimes can promote. Accordingly, we would argue against the notion that voluntary codes and other new forms of governance can (or should) be located along a simple public-private governance axis. Understanding the nature of such regimes thus demands, in our view, an approach that considers their attributes across several criteria. In the following section, we elaborate an approach that, we believe, provides a more nuanced way to conceive of governance regimes and then apply this approach to the FSC case.

### An Ecology of Governance

Despite the vastness of the scholarly output addressing the "new governance" phenomenon, progress toward developing a coherent taxonomy of governance remains very much in its infancy. Building on recent work by Treib, Bähr, and Falkner (2005), we contend that only a multi-dimensional approach is capable of capturing the full range and diversity of the phenomenon. In our view, three dimensions stand out in terms of developing such a taxonomy: the nature and identity of the parties and interests participating in such arrangements *(political dimension);* the nature of the instruments deployed, and the values sought to be advanced, under such arrangements *(regulatory dimension);* and the nature of the architecture within which such arrangements are housed *(institutional dimension).*

**The Political Dimension**
As discussed above, the focus of much governance scholarship has been on the identity of the participating parties and the interests such arrangements represent (or fail to represent). Thus, arrangements are typically classified as public or private depending on the range of participants and interests they include or exclude. As Freeman (2000) and Wood (2002-3) rightly underscore, scholars using this approach tend to understate the significant extent to which, in many such arrangements, both private and public values and interests are interwoven within complex interdependent networks of influence.

Throughout this volume, we have sought to elaborate the breadth and ambitiousness of the FSC's reach in terms of this political dimension. Whether, and to what extent, its reach exceeds its grasp is, of course, a matter of some debate. Despite its commitment to inclusiveness, the FSC's ability to achieve these aspirations on a global scale has been constrained by a variety of factors, including financial resources, management and administrative capacity issues (both within the FSC and among potential members), constitutional gaps and challenges, and lingering skepticism within the global forestry polity over the benefits of participation. Indeed, some representational constraints have been self-imposed. For example, as a professedly civil society organization, the FSC has itself implicitly endorsed the public-private dichotomy by adopting policies that constrain state participation in its affairs and governance.[4]

Even as it labours under these various limitations, we would contend that the FSC has succeeded in achieving a high level of inclusiveness in the political dimension, spanning differences in levels of economic development (North and South), spatial reach (provincial, national, global), and values and interests (economic, environmental, social, indigenous). In some ways, the negotiations that culminated in the FSC-BC final standard can be seen as a microcosm of the challenges inherent in the complex and murky politics of inclusiveness that FSC-AC confronts globally.

In contrast, as was elaborated in Chapter 2, other forest-certification schemes have typically adopted much less inclusive and ambitious approaches to political representation, although there are signs that this may be changing. Stakeholder participation in the ISO, a federation of national standard-setting bodies, has been and remains largely dependent on the inclinations and practices of its membership. Early on, a focus on technical standards meant that ISO members were often self-selected from leading companies in a designated industry with a specific interest in international standards development. Consequently, as Haufler (1999) notes, the resulting standards were largely developed by closely knit industrial elites from developed countries, a situation that mattered less when standards dealt with narrowly focused technical matters such as helmet safety or screw-thread diameters. It has never, however, been without its politics. Thus, Kuert (1997, 18) describes how even the founding of the ISO in 1946 in London was structured by "inch bloc" versus "metric bloc" politics, and Haufler (1999, 15-16) notes significant transatlantic tension between the United States and the European Union over the development of its 14000 series.

Likewise, the various national schemes falling under the umbrella of the PEFC (including the SFI and the CSA's sustainable forest management program) originated as industry-driven initiatives. For instance, as was chronicled in Chapter 2, the SFI was developed in-house by the American Forest and Paper Association (AF&PA) in consultation with its members, which consisted exclusively of US forest companies.

However, there is evidence that the FSC's main competitors are taking the challenge of inclusiveness more seriously (Meidinger 2006c). In recent years, the ISO has developed an action plan to promote participation of developing countries and NGOs in standards development, and individual members such as the American National Standards Institute (ANSI) and Standards Australia are likewise seeking to encourage more widespread stakeholder consultation (Haufler 1999; Cameron 2005). The SFI has also taken steps in this direction, most notably in 2000 when it established a Sustainable Forestry Board (SFB) with "tripartite" representation from industry, environmental groups, and the "broader forestry community" (SFB 2006a).[5] In spite of these moves, however, the SFI specifically and the PEFC more generally have been unable to obtain the support of mainstream environmental groups or of community and indigenous peoples' organizations.

The record of the FSC's sister organization, the Marine Stewardship Council, is similarly equivocal. During its early years it was taken to task for being too closely associated with the interests of two of its founding member organizations, Unilever and the WWF, at the expense of other interests including non-industrial fishers, women, and the South (Constance and Bonanno 2000, 131-32). In response to these criticisms, the MSC conducted a review of its governance arrangements in 2000 and established a technical advisory board (TAB), made up of fisheries and ecological scientists, and a stakeholder council to provide advice to the board of trustees about MSC's strategic direction and standards (Cummins 2004, 88-89). Although these initiatives have enhanced the inclusiveness of its board of trustees, the MSC remains vulnerable to criticism about the discretionary nature of representational arrangements.

In short, issues of inclusiveness and representation are likely to remain challenges for all certification organizations, particularly those that seek to realize these goals on a global scale. Although differences between the approach adopted by the FSC and those of other certification regimes have diminished somewhat over time, we would argue that, on balance, the FSC model remains at the forefront in this realm both in aspiration and in practice.

## The Regulatory Dimension
What we term here the "regulatory dimension" has been studied by other governance theorists (Jordan, Wurzel, and Zito, 2003, 2005; Kooiman 2003). Often, by giving firms more discretion over how they achieve regulatory compliance, the focus of such analyses is on the extent to which emerging governance arrangements rely on new policy instruments that depart from the putatively rigid and more coercive regulatory forms (commonly associated with traditional command-and-control-style regulation). However, we would argue that a more nuanced picture of the regulatory dimension begins to emerge when governance arrangements are assessed using multiple criteria. Three criteria, in our view, are particularly instructive: *bindingness* (whether regulation more closely resembles hard or soft law); *form* (whether it deploys management-, technology-, and/or performance-based standards); and *values* (the nature of the substantive and procedural values it seeks to promote).

### *The Hard Law–Soft Law Continuum*
While the hard law–soft law distinction originates in the realm of international law, increasingly it is also being deployed in the domestic realm (Kennett 1993; Kirton and Trebilcock 2004). Traditionally, it has been the task of legal scholars to define the boundary between these two legal forms. In this context, laws are considered "hard" to the extent that they impose generally applicable obligations, articulated in a relatively precise manner, that are in turn enforceable through the judicial authority delegated by the state. In contrast, "soft" law is typically conceived of as representing a weakening (or softening) of the relevant legal arrangements in terms of obligation, precision, and delegation (Abbott and Snidal 2000, 422; Tollefson 2004, 93).

Analyzed through this lens, FSC standards are doubtlessly soft law insofar as the obligations they create are self-imposed as a condition of membership. Notably, however, in many other respects the FSC model tilts "towards the hard law end of the legalization spectrum" (Tollefson 2004, 111), a conclusion fortified through an examination of the FSC-BC final standard in terms of the metrics of obligation, precision, and delegation. On one hand, as measured in terms of "generally applicable obligations," the final standard represents a softening when compared to prior versions of the standard, particularly in terms of the elimination of "major failures" (provisions that preclude certification where there is substantial non-compliance) and the elaboration of a stand-alone standard for small operators. Nonetheless, strong mandatory provisions remain, including an obligation on all members to uphold FSC principles.[6] In contrast, it can be argued that in terms of the criteria of "precision" (in particular, the detailed and specific nature of provisions contained in the final standard) and "delegation" (the extent to which member firms entrust continuing adjudicative authority to the FSC and its certifiers, subject only to the "exit" option), both the FSC-BC final standard and the FSC model more generally exhibit definite hard-law characteristics.

Indeed it bears noting, as we discussed in Part 1, that one of the main reasons the BC environmental movement invested so heavily in the development of the FSC-BC regional standard was a sense that provincial regulation of forest practices (notionally hard law) had, due to non-enforcement, regulatory amendments, and industry pressure, begun to resemble soft law in many relevant and practical respects (obligation, precision, and delegation among them).

In contrast, we would argue that most competitor regimes adopt a much more classically soft-law approach to regulation. Apart from a generic commitment to establish and maintain EMSs, the obligations imposed under the ISO and the PEFC are largely self-defined at the firm level. Indeed, under the ISO model, firms are the primary source of substantive standards (Meidinger 2000, 196). Some schemes that fall under the PEFC umbrella, such as the SFI and the CSA, do impose obligations to certain substantive principles relating to environmental management, but, when measured in terms of the "precision" metric, these are often relatively weak.[7] This weakness is due to the fact that such obligations are often highly qualitative (i.e., to employ sustainable forestry practices), leaving firms with a broad discretion to define and implement compliance strategies. Moreover, from the perspective of verification and monitoring, many competitor regimes entail relatively high levels of "delegation." Under the ISO, verification and monitoring are left almost entirely to firm discretion. Other regimes, such as the SFI, have begun to encourage voluntary third-party verification; however, firms continue to exercise considerable control over both the choice of verifier and the verification process (Meidinger 2000, 210).

### Form of the Standard Employed

In much of the governance literature, "government" is identified with heavy-handed command-and-control-style "prescriptive regulation" (see Chapter 1). As a result, it is often contended that a distinguishing feature of new governance arrangements is their deployment of more flexible, market-sensitive performance-based regulatory standards (Jordan, Wurzel, and Zito 2005). We would argue that this generalization is misleading in at least two main respects. At a conceptual level, as elaborated in Chapter 11, the oft-invoked dichotomy between prescriptive- and performance-based regulation is ill-conceived: prescriptiveness (defined in terms of the imposition of constraints on managerial discretion) is an inherent feature of virtually all forms of regulation. Thus, all forms of standard (management, technology, and performance based) entail prescription, albeit at different stages of the production process. Moreover, we would argue that the notion of performance-based regulation as something new is descriptively misleading. Historically, governments (particularly Canadian ones) have relied rather heavily on performance-based regulation in the environmental/resource management context and beyond. Although the growing popularity of this form of standard is arguably recent, it is erroneous to characterize reliance on such measures as a new form of governance.

This said, what does our examination of the FSC-BC case tell us about the form of the standard employed within this broader context? A leading scholar of forest certification has contended that "individual principles, standards and criteria used in *private* regulatory systems seem markedly different from the rules typically produced by *public* regulatory systems" (Meidinger 2000, 233, emphasis added). In support of this claim, he argues that private forest-certification systems are less precise and afford greater regulatory discretion than exists under comparable public rules.

Our research has led us, at least insofar as the FSC-BC final standard is concerned, to quite different conclusions. As chronicled in Chapter 11, our analysis of this standard reveals that it is highly detailed and precise in terms of imposing management, performance, and often hybrid management/performance obligations on firms seeking certification. Moreover, although it vests considerable discretion in both managers and certifiers, we would argue that, in this respect, it does not differ significantly from comparable public regulatory systems.

Where Meidinger's contention seems more compelling is in the context of an assessment of the FSC's rival certification regimes. Like the FSC, many of these regimes oblige companies to meet standards that are performance based. The difference, however, is that the obligations imposed under these rival standards are usually framed in much more open-ended, qualitative terms: for example, "to comply with applicable laws" and "to conserve ecosystem diversity." While similar qualitative performance standards are present in the FSC-BC final standard, they are usually accompanied by specific management requirements designed to facilitate independent monitoring and to bolster the probability of compliance.

*Range of Values Being Regulated*

Yet another criterion against which it is instructive to assess the FSC-BC final standard and the FSC model is the scope of the values they seek to advance. In substantive terms, the breadth of FSC aspirations in this regard is captured in its stated goals of promoting "environmentally appropriate, socially beneficial and economically viable management of the world's forests," which, in turn, are elaborated in its principles and criteria. As expansive as these goals are, they do not capture the full range of the organization's values, which include a commitment to a suite of process-related values such as democratic dialogue, interest representation, transparency, and stakeholder accountability (see Chapter 12).

Competitor certification regimes also express a commitment to environmentally appropriate forest management. Yet although such regimes are ostensibly in alignment with the FSC model, given the relatively open-ended and discretionary nature of the requirements they impose (which are typically framed in terms of obligations with respect to planning, management, and compliance with applicable laws), their ultimate impact in terms of environmental protection is, as Meidinger (2000, 202) puts it, "quantitatively indeterminate." Moreover, such regimes are deliberately designed around the core value of economic efficiency, leaving individual firms to determine how the deleterious environmental and social impact of forestry should be balanced off against the economic costs. Accordingly, under ISO and PEFC schemes a variety of values central to the FSC regime – including indigenous peoples' rights, worker safety, local employment, and community involvement – are left to firm discretion (208).

## The Institutional Dimension

Perhaps the least studied aspect of the new governance is what we would term the institutional dimension. Here the relevant inquiry addresses the formal institutional arrangements that structure decision making within the governance model in question.

On the basis of our analysis of the nature and operation of these arrangements within FSC-AC (see Chapter 12), we concluded that the FSC model exhibits many of the features and mechanisms of what we would term "democratic constitutionalism." Thus, like many democracies, the FSC has both formal and informal constitutional provisions that set out the overarching rights and responsibilities of its "citizens." It also seeks to provide broad representation to interests in the global forestry "polity" through a variety of institutional forms, including a chamber-based legislature (its General Assembly), a balanced executive (board and secretariat), and a judiciary (its Dispute Resolution Committee). Under the "unitary" form of governance the FSC has adopted, these arrangements are designed to balance the need for central leadership and coordination with the goal of giving national and regional initiatives the ability to define and interpret, principally through the standard-development process, how the FSC's principles and criteria are translated into nationally and regionally applicable regulation.

Although there is still room for improvement, the relative "thickness" of FSC's governance arrangements contrast rather starkly with the "thinness" of the arrangements in other schemes. The ISO has a complex governance arrangement, but its ISO 14001 series basically stands alone and operates without a substantive institutional architecture. All that exists, in effect, is the standard and associated documentation, which are interpreted by individual firms with periodic guidance from third-party auditors. Members of the larger forestry polity have few avenues of influence beyond complaint to the company and auditor if their practices do not conform to EMS standards.[8] The PEFC's arrangements vary from scheme to scheme. The CSA's institutional arrangements parallel the ISO's – there is a stand-alone scheme with no separate institutional architecture to monitor, enforce, and update the standard.[9] The SFI standard stands apart and bears a greater resemblance to the FSC, with the Sustainable Forestry Board established to oversee the implementation of SFI, update it in light of emerging forestry knowledge, and adjudicate alleged violations (see further discussion later in this chapter). Nonetheless, overall, the SFI regime neither aspires to nor reflects a commitment to broad democratic norms akin to those that remain a core tenet of the FSC model. Even the MSC, the FSC's fisheries counterpart, departs significantly from the FSC's innovative institutional structure by virtue of its board's structural isolation and lack of accountability to a global fisheries "polity."

Significantly, though, where authority is centralized within the MSC, it is decentralized in the ISO and PEFC schemes, despite formal constitutional provisions to the contrary. This seeming paradox is a function of the extent to which, as discussed earlier in the context of the hard law–soft law continuum, such regimes preserve broad discretion at the firm level to interpret and apply relevant standards. It is also a function of the relative weakness (if not absence) of central compliance monitoring.

## Summary
The foregoing analysis illustrates the utility of analyzing new governance arrangements using a multi-dimensional approach. It also underscores the limitations of received conceptual dichotomies such as private-public and state-market. Based on our analysis, we argue that the FSC model departs dramatically, both in aspiration and in practice, from its competitor schemes, as measured across political, regulatory, and institutional dimensions. In terms of the ecology of governance, it is sui generis. Although it could plausibly be contended that the FSC model tilts toward, or more closely resembles, a "public" as opposed to a "private" form of governance along all of these various dimensions, we believe that, in the final analysis, such a

characterization remains theoretically wanting. Accordingly, in the concluding section of this chapter, we offer and justify an alternative conceptual
vision of the mode of governance that the FSC embodies, which we term
"global democratic corporatism."

## Global Democratic Corporatism

Three organizational elements of the FSC underscore its uniqueness as a
mode of governance: its global vision of economically, ecologically, and
socially sustainable forestry; the democratic constitutionalism embedded in
its institutional arrangements; and the corporatist interest mediation system
it employs to balance the interests of its diverse membership. We turn now
to a closer consideration of these three defining elements.

### The Global Element

Geographically, forest-certification systems range from those that are primarily regional or national in scope to those that are decidedly global. The
FSC's vision is explicitly global, the aim being to make "environmentally
appropriate, socially beneficial, and economically viable" forestry available
to all forest managers around the world regardless of size of operation, nature of forest type (tropical, temperate, boreal), or intensity of management
(natural forest or plantation). The FSC's global vision emerged very early in
the organization's development and was first articulated by Hubert Kwisthout
in his original proposal for an international forest monitoring agency (see
Chapter 2). There were several rationales for a global approach: northern
environmentalists' desire to see improved practices in tropical developing
countries, the growing awareness of poor forest management in the temperate and boreal forests of the North, and the fear that schemes developed by
national governments and individual certifying bodies would lead to a race
to the bottom in terms of standards development.

There is little doubt that the FSC's vision of sustainable forest management
is global, but our analysis has also shown that there are significant differences of opinion within the organization about how this ambitious goal
should be realized. Whether one characterizes these differences in terms of
a dichotomy between "boutique" or "general store" approaches (as large
industry argues) or in terms of conflicting views on how "rigorous" standards
need to be to leverage real change in forest practices (as many environmentalist and social justice advocates prefer to frame it), it is clear that the evolution of the FSC-BC final standard underscores that the "vision" will remain
an issue for the organization for the foreseeable future. Of the competitor
schemes considered for this study, only the ISO proceeded at the outset
with a comparable global vision, which was in its case based on an EMS
model. The schemes that now collectively make up the PEFC (including

the SFI and the CSA) all originated as state-centred initiatives. It was only when the former Pan-European Forest Certification Council reinvented itself as the Programme for the Endorsement of Certification Council in 2002 (in large measure as a competitive response to the FSC), that national programs such as these had the opportunity to become part of a global network of domestically developed certification schemes.

Like the FSC, the MSC's vision at the outset was also global, its mission being to "safeguard the world's seafood supply by promoting the best environmental choice" (Cummins 2004, 86). The MSC was also motivated by many of the same ideals as the FSC, but it has encountered considerable difficulty in translating its global vision into practice. Early certifications were contentious and heavily criticized, leading to substantial institutional restructuring, although the global mission itself remained. Yet a global vision is one thing; a global reach, quite another. All organizations considered here have encountered difficulties realizing their respective visions, whether national or global. Given the original aspirations of the FSC and the MSC for sustainable forests and fisheries around the world, its members might justifiably feel disappointed, particularly with respect to the slow progress in redressing key issues in the South (Cashore et al. 2006).

## The Democratic Element

*Global Sectoral Polities*
A second defining element of the FSC as a governance regime is its commitment to democratic principles and practice. Although democracy is an "essentially contested" concept in the field of political theory, most analysts agree that at its core is the practice of "rule by and for the people."[10] The fundamental questions such a formulation raises are: Who are "the people"? And what does it mean to be ruled democratically? The conventional answer to the first question is those individuals, the *demos,* bound together through ties of blood, descent, culture, history, or some such amalgam. The people rule themselves by establishing a territorial political space – the *polis* – conceptualized in much political philosophy as occurring via the signing of a social contract that lends sovereignty to the state and provides it with the right to be recognized as such within the international state system. Ideally, the spaces occupied by the *demos* and the *polis* completely overlap to form the "nation-state," which was until recently viewed as the unrivalled unit of modernity and the only legitimate place for politics.[11] Within the nation-state, democratic rule is conceptualized by mainstream political realists as a competitive contest between two or more political parties that seek power via regular "free" and "fair" elections, supplemented by contests between relatively equally positioned interest groups for the ear of government with respect to any particular policy matter.[12]

Without disputing the verisimilitude of this long-standing *horizontal* image of politics in a world of relatively discrete *territorial* nation-states (each constituting a spatialized polity in its own right), we would argue that the FSC is a manifestation ·of an emerging *vertical* politics clustered around relatively discrete issue areas, each with the potential to constitute its own *sectoral polity*. In this conception, the FSC functions as a non-territorial, sectoral political space where global networks deliberate over the meaning of sustainable forest management and enforce the rights and obligations necessary to achieve this goal. In short, in this vision, the FSC represents the founding of a *global forestry polity* that may eventually challenge the authority of territorial states over forest management regulation.

This conception of an emerging global forestry polity builds on and elaborates some recent literature in this field, notably by Walby (2003) and Ferguson et al. (2000) on non-state polities, Ansell (2000) on networked polities, Blatter (2001) on multi-polity stems, and von Bernstorff (2003) on sectoral polities. Particularly instructive is Ferguson et al.'s definition of "polity," which, in their conception, consists of "those persons who identify with it, the resources it can command, the reach it has with respect to 'adherents' located in space in the broadest sense ... and issues. All polities are authorities and govern within their respective domain. Authority or governance in our definition is effective control or significant influence within a domain. Such authority need not be exclusive (it can be and often is shared), nor need it be regarded as legitimate, although legitimacy is an asset" (30).

Not only do Ferguson et al. escape the territorial trap by not equating polities with states, but their formulation also permits the formation of global sectoral polities when actors with resources gain rule-making authority over specific issue areas. In this book we have charted not only how the FSC played a major role in constituting this global forestry polity, but also how its actions mobilized others, notably disaffected members of industry, to respond by establishing the PEFC, which has had the effect of further consolidating the global forestry polity.[13] As the FSC sought to operationalize a new conception of global democracy to meet the governance challenges raised by this emerging global polity form, the PEFC adopted a conventional "multinational" approach to governance that locates authority in industry bodies within states overseen by an international board at the global level.

In developing our conception of a global sectoral polity, we are mindful of the pioneering work of Ansell, Blatter, and von Bernstorff (noted above), who derived similar conceptions from analyses of real world examples. Of these studies, von Bernstorff's is most akin to our own. He argues that the failed Internet Corporation of Assigned Names and Numbers (ICANN) initiative represented an attempt to establish "a democratic and 'a-centric' global polity to govern the 'Internet Community.'" In his view, this Internet community can be conceptualized as being composed of "all the citizens of the new virtual space, a community of those who have created and explored the new space and have been living together at the 'Cyberspace frontier'" (von Bernstorff 2003, 514). Von Bernstorff is highly critical of ICANN's claims to represent the interests of the Internet community, arguing that it is constitutionally unable to perform this role because it consolidated the hegemony of commercial over user and other interests. Based on our research, we would argue that the FSC suggests the potential for alternative and more democratic outcomes.

### The Global Forestry Polity and Democratic Rule

Insofar as the FSC represents a global forestry polity, "the people" to whom it is accountable thus become any and all who share its vision of sustainable forest management, including members of forest-dependent communities, forest workers, tourism operators, botanical forest product collectors, forest managers, foresters, indigenous communities, sawmillers, manufacturers, consultants, wholesalers, retailers, environmentalists, social justice activists, and final consumers.

Considerations of democratic practice with respect to this broadly conceived global forestry polity require that interests be mediated through a process that is accessible, accountable, deliberative, and responsive. A comparison of FSC's arrangements with its competitors reveals some interesting contrasts in this regard. Organizational accessibility and accountability to all current and potential members of this global forestry polity are core values of FSC's institutional ethic. Its membership provisions facilitate broad-based individual and organizational participation on the condition that prospective members commit to uphold its principles and criteria, pay a fee (which varies depending on North/South status and whether one is an individual, NGO, or business), and have their application supported by two existing members. Members are entitled to attend meetings of the General Assembly, put forward and vote on resolutions, and stand for and vote in elections for the board. Members are also entitled to participate in their country's national initiatives, and, as is documented by our account of the FSC-BC case, they played a key role at all levels, from regional to global, in the development of the FSC-BC final standard.

Although the FSC is accountable to a global forestry polity, it departs from the "one person, one vote" accountability of representative democracies by giving voice to organizational "persons" such as firms, community groups, civil society organizations, and unions organized into networks. Individuals may join the FSC, but their electoral weight can never be more than 10 percent of the total votes cast in any chamber, with 90 percent retained for organizational members within each of the six sub-chambers. This arrangement could seriously erode FSC's claims to be accountable should the number of individual members ever seriously outweigh the combined membership and/or customer base of its organizational members. However, this seems unlikely, given that the FSC currently has fewer than seven hundred members, the majority of which are organizations rather than individuals.

The FSC's arrangements differ substantially from those of its competitor schemes, which define the global forestry polity more narrowly and which, overall, can be considered both less accessible and less accountable. For example, the ISO is a federation of national standards organizations, and it admits to membership only a single body from each state that is "most representative of standardization in its country" (ISO 2006). ISO members range across a spectrum of institutional forms. The Standards Council of Canada is a statutory body that represents Canadian interests in the ISO; ANSI, the American National Standards Institute, is a civil society organization. Governmental and statutory bodies receive their mandates from the state and, as such, are subject to legally enforceable norms of accountability (though not necessarily accessibility) to the public at large, at least in democratic states governed by the rule of law. Where national standards bodies are civil society agencies, as in Australia and the United States, public accountability and accessibility is more attenuated. Given ISO's exclusive membership requirements, the participation of a large number of non- or proto-democratic states, and the variability in its members' organizational systems, it would be difficult to conclude that the ISO was especially accessible or accountable to "the people," broadly construed, that use its standards. This conclusion is reinforced by the observation of many analysts that the ISO has tended to be dominated by the interests of large industry (see especially Haufler 1999).

Like the ISO, the PEFC is an umbrella body that houses a diversity of schemes with varying institutional arrangements. The PEFC's conception of the global forestry polity is more limited than ISO's, with membership restricted to a single national governing body from each country. The PEFC does not prescribe how these national bodies are constituted, requiring only that they have the support of the "major forest owners of the country" (PEFC 1999). Consequently, PEFC members can have very different institutional structures, as a comparison of the CSA and the SFI makes clear. Moreover, the process of establishing national PEFC schemes invariably devolves to national forest industry interests, which enjoy a privileged position relative to other members of the global forestry polity. A case in point is Canada's CSA, which, as we noted earlier, was initiated by the Canadian Pulp and Paper Association and is a scheme that, consistent with ISO's EMS-based approach, has fragmented institutional arrangements for implementing, monitoring, or revising its standard. The use of the standard is promoted by the Canadian Sustainable Forestry Certification Coalition (CSFCC), which maintains a website with excellent information on current certifications, but actual formal responsibility for standard development remains with the

CSA. Moreover, any changes to the standard's forest management criteria and elements would come from an updating of the Canadian Council for Forest Ministers criteria and indicators for sustainable forest management, over which neither the CSFCC nor the CSA has authority. Meanwhile, the hierarchy for dispute resolution provides for a disputant to contact the certifying body in the first instance and the Standards Council of Canada in the second. This fragmentation of authority has the potential to benefit industry, as it allows different elements of the system to evade responsibility and accountability.

In contrast, the SFI is managed by the Sustainable Forestry Board (SFB), which was created in 2000 to enhance organizational accountability to US forestry stakeholders. The SFB is still largely dominated by industrial forestry interests, but its fifteen-person board is now explicitly corporatist in design, assigning five seats each to economic, environmental, and "broader forest" interests. Nonetheless, the SFB continues to suffer from a serious accountability deficit, in large measure due to its resistance to becoming a membership-based organization. At present, the SFB is run as a self-appointed board of volunteers that meets two to three times a year to conduct the organization's business, taking responsibility for the content of the SFI standard, the procedures, and the qualifications for audits. It has, nonetheless, evolved an elaborate institutional structure, with the majority of its work being done by an array of subcommittees, including the SFB Resources Committee (which recommends changes and/or enhancements to the SFI standard) and the External Review Panel (which provides advice on the implementation of the SFI and ensures that annual reports are objective and credible).

Indeed, in this regard there are some interesting parallels between the institutional structures of the SFI and the MSC. Both have eschewed the concept of open membership, adopted self-appointed boards, and, more recently, been forced to restructure following heavy internal and external criticism (the SFI establishing a more independent SFB with explicit tripartite representation, and the MSC creating a technical advisory body and a stakeholder council). However, although the MSC's restructuring has reduced its democratic deficit, it remains vulnerable to criticisms of executive dominance because appointments to its governing board remain an executive prerogative, constraining its accountability to the global fisheries polity.

Democratic governments are not only a form of rule by and for the people, they are also expected to be responsive to a diversity of interests and values and to safeguard minority groups against the "tyranny of the majority" when deliberating on policy matters. To ensure responsiveness, democracies have evolved constitutions that protect minorities from discrimination based on religion, race, ethnicity, and sexual orientation. Not infrequently, legislation is struck down as a consequence of high court deliberation on the constitutionality of one or more of its provisions on these and other grounds.

Certification schemes vary in the degree to which they aspire to achieve similar goals of responsiveness and deliberation. Our research suggests that, in terms of these metrics, the FSC rates highly both at the regional and national levels and on a global scale. The BC case, in particular, highlights the deliberative capacity of the FSC regime in the context of a highly complex, polycentric decision-making process. With deliberation occurring within and across six sub-chambers – as well as within networks of donors, certifiers, and indigenous peoples – it is a regime that endeavours strongly to ensure that all interests will be heard and that minority interests will be protected. Indeed, the fact that the FSC protects minority rights explains why it has been strongly endorsed by indigenous peoples and communities.

FSC's competitor schemes perform less well with respect to responsiveness and deliberation, although pressures to address these deficiencies are mounting. The ISO remains dominated by industry from OECD countries, although it has recognized the need for broader sectoral consultation within countries and the need to be more proactive with respect to developing countries. The momentum for change within the ISO is in part attributable to a growing appreciation that process standards (derived from the EMS model) raise broader and more complex social questions than the technical standards that formerly comprised the core of the ISO's work and, as such, need to be developed through more deliberative and consultative processes. Analogous changes are afoot within the PEFC, although given its status as an umbrella body housing a large number of nationally developed schemes, significant country-by-country variations are likely to persist. In some contexts, the CSA being a case in point, development of a more responsive and deliberative architecture has been thwarted by resource constraints. In other settings, pressures for reform have led to more thorough-going changes, as is illustrated by recent structural renovations undertaken by the SFI and the MSC.

### The Corporatist Element
Of the three terms we are using to describe the FSC, "corporatist" is likely to be the most normatively controversial. This is due to its past association with powerful tripartite alliances between government, business, and labour that have sometimes operated to the detriment of the broader public interest. We acknowledge the historical baggage that accompanies corporatism as a conceptual category, but we would argue that the form of corporatism embodied by the FSC not only minimizes the risks of abuse often conjured by the concept but also contributes to the organization's strengths. Among these strengths, we contend, is its capacity to promote stable governance, to mediate across groups with diverse and often conflicting interests, and to facilitate high-quality policy deliberation – all of which work toward fair,

balanced, and legitimate outcomes. Moreover, in contrast to direct, representative, and pluralist democratic forms, the FSC's variant of corporatism is not territorially limited, which allows it to articulate and mediate global interests more effectively than familiar place-based democratic arrangements are capable of doing.

The modern practice of corporatism has its roots in twentieth-century Europe. It is not inherently conservative in nature, but it took a particular nasty form in the 1930s in Germany, Spain, and Italy, where nationalism and racism were fused with corporatism to create fascism. In the aftermath of the Second World War, corporatism continued to be the subject of broad-based vilification that reached a zenith during the 1950s with the emergence of a pluralist vision of democracy that, particularly in North America, has retained normative ascendancy ever since. Despite this, as a model of democratic practice, corporatism retains significant influence and finds expression, to varying degrees, in the political systems of a number of OECD countries, including Sweden, Norway, Austria, Finland, the Netherlands, and Japan (Schmitter 1974).

Corporatism has been described as the inverse of pluralism: both are "systems of interest representation," but they differ in the way the "constituent units" are organized to represent interests (Schmitter 1974). In classical pluralism, a multitude of interest groups presumed to be relatively equal in power are viewed as competing against each other for influence over public policy, with government playing the neutral arbiter, brokering the necessary compromises in the public interest. In contrast, Schmitter defines corporatism as "a system of interest representation in which the constituent units are organised into a limited number of singular, compulsory, non-competitive, hierarchically ordered and functionally differentiated categories, recognised or licensed (if not created) by the state and granted a deliberate representational monopoly within their respective categories in exchange for observing certain controls on their selection of leaders and articulation of demands and supports" (13).

FSC-AC is founded on a tripartite corporatist structure of interest mediation that is reflected in the authority jointly vested in its three chambers; a structure that conceptually derives directly from the FSC's mission of promoting "environmentally appropriate, socially beneficial, and economically viable" management of the world's forests. Economic, environmental, and social interests in the North and South are aggregated within the FSC into six equally weighted sub-chambers. The assumption is that organizationally optimal standards will emerge from the deliberations of the individuals and networks represented within these sub-chambers; these standards will reflect and fulfill the vision, values, and interests of all groups, rather than provide special treatment to a single one. Although each group must compromise on its ideal vision of "sustainable forestry management" to achieve a workable outcome, they are each in a relatively powerful position to defend their bottom lines, thus ensuring that the final standard is sufficiently stringent to meet most groups' requirements.[14]

Our account of the negotiation of the FSC-BC final standard demonstrates the strengths of this corporatist form of interest intermediation in the deliberation that took place within and between chambers at the provincial, national, and international levels. Interest groups in British Columbia were organized into economic, social, environmental, and indigenous peoples chambers to undertake the negotiation of a standard for the province. At the end of an exhaustive three-year negotiating process, a standard emerged that had the full support of social, environmental, and First Nations chambers but only the partial support of the economic chamber. At this point, region-based interest intermediation gave way to a more complex process involving regional, national, and, especially, international processes, with the FSC-IC board and secretariat playing a leadership role in breaking the deadlock by identifying and promoting a policy of preliminary accreditation and certification. In 2005, this policy resulted in a standard that received unanimous approval from all chambers at every level.

The FSC's corporatist interest intermediation arrangements contrast with those adopted by most other standards institutions within and beyond the forest sector. The ISO, for example, generally supports and adopts a "stakeholder" approach to interest mediation, which reflects its pluralist-influenced conception of decision making. The pluralist form of interest mediation works well, however, only when stakeholders have roughly equal levels of negotiating power, a situation that prevents any single group from gaining control of the agenda and dominating negotiations. In the ISO's case, however, commentators concur that the organization remains dominated by those large industries that have the most to gain or lose from the promulgation of a standard. This matters less when negotiations are about narrow, product-based technical standards; it becomes an important public issue when standards concern product safety, human health, and the environment. Ultimately, because it is a devolved system, the ISO depends on the organizational structures of national standards-setting bodies. These bodies have tended to adopt a narrow stakeholder model to date – consulting industry members over standards development – but there appears to be growing interest in expanding the level of consultation to embrace a larger number of stakeholders. This is a positive development, but it does not address the fundamental imbalance in power between industry and other interests, nor does it address ISO's domination by OECD countries to the detriment of those in the Third World.

The PEFC has developed a hybrid stakeholder-/chamber-based approach to standard setting (PEFC 2005). Under its procedures, lead responsibility for standard setting devolves to the forestry organization that has the support of most forest owners, as determined by a "balanced" multi-stakeholder forum. The PEFC's corporatist approach is evident in its idea that the negotiating forum should seek to have balanced representation from "producers, buyers, consumers, etc." This arrangement underscores the fact that the PEFC views interests as being organized into groups; however, it considers these groups to be mostly commercial and reflective of the forest product chain, rather than being more broadly political and reflective of worker, community, indigenous, and environmental interests. In consequence, schemes recognized by the PEFC tend to be industry driven and exclude the range of relevant social and environmental interests. Thus, for example, SFI's standard is the product of an industry-dominated process driven by the AF&PA, and it has been unable to transcend its foundational limitations despite the establishment of the SFB and an effort to reach out to other interests over the years. Similarly, although CSA's negotiation arrangements were more open and transparent at the outset, the explicitly stakeholder model of interest representation guaranteed industry a dominant position.

The MSC departs most significantly from the FSC with respect to its interest intermediation arrangements. As noted, the MSC is governed by an appointed board of directors comprising fifteen members. Although it has made an effort to include board members from diverse sectors, it appears that indigenous peoples, communities, small fishers, and developing countries remain under-represented. Commercial interests appear to dominate – and could certainly come to dominate in the future because there is no built-in mechanism to secure relatively equal bargaining power across interests. The MSC has moved to curb criticism of its institutional arrangements by establishing a technical advisory board and a stakeholder council, but neither is structured to give explicit voice to the wide range of interest networks insufficiently represented on the board. And, of course, the board remains in control and can accept, reject, or transform the advice it receives. In contrast to the FSC, therefore, we would argue that the MSC is characterized by a largely elitist, pluralist model of interest intermediation. Even though the FSC and the MSC are ostensibly sister organizations, it is difficult to sustain the argument that there is, as yet, a single coherent "stewardship council" certification model.[15]

Those critical of corporatism argue that, historically, it has been an institutional form used to silence dissent as often as it has fostered collaborative public policy.[16] However, these criticisms are more applicable to coercive, state-sponsored corporatist arrangements than to societal corporatism, especially that which occurs in a body that does not have recourse to coercion and force. The FSC differs from spatially organized corporatist arrangements, such as those that exist in a number of European states, in Mexico under the PRI, and, in a truncated form, in Japan and Korea. When corporatism dominates a single national space, there are clear dangers that it will become institutionalized and resistant to participation by new voices. Unlike states, however, the FSC does not claim a totalizing mandate to make laws within a specific geographic space across a wide variety of sectors subject only to the provisions of a constitution. FSC's mandate is restricted to forest management, and it seeks regulatory authority only over a polity defined in sectoral terms. Nor has the FSC-style corporatism operated in practice to exclude interests from making the case for recognition as equal corporate partners. As we noted at the end of Chapter 12, the FSC has proved itself to be a dynamic learning organization. Not only was its original fused social/ environmental chamber divided into two separate chambers in 1996, but other initiatives at the national level, such as the creation of a fourth First Nations chamber in Canada, illustrate an admirable institutional capacity for adaptive management.

## Conclusion

In this chapter we have sought to demonstrate the utility of analyzing new governance arrangements across multiple dimensions – including political, regulatory, and institutional – as a means of capturing the richness and nuance embedded in these arrangements. As a corollary, in this regard we have argued that orthodox conceptual dichotomies, particularly the public-private distinction, should be employed cautiously (if at all), given their tendency to circumscribe the analytic field of view and potentially disguise this richness and nuance. In offering this multi-dimensional analysis of the FSC as a governance arrangement, we have deployed a somewhat eclectic variety of conceptual tools and distinctions, including political network analysis, attention to forms of standard and law, and comparative constitutionalism. In doing so, we readily recognize that both our methodological choices and the dimensional elements we have chosen to study have, to some extent, been driven by the subject matter of our research and our respective disciplinary interests. As such, doubtlessly alternative methodologies and research foci aimed at elaborating these dimensions of governance in this and other contexts could usefully be deployed.

Drawing on our analysis along these dimensions, we have argued that the FSC represents an emerging new governance form: global democratic corporatism. Under the auspices of this new governance form, social, economic, and environmental interests are organized into six sub-chambers under decision-making rules that prevent any single interest from dominating. Moreover, interests are mediated not only through FSC-established chambers, but also through the homogeneous interest networks that permeate them. Thus, we have argued that the FSC creates a political space where actors organized into networks and disciplined by chambers engage in deliberation over the meaning, practice, and verification of sustainable forest management.

The FSC's global democratic corporatist governance system has been put to the test several times in the past decade. Indeed, the BC case was arguably its toughest challenge yet. But while the system has strained at times to achieve acceptable solutions, it has not buckled. And although the FSC-BC case demonstrates how traumatic standards negotiations via this process can be for the individuals and networks concerned, it also highlights how the checks and balances built into FSC's governance arrangements operated to produce a high-quality standard that respects the values of core constituencies, in the process giving new meaning to the concept of sustainable forest management in British Columbia.

# 14
# Reflections on the Nature and Significance of the FSC-BC Case

The emergence of the FSC as a global regulatory and governance institution crystallizes two interwoven and, at first blush, contradictory themes. At one level, the rise of the FSC reflects a strategic decision on the part of a diverse array of civil society organizations to forgo hard law and conventional state-based vehicles of international diplomacy in favour of market forces to leverage progress toward sustainable forest management broadly conceived. Yet, although market forces will, in part, decide the fate of this venture, perhaps what is most intriguing about the FSC is not its status as a market-based mechanism but the bold, arguably unparalleled, nature of its foray into global governance and regulation. Indeed, to harness these market forces, the FSC's founders were challenged to imagine and create a new social space, an "ideal" governance and regulatory institution that embodied the virtues and the values that, in their view, were so lamentably absent from existing state and supra-national forestry institutions.

In this book, we have sought to explore, in detail, governance and regulation within the FSC – employing the vehicle of the FSC-BC standard-development process – from a variety of perspectives: "grassroots" and "head office," process and substance, practice and theory. In Part 1, a principal concern was to understand the complex dynamics of the standard-negotiating process in British Columbia and how that process compares to the experience of analogous FSC jurisdictions. Part 2 elaborated further upon this comparative analysis, through an in-depth examination of the substance and implications of BC's final standard against a backdrop of standards developed in these same comparator jurisdictions. Finally, in Part 3, building on this procedural and substantive analysis, we sought to draw some conclusions about the nature and implications of the FSC regime for emerging theories of governance and regulation.

In this conclusion we reprise our findings, offering some final reflections on our research that address the following issues: the various barriers that nearly thwarted the development of an FSC-BC final standard; how, in the process of confronting these daunting barriers, a made-in-British Columbia consensus standard did emerge; the nature and significance of this final standard both on its own merits and in a broader comparative context; and what the FSC-BC case tells us about the future of the FSC, in particular the challenges and priorities that it must confront as an experiment in global forestry governance and regulation.

## Barriers to Developing the FSC-BC Standard

The reasons the FSC-BC standard-setting process was so challenging and came so perilously close to foundering are, we would argue, perhaps more apparent than the reasons why, at the eleventh hour, it succeeded.

We would argue that the barriers to achieving consensus in the FSC-BC standard-setting process can be understood as emanating from some of the key challenges inherent in governance, particularly the need to mediate across diverse interests to achieve legitimate, consensual outcomes. Some of these challenges arose out of what participants brought to the process (i.e., visions, values, negotiating strategies and goals, leadership); others relate to the structures and rules under which the negotiations occurred and the ensuing standard was ratified (i.e., norms and rules of representation and accountability, clarity of lines of authority and responsibility, and organizational roles). We will explore these various barriers under the headings of vision and values, democracy, and leadership.

### Vision and Values

Any exercise in governance necessitates the mediation of competing interests. As the FSC-BC experience underscores, moving from a shared statement of values – articulated, in the FSC's case, in the overarching objective of promoting "environmentally appropriate, socially beneficial, and economically viable" forestry – to a shared elaboration of how these values translate into on-the-ground regulation can be a daunting and divisive process. The debate within the FSC-BC standard-development process, and one that continues to reverberate throughout the FSC as a whole, was framed by industry as a choice between a "mainstream" or a "boutique" standard. For large industry especially, the goal was to secure the former: a standard that was broadly achievable and created incentives for a wide range of forest managers to secure certification at relatively low cost. Industry representatives argued that the higher the bar was set, the higher the cost to industry and the price to consumers, with negative effects in the form of low industry take-up, low volumes of certified timber in the marketplace, and a "boutique standard" endorsed by a small number of highly efficient and/or deeply committed operators.

Environmental, social, and First Nations participants in the BC process were mindful of these market concerns, yet they were determined to have an FSC-BC standard that went well beyond current practice and legislated requirements. Environmentalists argued forcefully in favour of replacing industry's "sustainable forestry" approach with a more robust "ecosystem-based forestry" approach that gave much more weight to maintaining a range of forest attributes over time through landscape-level planning, protection of natural forest cover and biodiversity, and careful management of riparian zones and high conservation value forests. First Nations representatives were determined to achieve a much higher degree of consultation than existed under provincial law over whether, when, where, and how logging would occur on their lands, and they were determined to secure tangible benefits from such activity. Social and community activists were committed to ensuring that any FSC-BC standard should translate the FSC's social principles – sometimes treated rather cursorily by FSC certifiers – into meaningful practices on the ground that generated tangible benefits to communities and workers. Their vision was of a rigorous standard that protected biodiversity, old-growth forests, and the livelihoods of workers and communities.

During the early stages of standard development in British Columbia, these differences in vision remained latent. Given the differences in interests and values that separated the parties at the outset – fuelled by a bitter ongoing struggle over protected areas in the province – remarkable progress was made toward a shared vision of what "sustainable ecosystem-based forestry" might mean in British Columbia. Later in the process, however, the "vision thing" came to be an important barrier, preventing the parties from pushing negotiations that extra step. Large industry refused to sign on to the third draft (D3) because they viewed it as setting the bar too high, creating a "boutique standard" that would have little impact in the market. Environmental, First Nations, and social representatives, on the other hand, viewed D3 – and the non-support of large industry – as the price to be paid to achieve an ecosystem-based forestry standard that adequately protected biodiversity, First Nations, and communities. Another three years of deliberation were required, together with on-the-ground experience of the standard in operation, before all parties could be convinced to come back to the table and forge a final standard.

If the "vision thing" had been more adequately addressed at the outset, could the later negotiations have progressed more smoothly? In this regard, the FSC-Boreal negotiations are instructive. There is considerable evidence – both in the way the Boreal negotiations were organized and from interviews with key informants – of organizational learning from the FSC-Maritimes and FSC-BC processes. Prior to negotiations, organizers circulated a draft vision statement that was later approved and consolidated into three simple paragraphs, of which the third is the most revealing. It mandated the negotiators to "develop a feasible and widely adopted certification system. If FSC-Canada is to be successful in its endeavours then it must develop a standard that is actually implemented. It must be practical for large as well as small-scale operations, and must confer advantages that outweigh the costs of implementation and auditing" (FSC-CAN 2002a, 2).

The statement articulates the view that the FSC should strive to be a "mainstream" standard, one that is "widely adopted" and "actually implemented" and that contains "advantages that outweigh the costs of implementation and auditing." This elaboration of the meaning of "environmentally appropriate, socially beneficial, and economically viable" allowed drafters of the standard to reflect not only on the substance of each indicator as it was negotiated, but also, as the standard neared completion, on whether they had achieved an adequate balance between the interests of the various parties and, in particular, on whether the standard provided adequate incentives for industry take-up. Had such a vision statement been in place in the BC process, some of the problems encountered when D3 was being finalized could potentially have been averted by providing the parties with a framework within which to continue negotiations, however difficult, in the search for a fully consensual outcome.

### Democratic Representation and Accountability

As elaborated in the preceding chapter, corporatism is a key element of the FSC governance model, a reflection of the practical obstacles to implementing a traditional, liberal democratic, "one person, one vote" approach on a global scale and of the desire to give equal representation to a range of interests that such traditional arrangements do not often succeed in balancing. The consensus that finally emerged in support of the BC standard provides reason for optimism about the viability of a global democratic corporatist approach, but our FSC-BC case study also highlights some of the challenges inherent in a governance model of this kind.

Key among these are issues of representation and accountability. In stateled corporatism, governments select their own compliant organizations to serve as sectoral representatives. In this context, for example, the voice of workers is only recognized when expressed through "official unions," whose role is often less to represent labour interests than to ensure industrial peace. Representation and accountability issues assume a somewhat different guise in civil-society-led, "grassroots" corporatism. In this context, the challenge is to ensure that the leadership that emerges to speak on behalf of sectoral interests both represents and is accountable to the interests on whose behalf it purports to speak. The FSC-BC case reveals a significant sectoral variation in the extent to which this objective was realized.

By far the most robustly representative and accountable chamber in the BC case was that which represented the environmental sector. The BC environmental chamber contained a large number of members representing a veritable who's who of the BC ECSOs, including Greenpeace, Sierra

Club of BC, and the David Suzuki Foundation. To determine who should represent the environmental movement on the FSC-BC steering committee (SC), detailed constituency consultations were organized, nominees were submitted, and elections were held. There was no doubt in the minds of the environmental members on the SC about who they represented or the concomitant need to be accountable to their constituents about the progress of negotiations.

Representation and accountability were much bigger challenges for the other three chambers. Due to their, at best, lukewarm attitude to the FSC, most major forest companies did not participate in the BC standard-development process. As a result, their ostensible representative in the process acted alone for the most part, consulting informally with a few like-minded colleagues but relying mainly on his knowledge of the industry and what was acceptable to his company. Similarly, the small number of social chamber representatives – and the non-membership and active hostility of the IWA – meant that these appointees tended to act in a personal capacity and were less formally accountable to a larger constituency than were their environmental chamber colleagues. Achieving representation of and accountability to First Nations also proved challenging because of the diversity of the interests to be represented and the logistics of engaging with this broad constituency, which was preoccupied by a range of other pressing issues.

The representation of interests in the FSC-BC process – both within and across chambers – was far from ideal, but it went significantly above and beyond most previous provincial government stakeholder processes and those of competitor schemes. For all its flaws, the various homogenous networks assembled within FSC's chamber-based system of interest mediation were able to coordinate their actions internally to secure the interests of a broad range of groups – large and small industry, communities, environmentalists, First Nations, and forest workers. Indeed, the openness, inclusiveness, and level of participation that occurred within and between networks in the FSC-BC process was recognized as superior to that of many other FSC regional-standard negotiations.

Although the FSC could potentially have done more to address these issues, as a voluntary membership-based organization its role in managing matters of intra-chamber representation is limited. Thus, while it can reach out to non-members (by hosting stakeholder forums and the like), in the end it is accountable to a global forestry polity of members that share its vision. As long as the various networks remain open to new members and adopt transparent and accountable processes to ensure all members are informed and able to participate, we would contend that each chamber should be left to determine its own modus operandi.

To fully secure such openness, transparency, and accountability, we have argued that the FSC should adopt a "bill of rights" based on the "International non governmental organisations accountability charter" described in Chapter 12. Such a charter would formalize and constitutionalize existing arrangements, ensuring that all players within the FSC system fully appreciate the dangers inherent in processes that are closed, elite-driven, and non-transparent. By adopting a charter of this kind, the FSC will not only safeguard member interests but also help to ensure that it remains at the forefront of democratic practice.

## Leadership

A third area where weaknesses in the FSC's organizational model were clearly evident relate to the absence of leadership at crucial junctures in the development of the BC standard. This absence of leadership was the outcome of structural difficulties within the FSC system rather than the failings of any of the individuals involved. For example, as we have noted at various points throughout this volume, the FSC-BC process commenced in 1996, only three years after the FSC's formal founding in Toronto and around the same time that FSC-AC became fully operational. By 1999, when FSC-BC was firmly established with staff, a growing budget, an elected steering committee, and a first draft of the standard, FSC-Canada was still finding its feet. Those who exercised leadership in British Columbia and moved the FSC process forward were working in something of a vacuum, relying heavily on a draft "National initiatives manual," which was useful but failed to provide detailed guidance on exact steps and processes. As a consequence, the steering committee was often forced to improvise, which contributed to some errors in judgment that, with the benefit of hindsight, could have been avoided.[1]

By 2002, FSC-Canada was up and running. However, although it had formal authority to refuse approval of the D3 standard, it lacked the moral authority to do so. Efforts to rein FSC-BC in were fraught with difficulties – not only organizational but also geographical, as such a move would have roused the traditional western Canadian fear of domination by eastern Canada. Even those who may have wished to block FSC-Canada's approval of D3 encountered a structural problem with FSC-Canada's statutes, which permitted a decision to be made by majority vote, without specifying the need for a minimum of 50 percent support from each chamber. If such a provision had existed, FSC-Canada would have been obliged to refer D3 back to FSC-BC.

Of course, FSC-Canada can justifiably claim that it was also operating in something of a leadership vacuum. Although it was a national body, it received relatively little support from FSC-AC and remarkably little guidance on how to establish and run its operation. Moreover, on becoming established, it had had to deal with one of the most public and acrimonious disputes affecting the FSC system – the outcry over the FSC-Maritimes standard (described in Chapter 5). Soon after, FSC-Canada found itself embroiled in a dispute over BC's D3 standard. It is hard to fault the FSC-Canada board for handing D3 on to FSC-AC, but there was an opportunity for the organization to take stock of its decision, even after it had been made. Had this been done, it would have been clear that, although proponents of D3 were formally entitled to insist that D3 be forwarded to FSC-AC, the decision was in fact against the spirit of deliberative compromise for which the FSC stands. The simple fact is that, within the FSC, all interests have to be accommodated to some degree, and this had not been achieved at the FSC-Canada level.

The buck stops, of course, at the FSC-AC level, with its board of directors and executive director. Could FSC-IC have shown more leadership? Perhaps. But such a judgment tends to underestimate the magnitude of the challenge associated with setting up a new, innovative, global organizational structure on a shoestring budget with a voluntary board operating in Mexico, at a distance from the action. With healthier finances, FSC-AC could have hired more staff and devoted more resources to policy formulation, national organization scrutiny, and board and staff training. As it was, it had a single officer in charge of policy development who was responding to numerous requests for policy across a huge range of issues. In short, it would be unfair to target him, or any of the other personnel in FSC-AC, for failing to show leadership. Rather, the absence of leadership was endemic in the organization because of problems associated with its unique structure, recent establishment, lack of resources, and distance from the action. One of the reassuring lessons from the BC case is that, by 2002, when D3 was forwarded to FSC-AC for its consideration and accreditation, the organization was well on the road to addressing many of its structural leadership deficiencies, as its handling of the ensuing preliminary accreditation process underscores.

### Explaining the Improbable: Understanding the Emergence of the Final Standard

Confronted by these various barriers, the fact that a consensus emerged from the BC standard-setting process is no small achievement. We would argue that a complex array of factors explains this somewhat improbable result.

In terms of realpolitik, we would argue that negotiation theory offers insights into why, particularly during the critical late stages of the process, key FSC-BC members continued to invest energy in the standard-setting process and were willing to make compromises in order to secure a deal. Environmental and First Nations groups had invested considerable time and energy into negotiating the D3 standard and had much to gain from its approval. Both chambers were aware that, with D3, they had achieved a good deal of what they wanted: environmentalists had secured substantial old-growth and biodiversity protection and First Nations had achieved significant recognition of indigenous rights and title interests. Their BATNA (best alternative to a negotiated agreement) (Fisher and Ury 1981) was to continue to pressure industry and governments on these fronts as before, but they would have had to do this without the benefit of a fully accredited FSC-BC standard. Large industry, which at this point consisted exclusively of Tembec, was in a more favourable position. It was interested in being FSC certified, but its BATNA was to seek certification under existing generic FSC standards or, failing that, to opt for SFI or CSA certification

However, we believe that the lessons for governance from the FSC-BC experience are potentially much more far-reaching and intriguing than those that emerge from a conventional analysis of bargaining positions and strategies. For one, we would argue that the spirit of compromise and flexibility that was exhibited as the lengthy and hands-on negotiation process wore on, and which was critical to the compromise that was forged, underscored the strength of the process in terms of social learning and trust building. By bringing representatives from diverse interest groups and backgrounds together in a shared, equitable, discursive environment to work toward a common goal, the negotiations enabled participants to learn from one another, facilitated the development of personal relationships that helped to bridge value differences, and (to borrow again from negotiation theory) allowed creative problem-centred solutions to be identified based on underlying "interests" rather than competing "positions."

One example of the benefits of this approach is the embedding of the "range of natural variability" (RONV) concept within the FSC-BC standard. RONV constitutes a progressive resolution of the potentially divisive positional debate between environmentalists and industry over the issue of clear-cut logging. RONV shifts attention to the underlying variation over time in a forest's natural condition, requiring management practices that fall within designated norms. In some cases, clear cutting will be acceptable because it will result in forest regrowth within the range of natural variation. In other cases, however, clear-cut logging will be substantially modified or replaced with various forms of selective, patch, and contour logging that preserve more of the forest stand and, ultimately, the treed landscape in order to be consistent with historical biological norms.

Another set of governance-related factors that worked in favour of reaching a settlement was the participants' sense of membership in the wider FSC collectivity: what we have earlier described as an emerging global forestry polity. Without these allegiances and the sense of belonging to something larger than oneself or one's organization, the corresponding desire to stick with the organization through its growing pains would not have been instilled, and the demands and pressures of the negotiation process could have easily prompted many of the key players to walk away. Indeed, in the wake of the Accreditation Business Unit's critical report on the preliminary standard, many participants did step back from the process, only to return later to "finish the job," a development that was in no small measure due to a sense of organizational loyalty and a recognition of the validity of (at least some of) the concerns that had been raised at the FSC-IC level.

Some might be tempted to discount the virtues we have just identified in the FSC-BC process and in the approach adopted by some of the negotiators (i.e., commitment, creativity, adaptiveness, pragmatism, and problem solving) as being borne of necessity, materializing only when it became apparent that the D3 standard would not be approved by FSC-AC. We do not entirely dismiss the notion that an assessment of the stakes influenced the final bargain, but we believe that the FSC-BC process provides an exemplar of a broader phenomenon in the environmental and natural resource management context.

There is strong evidence to suggest, particularly in Western liberal democracies, that the appetite to experiment with new approaches to long-standing and highly complex environmental and resource management challenges has never been greater. The intensity of the ideological war between defenders of traditional regulatory approaches and proponents of market-based alternatives has been subsiding, and there appears to be a growing sense that solutions must emerge from, and be driven by, a fuller understanding of the unique contextual features and dimensions of the problem at hand. Corollaries of this emerging problem-based approach to regulation and governance include recognition of the need for institutional forms to evolve to fit new functions, and of the need for a robust commitment to promote a cluster of institutional design characteristics, including adaptive management, social learning, transparency, inclusiveness, equity, and accountability.

Extrapolating from these trends, Freeman and Farber (2005) argue for a "modular" conception of regulation and resource management. This "modularity" metaphor is intended to convey the notion that regulatory arrangements – tools, structures, and relationships – should be provisional and rearranged periodically to suit evolving functional needs. In their words, "in its most idealized formulation, modularity supposes that both the tools and the governance structures with which we approach environmental regulation and resource management can be built, unbuilt and rebuilt" (798). According to the authors, the "theoretical payoff" of their modularity concept is that "it captures a moment of maturation in both administrative law and environmental law which has yet to be named and fully described. In this moment traditional forms of action and institutional structures are giving way to a 'problem focus' that calls for new arrangements, new strategies and new capacities" (804-5).

We contend that this modularity metaphor, though developed by Freeman and Farber to guide reform of state regulatory institutions and practices, is one that, appropriately interpreted, resonates and provides important insights into the FSC-BC case.[2] Although the FSC as a whole can be regarded as a governance "experiment," FSC-BC has a well-earned reputation for pushing the envelope. Few regional or national standard-setting initiatives rival that of FSC-BC, whether they are measured in terms of innovation, comprehensiveness, duration, controversy, or cost.

To the extent that a key metric of modularity is an ability to adapt governance and regulatory forms to requisite functions, FSC-BC deserves high marks. It is no coincidence that it was FSC-BC that pioneered the concept of a fourth, First Nations chamber and that several of its more prominent Aboriginal members have subsequently played a leading role within FSC-IC on indigenous issues. Likewise, more than most regional or national programs, FSC-BC demonstrated an appetite and capacity for innovation in standard setting, both *institutionally* (e.g., the creation of the Technical Advisory Team, which had responsibility for negotiating and crafting the language of much of D2, and the standards team, which acted as advisors on D3) and *conceptually* (e.g., developing and refining the concept of RONV, pioneering the use of "satisfaction" tests under P3, and the use of hybrid management/process standards under P6 and P9). Moreover, we would argue, in terms of demonstrating "maturity" and a pragmatic ability to focus on the problem at hand, there can be little doubt that, in the final analysis, at the eleventh hour, FSC-BC met the challenge.

As the senior level of management, FSC-AC also deserves some credit. We would argue that, despite significant structural constraints related to inexperience, lack of resources, and distance, FSC-AC has demonstrated an admirable organizational maturity and a capacity for creative, modular praxis. Two examples of this are its recent leadership efforts to develop and articulate a policy on the form and content of standards and its management (via preliminary accreditation) of the potentially explosive showdown with FSC-BC over approval of the D3 standard. Experience has provided FSC-AC with a body of evidence of what works and what does not, and that experience must now be interpreted and embedded in policies that guide national and regional standards development around the world. Building on this experience, the FSC system appears to be ready for further institutional innovation in the form of increased devolution, to the national level, of responsibility for membership, standards development, certification oversight, and fundraising. This devolution will occur under a more clearly articulated set of requirements for constituting such operations. Furthermore, and in response to criticisms from its members in the global South where conditions are so far much less favourable to FSC certification (Cashore et al. 2006), the organization is now examining a range of alternative and innovative arrangements for certification, including the possibility of "step wise" certification that would enable forest management operations to move through a graduated approach from adequate to good to appropriate forest management practices.

**The FSC-BC Final Standard in Comparative Context**

When we embarked on this project, one of our goals was to assess the substantive content of the FSC-BC final standard both on its own terms and in comparative context. Widely touted to be the most comprehensive and rigorous FSC standard of its kind, we set out to consider whether and to what extent this portrayal was justified.

As is discussed in detail in Chapter 11, an initial priority is to be clear about the nature of the measures being deployed in such an assessment. To this end, we contended that a threshold distinction must be made between three forms of standard: management, technology, and performance. Each of these forms of standard, in turn, represent interventions at different stages of the production process that are aimed at influencing firm behaviour. All three of these forms of standard, we would argue, can therefore be regarded as *prescriptive* to the extent that they fetter or constrain action at the firm level. It is difficult, if not impossible, to measure how and to what extent the standards will trigger alterations in firm behaviour. There are a number of reasons for this uncertainty. It can be hard to estimate the extent to which new, standard-driven practices will depart from existing practices dictated by extant legislated requirements. All firms have different operating practices, and some firms' current practices may be closer to those demanded by the standard than others. As well, firm behaviour can be influenced by factors relating to the precision of, and discretion associated with, the standard, which in turn depend on whether a standard is framed in quantitative or qualitative terms. For these and other reasons, no simple correlation can be drawn between the prescriptiveness of the standard and its cost implications for particular firms.

Mindful of these realities, we chose, in our comparative analysis, to eschew offering generalized conclusions about the relative prescriptiveness of the FSC-BC final standard versus its counterparts. Instead, our analysis of the standards examined in *Setting the Standard* focuses on whether and to what extent they provide clear and viable guidance to managers and on the form (management-, technology-, or performance-based) that this guidance takes. We will now offer some concluding thoughts on these themes in relation to the four subject-matter areas addressed in our analysis in Part 2.

## Tenure and Use Rights

Within this category, two issues stand out as particularly thorny in the development of the FSC-BC final standard, both of which are related to the central role played by the Crown in provincial forest management. The first concerned how to reconcile the BC system's unique reliance on volume-based (as opposed to area-based) tenure rights with the FSC requirement that forest managers demonstrate long-term rights to forestlands for which they are seeking certification. The second arose as a consequence of the key role

British Columbia's chief forester plays in establishing the allowable annual cut (AAC) across the vast majority of the province's forested land base.

In British Columbia, volume-based tenures constitute almost 50 percent of the province's AAC in a jurisdiction where the Crown owns approximately 95 percent of the forestland. The significance of volume-based tenures to the province's forest industry, coupled with the absence of a clear connection between this form of tenure and a definable geographic land base, presented drafters of the BC standard with a difficult challenge that was not confronted in other jurisdictions where area-based licences and private holdings are the predominant forms of tenure (which is the case with our comparator jurisdictions). As is recounted in Chapter 6, the drafters considered formally excluding volume-based tenures from FSC certification, but this approach was rejected for two reasons: it would have unduly restricted the FSC's potential growth within the province, and it would have had a particularly adverse impact on small operators. In the final analysis, however, we would argue that the proposed solution (which requires direct provincial government involvement in volume-based certifications) falls short of ensuring against either of these outcomes and is likely to remain a vexing issue for FSC-BC for the foreseeable future.

The second issue, related to AAC, had much broader relevance to other FSC regions. In British Columbia and other jurisdictions where timber is harvested on public lands, the rate of cut is established by the government; on private lands (in British Columbia and elsewhere), this determination rests with the forest manager. In both contexts, FSC Criterion 5.6 specifies that the "rate of harvest ... shall not exceed levels that can be permanently sustained." Predictably, how and to what extent this criterion should be addressed in the BC standard was a source of considerable contention. In grappling with this task in jurisdictions where the AAC is set by government, a central question is one commonly confronted in voluntary standard setting: how do you bind government to an arrangement to which it is not (and does not wish to be) a party?

The FSC-BC final standard and most of our comparator standards tackle this dilemma by requiring the forest manager to justify the government-established cut rate, essentially employing the same methodology and techniques used by those responsible for AAC calculations in government. Thus, like its comparators, the FSC-BC final standard adopts a management-based form of standard. However, it goes further than its counterparts in one key respect by requiring the manager to demonstrate that "present and projected annual timber harvest rates ... do not exceed the projected long-term harvest rate." This performance-based standard, although it is not unique amongst our comparators, is framed in more precise and auditable terms than elsewhere (see Table 6.3). And, as we note in Chapter 6, it is in direct conflict with prevailing provincial government policy and in some management areas could require an immediate drop in AAC of as much as 25 percent. How rigorously FSC certifiers will interpret and apply this obligation, and whether and to what extent the chief forester will seek to accommodate managers confronting this issue, remains to be seen.

### Community and Workers' Rights

Under community and workers' rights, the task confronting those charged with developing the FSC-BC final standard was to give meaning to FSC Principle 4, which provides that "forest management and operations shall maintain or enhance the long-term social and economic well being of forest workers and local communities." This broad obligation is expanded by criteria that require managers to give "local" communities opportunities to access employment, training, and other services; to incorporate social impact assessment in their planning processes; and to consult with people and groups "directly affected" by their operations (Criteria 4.1 and 4.4).

As is chronicled in Chapter 7, debate within FSC-BC during the standard-development process crystallized around two sets of issues: *definitions* (in particular, giving meaning to the adjectives "local" and "directly affected") and *entitlements* (in particular, determining how precisely worker and community rights should be specified).

For the most part, the comparator standards in our study largely eschew the definitional challenges inherent in Principle 4 and its associated criteria – particularly as they define the concept of "local worker" and persons "directly affected" by forest operations – and leave such interpretive judgments to the discretion of FSC certifiers.[3] In contrast, after much debate, the FSC-BC final standard addresses these definitional uncertainties directly. Intriguingly, "local worker" is defined using an objective standard (workers who reside within daily commuting distance of the forest operation in question) and "directly affected" is defined subjectively (persons who consider themselves to be directly affected).

The FSC-BC final standard likewise departs from our comparators in terms of the detail with which it articulates the entitlements of local workers, communities, and directly affected persons. For the most part, the comparators contain little more than a recapitulation of the language of the relevant principle and criteria, once again, in this context, leaving broad interpretive discretion to certifiers. In contrast, the FSC-BC final standard (and, to a lesser extent, the FSC-Boreal standard) imposes a range of largely performance-based obligations on managers, governing their relationships with both local workers and directly affected persons. Although these obligations are, for the most part, framed in terms that preserve the broad interpretive discretion of certifiers, the FSC-BC approach does set a more rigorous standard than its peers, particularly with respect to dispute resolution in relation to directly affected interests.

## Indigenous Peoples' Rights

We have argued that, in various respects, the context within which standard development occurred in British Columbia was unique; this is particularly so with respect to indigenous rights. As Chapter 8 discusses in detail, British Columbia stands alone in Canada and, indeed, the Commonwealth in the uniqueness of the challenge it faces to achieve a meaningful and lasting reconciliation of what are still largely undefined and indeterminate Aboriginal rights and title. Arguably, no issue looms larger and is more important to the province's future.

Across a range of issues, the FSC-BC final standard reflects a broad-based recognition of the desirability of vesting in First Nations the ability to control whether and how forest development occurs on their traditional lands. In this respect, the made-in-BC standard, more than any other comparable standard, literally translates the unequivocal language in Criterion 3.1 that "indigenous peoples shall control forest management on their lands and territories unless they delegate control with free and informed consent." Moreover, it seeks to achieve this result through a relatively sparse yet highly precise set of performance-based standards that impose on managers the obligation to provide evidence that potentially affected First Nations are satisfied that their Aboriginal rights have been both recognized and respected.

Of the comparator standards examined, only the Boreal standard approaches BC's rigour in terms of the evidence of indigenous support for proposed forestry operations required as a prerequisite to certification. Likewise, only the Boreal and the BC standards require managers to take steps to promote indigenous participation in forest management and development.

Perhaps the most notable feature of the evolution of this aspect of the BC standard is the extent to which, through the iterative drafting process, it moved away from a management-based standards model (prescribing in detail the procedures managers should employ in their dealings with First Nations) to a performance-based one that is focused on outcomes (ensuring affected First Nations' satisfaction with proposed forestry activities).

## Environmental Values

By far the most controversial and daunting challenge confronted by the drafters of the FSC-BC standard was achieving consensus on provisions relating to environmental values. It was in this context that the most heated and divisive debates took place between industry interests and environmentalists. It was also in this context that we can observe the biggest differences between the final standard and its D3 predecessor, most notably in the elimination of the twenty-nine major failures that were a core component of large industry's critique of the D3 text.

Relative to our comparators, we concluded that the BC final standard is more comprehensive in scope, articulates more explicit performance expectations, requires more elaborate consultative procedures, and demands significantly more data collection on the part of managers (Chapter 9). At the same time, it is worth remembering that the FSC standards developed in our two US comparator jurisdictions (the Pacific and Rocky Mountain standards) were designed primarily to be employed on private lands. Given the predominant role of public forests in the BC context, and the prevalence of high conservation values, particularly in BC coastal forests, a significant level of deviation between a BC standard and its US counterparts would be expected.

Also worth underscoring is the extent to which the final standard departs from its predecessor versions. Compared to D3 and earlier drafts, the final standard is considerably more concise and flexible. Indeed, we anticipate that managers will welcome its generally qualitative as opposed to quantitative approach to establishing performance measures. It also rates highly for the innovative manner in which it employs hybrid management-/ performance-based forms of standard and for its development and deployment of new approaches and concepts, most notably RONV, which aim to preserve managerial autonomy while ensuring accountability for outcomes (Chapter 11).

In short, although the FSC-BC final standard is the most comprehensive articulation of an ecosystem-based model of forest management – not only relative to our comparators but also within the broader FSC "family" – in the final analysis its approach to leveraging change to forest practices must, perhaps to the disappointment of some of its earliest and most fervent supporters, be judged incrementalist as opposed to revolutionary.

## Looking to the Future: Challenges and Priorities for the FSC in the Wake of the FSC-BC Saga

We conclude by reflecting on the implications of the BC case for the FSC as an experiment in global democratic corporatism. Earlier in this chapter, we contended that one of the reasons a consensus standard finally emerged in British Columbia was the capacity of the organization to engage in a modular, non-ideological approach to governance and regulation. In this concluding section, in keeping with the modularity metaphor, we offer some thoughts on the principles and values that the FSC must enshrine to ensure its continuing capacity to adapt and thrive in the face of complex, ever-evolving external and internal demands and pressures.

The FSC has invested heavily in developing sophisticated governance arrangements, but it has not made a comparable investment on the regulatory front. We would argue, therefore, that a key priority for the FSC as it moves forward is the need to focus more squarely on its regulatory mission. This need is particularly compelling in light of the ongoing work associated with harmonization, the need to adapt the FSC's approach to make it fully workable in developing countries, and the expectation that regional and national standards will be revised on a rolling, five-year basis.

In tackling this challenge, the FSC has the opportunity to benefit from a close examination of standards development in British Columbia. There is no question that this was a long and arduous process. We have chronicled and analyzed many of the reasons why this was the case. Several of these factors relate to the novelty of the challenge confronted by drafters of the standard and to the competing normative visions of what the standard should promote. We would argue, however, that the process was unnecessarily prolonged and complicated by the FSC's inability to provide clearer up-front guidance on standard setting.

To be effective, standard-setting processes must move forward on the basis of a shared conceptual framework and vocabulary. To a large extent, this was absent during the FSC-BC standard-development process. As a result, debates about the standard tended to be refracted through the distorting lens of the supposed dichotomy between prescriptive- and performance-based regulation.

Standard setting, whether it takes place under the auspices of the state or a voluntary regime, invariably represents a prescriptive intervention in the process of production. All three forms of standard (management-, technology-, and performance-based) entail elements of prescription. What distinguishes these differing forms of standard is the stage in the production process at which they intervene. Likewise, although the degree to which a standard constrains managerial discretion can vary, all three forms of standard have cost implications for business, and it is not possible to divine the extent of these costs in the abstract. Thus, in general, the form of a standard does not constitute a principled basis for preferring one type over another.

Standard setters must approach the task of determining what form of standard is most appropriate in any given context with an open mind. In general, in a complex, fast-changing, and highly differentiated environment, we would argue that management- and/or performance-based standards will be more suitable than technology-based standards. Beyond this general observation, however, it is counterproductive to entertain any preconceptions about the relative merits or suitability of a particular form of standard. In keeping with the modular approach to regulation and governance discussed earlier, effective standard setting is a creative enterprise that involves adapting form to function in a flexible, context-driven fashion.

Encouragingly, albeit belatedly, it would appear that the FSC is realizing the importance of showing leadership on this front. In its 2004 "Policy on the structure and content of standards," there is no reference to what we would deem the analytically unhelpful concept of "prescriptiveness" that figured so prominently in the Accreditation Business Unit's critique of the FSC-BC

D3 standard. As we discussed in Chapter 9, the policy evinces a preference for performance-based standards, but it recognizes that management-based standards can perform a useful regulatory role when they are linked to "an expected outcome" and "evidence of implementation." This said, we would argue that much work remains to be done at the FSC-AC level. The policy provides a starting point for the FSC discussion of these critical issues to ensure consistency and coherence throughout the FSC on a global scale, but a comprehensive elaboration of the rationale and nature of the organization's approach to standard setting, together with an implementation strategy that addresses logistical questions (including harmonization and standards review), are long overdue.

In the governance arena, there is much the FSC – and many others – can learn from the BC case. Overall, the systems put in place by the FSC in 1993 withstood the test imposed on them by the FSC-BC imbroglio. Participants in various networks – environmental, social, indigenous, economic – with the assistance of national and head-office staff, were able to prevent a complete breakdown in negotiations and, with preliminary accreditation, found a way to continue to involve large industry without alienating other interests. To date, preliminary accreditation remains a "made-in-BC" solution – the Brazilian and Papua New Guinean standards that were cited as possible candidates for the new procedure have not taken it up – but it constitutes a useful tool in the FSC armoury. As well, the process by which it was negotiated highlights the strengths of FSC's formal and informal governance arrangements. These are, as we have underscored, unique in the global system and constitute a very different and intriguing vision of the possibilities of global democratic governance in which duly constituted civil society organizations like the FSC aggregate relevant interests in the economic, environmental, social, and indigenous peoples spheres and negotiate the standards of production by which they agree, voluntarily, to be bound. There are many other sectors where such organizations could make an important contribution – almost all commodities appear to be governable via some form of stewardship council – and many other economic sectors are likewise amenable to such measures (Meidinger 2006b).

Indeed, in several sectors beyond forestry and fisheries, new stewardship schemes are emerging. One sector that has made significant progress down this path is mining, where WWF Australia has taken the lead in bringing together a diverse range of companies under its Mining Certification Evaluation Project (MCEP). In its final report (MCEP 2006), MCEP concluded there was "scope for a scheme for third-party certification of mine sites" and identified the need for "appropriate governance" arrangements in several areas, including structures and procedures, standards and assessment, and dispute resolution (47). Subsequently, the parties established an Initiative for Responsible Mining Assurance, which is now exploring the possibility of establishing and institutionalizing "a voluntary system to independently verify compliance with environmental, human rights and social standards for mining operations" (IRMA 2007). Tourism is another sector where the stewardship model shows promise. For some time, the Rainforest Alliance has been working to develop a Sustainable Tourism Stewardship Council (STSC) "to promote globally recognized, high-quality certification programs for sustainable tourism and ecotourism through a process of information sharing, marketing, and assessment of standards" (Rainforest Alliance 2003). In its 2003 final report, the Rainforest Alliance envisioned a three-step process to tourism certification: establishment of an STSC network, followed by an STSC association, and finally an STSC accreditation system. Finally, discussions continue within the coffee sector on the desirability of uniting the various existing schemes within a single Coffee Stewardship Council. At a 2006 "brainstorming workshop" organized by the United Nations Conference on Trade and Development (UNCTAD) and the International Institute for Sustainable Development (IISD), participants concluded that one proposal holding "particular promise as a means for improving sustainability in the sector" was a suggestion to establish a multi-stakeholder process for developing a broad-based international strategy, and potentially a standard, for sustainability across the coffee sector (UNCTAD/IISD 2006).

The BC case can be interpreted as highlighting some of the strengths of FSC's democratic institutional arrangements; yet, it is also true that national organizations could usefully have greater guidance on their establishment and on the relationship between national and sub-national bodies. In 1998, all that existed to guide FSC-BC and FSC-Canada was the draft "National initiatives manual," which was silent on many of the critical organizational issues related to statutes, finances, and setting up sub-national chapters. It also contained a frankly confusing account – in part drawn from FSC-IC's own statutes – of the relationship between an FSC contact person, working group, board, and national initiative. In addition, although the formal governance structure of the FSC is relatively clear, the organization should pay more attention to its informal governing networks, especially those for certifying bodies and donors. We hope that our account of the FSC-BC case will put to rest some of the worst of the conspiracy theories – i.e., that the organization is run by either certifiers or donors – but these groups do emerge as bodies with considerable influence, and greater attention should be paid to their accountability to both FSC members and the board.

It is of some concern to see pressures growing within the FSC system to constrain its global democratic corporatist ethos. Increasingly, it is being contended that the FSC is "in the business" of providing certification services to the forest industry and is therefore competing with other certification service providers such as the ISO, the PEFC, and, at the national level, the CSA, the SFI, and other country-level schemes.[4] This view is true up to a point, but there is a danger that the managerial perspective underlying it will produce an exaggerated focus on FSC's service delivery mission and ignore its fundamental governance mandate of ensuring the equitable brokering of diverse social values in the interests of the global forestry polity as a whole. In contrast, we consider that the FSC needs to embrace global democratic corporatism even more wholeheartedly by establishing a larger "legislative" board, composed of between twenty-four and thirty elected representatives, that devolves day-to-day decision making to a nine-member executive committee. There are costs associated with this recommendation, but it is a truism that democracy requires resources – people, time, and money – and a variety of new constituencies may be persuaded to provide these resources if they are confident that the FSC is both democratic and accountable. Signing the "International NGOs accountability charter" and integrating it into a constitutional "bill of rights" would significantly bolster this confidence.

On 30 March 2006, a gala event was held in Vancouver to celebrate the accreditation of the FSC-BC final standard. A wide range of speakers was brought in to address the celebrants, including Brent Rabik of Alberta-Pacific Forest Industries, Lisa Matthaus of the Sierra Club of BC, Jim Carr of the Pulp, Paper and Woodworkers of Canada, and Chief Mike Retasket of the Union of BC Indian Chiefs. In his speech, Rabik observed, "Bringing together such a diverse group of interests to collectively develop and agree upon the B.C. standards is a major achievement and all participants should be very proud of their accomplishment. Forest companies now have a made-in-BC international FSC-accredited certification to demonstrate to customers globally that their forests are responsibly managed to the highest standard" (FSC-CAN 2006).

If the FSC can take root in a jurisdiction as fraught with challenges and complexity as British Columbia, it bodes well for its prospects elsewhere. Likewise, if the FSC can succeed in the forest sector, this suggests that the governance model on which it is based is robust enough to support analogous initiatives in a range of other tradable goods and services. However, the ultimate reach and implications of this governance model remain uncertain, depending not only on the fate of the FSC but also on whether a core premise of its global vision – that purchasers of products are, in fact, willing to act on their latent preferences – comes to fruition. Time will tell. Meanwhile, those interested in influencing the course of events have an opportunity to do so by voting, in the marketplace, for an ambitious and novel form of global democratic corporatist governance.

# Appendix

The accredited FSC-BC final standard and related documents can be accessed at http://fsccanada.org/BritishColumbia.htm.

## FSC INTERNATIONAL STANDARD
## PRINCIPLES AND CRITERIA FOR FOREST STEWARDSHIP

These principles and criteria of the FSC have been extracted from document FSC-STD-01-001 (version 4-0) EN. For the complete set of standards, including the introduction and glossary, see: http://www.fsc.org/en/about/policy_standards/princ_criteria.

### Principle 1: Compliance with laws and FSC Principles

Forest management shall respect all applicable laws of the country in which they occur, and international treaties and agreements to which the country is a signatory, and comply with all FSC Principles and Criteria.

1.1 Forest management shall respect all national and local laws and administrative requirements.

1.2 All applicable and legally prescribed fees, royalties, taxes and other charges shall be paid.

1.3 In signatory countries, the provisions of all binding international agreements such as CITES, ILO Conventions, ITTA, and Convention on Biological Diversity, shall be respected.

1.4 Conflicts between laws, regulations and the FSC Principles and Criteria shall be evaluated for the purposes of certification, on a case by case basis, by the certifiers and the involved or affected parties.

1.5 Forest management areas should be protected from illegal harvesting, settlement and other unauthorized activities.

1.6 Forest managers shall demonstrate a long-term commitment to adhere to the FSC Principles and Criteria.

**Principle 2: Tenure and use rights and responsibilities**
Long-term tenure and use rights to the land and forest resources shall be clearly defined, documented and legally established.

2.1 Clear evidence of long-term forest use rights to the land (e.g. land title, customary rights, or lease agreements) shall be demonstrated.

2.2 Local communities with legal or customary tenure or use rights shall maintain control, to the extent necessary to protect their rights or resources, over forest operations unless they delegate control with free and informed consent to other agencies.

2.3 Appropriate mechanisms shall be employed to resolve disputes over tenure claims and use rights. The circumstances and status of any outstanding disputes will be explicitly considered in the certification evaluation. Disputes of substantial magnitude involving a significant number of interests will normally disqualify an operation from being certified.

**Principle 3: Indigenous peoples' rights**
The legal and customary rights of indigenous peoples to own, use and manage their lands, territories, and resources shall be recognized and respected.

3.1 Indigenous peoples shall control forest management on their lands and territories unless they delegate control with free and informed consent to other agencies.

3.2 Forest management shall not threaten or diminish, either directly or indirectly, the resources or tenure rights of indigenous peoples.

3.3 Sites of special cultural, ecological, economic or religious significance to indigenous peoples shall be clearly identified in cooperation with such peoples, and recognized and protected by forest managers.

3.4 Indigenous peoples shall be compensated for the application of their traditional knowledge regarding the use of forest species or management systems in forest operations. This compensation shall be formally agreed upon with their free and informed consent before forest operations commence.

**Principle 4: Community relations and worker's rights**
Forest management operations shall maintain or enhance the long-term social and economic well-being of forest workers and local communities.

4.1 The communities within, or adjacent to, the forest management area should be given opportunities for employment, training, and other services.

4.2 Forest management should meet or exceed all applicable laws and/ or regulations covering health and safety of employees and their families.

4.3 The rights of workers to organize and voluntarily negotiate with their employers shall be guaranteed as outlined in Conventions 87 and 98 of the International Labour Organisation (ILO).

4.4 Management planning and operations shall incorporate the results of evaluations of social impact. Consultations shall be maintained with people and groups (both men and women) directly affected by management operations.

4.5 Appropriate mechanisms shall be employed for resolving grievances and for providing fair compensation in the case of loss or damage affecting the legal or customary rights, property, resources, or livelihoods of local peoples. Measures shall be taken to avoid such loss or damage.

## Principle 5: Benefits from the forest

Forest management operations shall encourage the efficient use of the forest's multiple products and services to ensure economic viability and a wide range of environmental and social benefits.

5.1 Forest management should strive toward economic viability, while taking into account the full environmental, social, and operational costs of production, and ensuring the investments necessary to maintain the ecological productivity of the forest.

5.2 Forest management and marketing operations should encourage the optimal use and local processing of the forest's diversity of products.

5.3 Forest management should minimize waste associated with harvesting and on-site processing operations and avoid damage to other forest resources.

5.4 Forest management should strive to strengthen and diversify the local economy, avoiding dependence on a single forest product.

5.5 Forest management operations shall recognize, maintain, and, where appropriate, enhance the value of forest services and resources such as watersheds and fisheries.

5.6 The rate of harvest of forest products shall not exceed levels which can be permanently sustained.

## Principle 6: Environmental impact

Forest management shall conserve biological diversity and its associated values, water resources, soils, and unique and fragile ecosystems and landscapes, and, by so doing, maintain the ecological functions and the integrity of the forest.

6.1   Assessment of environmental impacts shall be completed – appropriate to the scale, intensity of forest management and the uniqueness of the affected resources – and adequately integrated into management systems. Assessments shall include landscape level considerations as well as the impacts of on-site processing facilities. Environmental impacts shall be assessed prior to commencement of site-disturbing operations.

6.2   Safeguards shall exist which protect rare, threatened and endangered species and their habitats (e.g., nesting and feeding areas). Conservation zones and protection areas shall be established, appropriate to the scale and intensity of forest management and the uniqueness of the affected resources. Inappropriate hunting, fishing, trapping and collecting shall be controlled.

6.3   Ecological functions and values shall be maintained intact, enhanced, or restored, including:

   a)  Forest regeneration and succession.
   b)  Genetic, species, and ecosystem diversity.
   c)  Natural cycles that affect the productivity of the forest ecosystem.

6.4   Representative samples of existing ecosystems within the landscape shall be protected in their natural state and recorded on maps, appropriate to the scale and intensity of operations and the uniqueness of the affected resources.

6.5   Written guidelines shall be prepared and implemented to: control erosion; minimize forest damage during harvesting, road construction, and all other mechanical disturbances; and protect water resources.

6.6   Management systems shall promote the development and adoption of environmentally friendly non-chemical methods of pest management and strive to avoid the use of chemical pesticides. World Health Organization Type 1A and 1B and chlorinated hydrocarbon pesticides; pesticides that are persistent, toxic or whose derivatives remain biologically active and accumulate in the food chain beyond their intended use; as well as any pesticides banned by international agreement, shall be prohibited. If chemicals are used, proper equipment and training shall be provided to minimize health and environmental risks.

6.7   Chemicals, containers, liquid and solid non-organic wastes including fuel and oil shall be disposed of in an environmentally appropriate manner at off-site locations.

6.8   Use of biological control agents shall be documented, minimized, monitored and strictly controlled in accordance with national laws and internationally accepted scientific protocols. Use of genetically modified organisms shall be prohibited.

6.9   The use of exotic species shall be carefully controlled and actively monitored to avoid adverse ecological impacts.

6.10 Forest conversion to plantations or non-forest land uses shall not occur, except in circumstances where conversion:

a) entails a very limited portion of the forest management unit; and

b) does not occur on high conservation value forest areas; and

c) will enable clear, substantial, additional, secure, long term conservation benefits across the forest management unit.

## Principle 7: Management plan

A management plan – appropriate to the scale and intensity of the operations – shall be written, implemented, and kept up to date. The long term objectives of management, and the means of achieving them, shall be clearly stated.

7.1 The management plan and supporting documents shall provide:

a) Management objectives.

b) Description of the forest resources to be managed, environmental limitations, land use and ownership status, socio-economic conditions, and a profile of adjacent lands.

c) Description of silvicultural and/or other management system, based on the ecology of the forest in question and information gathered through resource inventories.

d) Rationale for rate of annual harvest and species selection.

e) Provisions for monitoring of forest growth and dynamics.

f) Environmental safeguards based on environmental assessments.

g) Plans for the identification and protection of rare, threatened, and endangered species.

h) Maps describing the forest resource base including protected areas, planned management activities and land ownership.

i) Description and justification of harvesting techniques and equipment to be used.

7.2 The management plan shall be periodically revised to incorporate the results of monitoring or new scientific and technical information, as well as to respond to changing environmental, social and economic circumstances.

7.3 Forest workers shall receive adequate training and supervision to ensure proper implementation of the management plan.

7.4 While respecting the confidentiality of information, forest managers shall make publicly available a summary of the primary elements of the management plan, including those listed in Criterion 7.1.

### Principle 8: Monitoring and assessment

Monitoring shall be conducted – appropriate to the scale and intensity of forest management – to assess the condition of the forest, yields of forest products, chain of custody, management activities and their social and environmental impacts.

8.1 The frequency and intensity of monitoring should be determined by the scale and intensity of forest management operations as well as the relative complexity and fragility of the affected environment. Monitoring procedures should be consistent and replicable over time to allow comparison of results and assessment of change.

8.2 Forest management should include the research and data collection needed to monitor, at a minimum, the following indicators:

a) Yield of all forest products harvested.
b) Growth rates, regeneration and condition of the forest.
c) Composition and observed changes in the flora and fauna.
d) Environmental and social impacts of harvesting and other operations.
e) Costs, productivity, and efficiency of forest management.

8.3 Documentation shall be provided by the forest manager to enable monitoring and certifying organizations to trace each forest product from its origin, a process known as the "chain of custody."

8.4 The results of monitoring shall be incorporated into the implementation and revision of the management plan.

8.5 While respecting the confidentiality of information, forest managers shall make publicly available a summary of the results of monitoring indicators, including those listed in Criterion 8.2.

### Principle 9: Maintenance of high conservation value forests

Management activities in high conservation value forests shall maintain or enhance the attributes which define such forests. Decisions regarding high conservation value forests shall always be considered in the context of a precautionary approach.

9.1 Assessment to determine the presence of the attributes consistent with High Conservation Value Forests will be completed, appropriate to scale and intensity of forest management.

9.2 The consultative portion of the certification process must place emphasis on the identified conservation attributes, and options for the maintenance thereof.

9.3 The management plan shall include and implement specific measures that ensure the maintenance and/or enhancement of the applicable conservation attributes consistent with the precautionary approach.

These measures shall be specifically included in the publicly available management plan summary.

9.4 Annual monitoring shall be conducted to assess the effectiveness of the measures employed to maintain or enhance the applicable conservation attributes.

## Principle 10: Plantations

Plantations shall be planned and managed in accordance with Principles and Criteria 1-9, and Principle 10 and its Criteria. While plantations can provide an array of social and economic benefits, and can contribute to satisfying the world's needs for forest products, they should complement the management of, reduce pressures on, and promote the restoration and conservation of natural forests.

10.1 The management objectives of the plantation, including natural forest conservation and restoration objectives, shall be explicitly stated in the management plan, and clearly demonstrated in the implementation of the plan.

10.2 The design and layout of plantations should promote the protection, restoration and conservation of natural forests, and not increase pressures on natural forests. Wildlife corridors, streamside zones and a mosaic of stands of different ages and rotation periods, shall be used in the layout of the plantation, consistent with the scale of the operation. The scale and layout of plantation blocks shall be consistent with the patterns of forest stands found within the natural landscape.

10.3 Diversity in the composition of plantations is preferred, so as to enhance economic, ecological and social stability. Such diversity may include the size and spatial distribution of management units within the landscape, number and genetic composition of species, age classes and structures.

10.4 The selection of species for planting shall be based on their overall suitability for the site and their appropriateness to the management objectives. In order to enhance the conservation of biological diversity, native species are preferred over exotic species in the establishment of plantations and the restoration of degraded ecosystems. Exotic species, which shall be used only when their performance is greater than that of native species, shall be carefully monitored to detect unusual mortality, disease, or insect outbreaks and adverse ecological impacts.

10.5 A proportion of the overall forest management area, appropriate to the scale of the plantation and to be determined in regional standards, shall be managed so as to restore the site to a natural forest cover.

10.6 Measures shall be taken to maintain or improve soil structure, fertility, and biological activity. The techniques and rate of harvesting, road and

trail construction and maintenance, and the choice of species shall
not result in long term soil degradation or adverse impacts on water
quality, quantity or substantial deviation from stream course drainage
patterns.

10.7 Measures shall be taken to prevent and minimize outbreaks of pests,
diseases, fire and invasive plant introductions. Integrated pest management shall form an essential part of the management plan, with
primary reliance on prevention and biological control methods rather
than chemical pesticides and fertilizers. Plantation management should
make every effort to move away from chemical pesticides and fertilizers,
including their use in nurseries. The use of chemicals is also covered in
Criteria 6.6 and 6.7.

10.8 Appropriate to the scale and diversity of the operation, monitoring of
plantations shall include regular assessment of potential on-site and
off-site ecological and social impacts (e.g. natural regeneration, effects
on water resources and soil fertility, and impacts on local welfare and
social well-being), in addition to those elements addressed in principles 8, 6 and 4. No species should be planted on a large scale until
local trials and/or experience have shown that they are ecologically
well-adapted to the site, are not invasive, and do not have significant
negative ecological impacts on other ecosystems. Special attention will
be paid to social issues of land acquisition for plantations, especially
the protection of local rights of ownership, use or access.

10.9 Plantations established in areas converted from natural forests after
November 1994 normally shall not qualify for certification. Certification
may be allowed in circumstances where sufficient evidence is submitted to the certification body that the manager/owner is not responsible
directly or indirectly of such conversion.

# Notes

## Chapter 1: Introduction

1 For an instructive discussion of the distinction between hard and soft law, see Kirton and Trebilcock (2004). To date, this distinction has enjoyed its greatest currency in the realm of international law. In this setting, laws are typically said to be "hard" to the extent that they impose generally applicable obligations, articulated with relative precision, that are in turn judicially enforceable under powers delegated by the state. Soft law is thus seen as a weakening (or softening) of the legal arrangements along the three dimensions of obligation, precision, and delegation: see Abbott and Snidal (2000, 422); Tollefson (2004, 93), and further discussion in Chapter 13.

2 By this juncture, it was generally believed that international efforts to improve forest management in the tropics – via the Tropical Forestry Action Plan and the International Tropical Timber Organization – had, in large measure, been a failure. Meanwhile, negotiations to establish a forest convention at the United Nations Conference on Environment and Development (UNCED) had yielded only a rather weak set of voluntary "forest principles" (formally known as the "Non-legally binding authoritative statement on principles for a global consensus on the management, conservation and sustainable development of all types of forest"). The very modest achievements of subsequent initiatives like the Intergovernmental Panel on Forests (1995-97), the Intergovernmental Forum on Forests (1997-2000), and the United Nations Forum on Forests (2000 to present) have further consolidated this view. For a recent review of global forest institutions, see Humphreys (2006).

3 Although timber boycott campaigns were highly effective in motivating retailers and consumers to consider alternative products, some ECSOs, including the World Wide Fund for Nature (WWF) and the World Conservation Union (IUCN), became increasingly concerned about their potential perverse consequences. By depreciating the price of timber and the value of the land on which it was grown, boycotts created a potential incentive to convert forestland to cash crop production.

4 For example, FSC Canada's by-laws provide for a board of between eight and twenty members drawn from four constituencies: Aboriginal, economic, environmental, and social (para. 9).

5 About 21 million hectares of this total are located in seven Canadian provinces. Currently, only about 1 percent of British Columbia's forestlands are FSC certified (just over 577 thousand ha). Significantly more forestlands are certified in other provinces, including Ontario (10.5 million ha), Alberta (5.5 million ha), and Quebec (4.5 million ha).

6 See Canada, External Advisory Committee on Smart Regulation (2004). This report later formed the basis for an ambitious smart regulation strategy announced by Canada's federal government in 2005. See also Coglianese, Nash, and Olmstead (2003).

7 For a seminal article on the need for such analyses, see Kersbergen and Waarden (2004).

8 See Tollefson (2004); Bernstein and Cashore (2004); Rhone, Clarke, and Webb (2004).

9 The FSC-BC final standard can be found at the following URL: http://www.fsccanada.org/SiteCM/U/D/48B4F585905BF469.pdf.

10  The terms "political dimension," "regulatory dimension," and "institutional dimension" are our own. Treib, Bähr, and Falkner (2005) employ parallel terminology that captures analogous phenomena: "politics," "policy," and "polity."
11  A full exposition of the nature and significance of this conception is provided in Chapter 13.
12  A new Swedish standard has since been negotiated (2005) to replace the 1998 standard, but it has not yet been accredited by FSC-IC.
13  In all cases, the versions of the standards selected were the first final versions negotiated. In the United States, a process was established in 2005 to review all US regional standards, with substantive changes likely in the provisions of the Rocky Mountain and Pacific Coast standards. Further details on comparators are provided in Chapter 5.
14  We conducted in-depth, semi-structured interviews with individuals intimately familiar with or directly involved in the FSC-BC standard-development process at all levels of the FSC-AC (see References for a list of interviewees).

### Chapter 2: The Rise and Rise of Forest Certification

1  Market-based instruments benefit those companies engaged in environmentally sustainable production, but other companies experience MBIs as disincentives or punishments. For example, because it increases costs relative to other, more sustainable, competitors, a carbon tax is experienced as a punishment by those companies that generate large carbon emissions.
2  Founding members of WARP were Scott Landis (writer), John Curtis (Luthiers Mercantile), John Shipstad (secretary/treasurer), Andrew Poynter (A&M Wood Specialties), Lew Lorini (Woodshop News), Dick Boak (ASIA Newsletter, C.F. Martin), Dick Jagels (University of Maine, Dept. of Forest Biology, and frequent contributor to *Wooden Boat*), Leonard Lee (Lee Valley Tools), Gary Hartshorn (director of science for World Wildlife Fund), David Ellsworth (president of the Association of Wood Turners), Ken Kupsche (Woodcraft Supply), Silas Kopf (furniture maker), Ivan Ussach (Rainforest Alliance), and Fiona Wilson (*American Woodworker* magazine) (see Luthiers Mercantile Catalogue 1993).
3  Members of the FSC interim board were Julio Centeno, Chris Elliott, Debbie Hammel, Dagoberto Irias, Dominique Irvine, Alan Knight, and Andrew Poynter. The interim board was broadly representative of what later came to be FSC's three chambers, with Centeno and Irias representing the South, Irvine representing indigenous people through her role in Cultural Survival Inc., Knight representing industry via B&Q (a major UK building supplies retailer), and Elliott the environmental movement through his role as a forest officer with WWF International. The interim board was assisted by Alan Pierce and Jamie Ervin and was run out of Burlington, Vermont, in the northeastern United States.
4  For details of the Enron and WorldCom collapses, see Unerman and O'Dwyer 2004.
5  Questionable certifications are subject to formal FSC investigation and decertification, as happened to companies like Perum Perhutani in Indonesia and Leroy Gabon in Gabon: www.FSC-watch.org.
6  To take a trivial example, consider how difficult it would be to use an electrical appliance if there were no national standards governing the size, shape, orientation, and voltage of electrical sockets. Consider also the advantages realized by a company that manages to establish its own product's specifications as the national or global standard.
7  The original eleven representatives were from Austria, Belgium, Czech Republic, France, Finland, Ireland, Norway, Portugal, Spain, Sweden, and Switzerland (PEFC 2006a).
8  George Fenwick of the American Bird Conservancy and Stephen McCormick of the Nature Conservancy.
9  Early members of the CSFCC included the BC Pulp and Paper Association, the Council of Forest Industries, and the Interior Lumber Manufacturers' Association (Cashore, Auld, and Newsom 2004, 268n8). Today, CSFCC has about twenty members, including most provincial forest industry associations (CSFCC 2007).
10  According to the CSA standard, "In this standard, use of the CCFM SFM criteria and elements as a framework for value identification provides vital links between local-level SFM and national- and provincial-scale forest policy, as well as a strong measure of consistency in identification of local forest values across Canada" (CSA 2003, 2).

11 These obligations are found in three different places in the main document: under public participation (Section 3), sustainable forest management (Section 6), and "continual improvement" (Section 7).
12 Such groups dispute the claims in CSA documentation that the 1996 version of the standard was developed using "an open and inclusive process" with a CSA SFM Technical Committee consisting of "timber producers, including woodlot owners, ... scientists, academics, and representatives of provincial and federal governments, as well as environmental, consumer, union and Aboriginal representatives" (CSA 2003, 1). They view these claims as half-truths, as major environmental groups effectively boycotted the process, while Aboriginal peoples were seriously under-represented. See Elliott (1999) for a detailed account of the process leading up to the establishment of the first CSA standard.
13 The CSA is embedded in Canada's National Standards System, two key goals of which are to "improve access to existing and new markets for Canadian goods and services" (SCC 2005, 5) and "build competitive advantage through technology and information transfer and global market intelligence" (6).
14 A 2003 report contains a description of the Fort St. John timber supply area (TSA) certification. The report notes that "the Fort St. John certification is unique in that the certificate will be issued to a group of participants operating in the TSA including: BC Timber Sales, Cameron River Logging Ltd., Canadian Forest Products Ltd., Louisiana-Pacific Canada Ltd., Slocan Forest Products Ltd., and Tembec Inc. ... The Fort St. John TSA CSA certification is being put in place by the participants of a results-based pilot project. The intent of the pilot project is to test ways to improve the administrative and regulatory framework of forest practices while maintaining high environmental standards" (CanFor 2003, 1).
15 According to Synnott, "staff of Coopers and Lybrand attended as observers, gathering material for the design of the Marine Stewardship Council. Following their recommendation, to avoid what were perceived as FSC's weaknesses, the MSC was established quite differently, not as an Association of members" (2005, 25).
16 Current certified fisheries include Alaskan salmon, Bury Inlet cockles, Loch Torridon nephrops, Mexican and Baja California spiny lobster, New Zealand hoki, Patagonian scallop, South African hake, South Georgia toothfish, South West mackerel (handline), Thames herring, and western Australia rock lobster. Fisheries currently undergoing assessment include British Columbia salmon, California chinook salmon, Gulf of California sardines, and Chilean hake (MSC 2007).

## Chapter 3: The BC Forest Policy Context

1 This extraordinary majority was massively reduced after the May 2005 election, when the Liberal Party of BC was returned to power with forty-six seats to the NDP's thirty-three.
2 A "normal" forest is one with a distribution of age classes that will allow an equal volume of timber to be harvested annually or periodically in perpetuity. The concept, which evolved in Europe, was embraced by North American foresters at the beginning of the twentieth century and became a cornerstone of "sustained-yield" forest management across the continent.
3 These realities are strikingly underscored by a recent decision of the BC Supreme Court: see *Tsilhqot'in Nation v. British Columbia*, 2007 BCSC 1700. In this decision, after one of the longest and most complex trials in provincial history, Justice Vickers opined that the Tsilhqot'in Nation had satisfied him that it had Aboriginal title to a large area in northern BC and that, in respect of these lands, the provincial Forest Act was of no force and effect: see http://www.courts.gov.bc.ca/Jdb-txt/SC/07/17/2007BCSC1700.pdf.
4 *Forest Act*, R.S.B.C 1996, c. 157.
5 Unlike area-based tenures, volume-based tenures do not delineate an area of land within which timber-harvesting rights must be exercised; rather, they identify a timber supply area within which licencees may harvest a specified volume of Crown timber annually.
6 Reforms introduced by the Liberal government under the Forest Revitalization Plan in March 2003 took back 20 percent of the AAC from all Crown forest tenure with an AAC over 200,000 cubic metres. Once this process has been fully implemented, the two major licence types will account for a much smaller proportion of the AAC – about 65 percent.

7  The Small Business Forest Enterprise Program originated in 1979 when a new *Forest Act* provided that a proportion of the allowable annual cut be made available for sale to small businesses, as defined by the act, through competitive auctions. The program was expanded in 1987, and by 2002 it accounted for 12 to 13 percent of the total provincial timber harvest. In 2003, the provincial government announced the introduction of the Market-Based Timber Pricing System. Under this initiative, the SBFEP was discontinued and the timber formerly sold under the program is now administered by BC Timber Sales and made available to all parties.

8  *Forest Act*, R.S.B.C. 1936, c. 102.

9  *Forest Act*, R.S.B.C. 1979, c. 140; *Ministry of Forests Act*, R.S.B.C. 1979, c. 240.

10  *Calder v. British Columbia (Attorney General)*, [1973] S.C.R. 313 *[Calder]*.

11  The protected areas strategy was an interagency initiative to develop a provincial system of areas protecting conservation, recreation, and cultural heritage values with the goal of designating 12 percent of BC as protected area by the year 2000 (Wilson 2001).

12  British Columbia's Forest Practices Board defines the landscape unit as "an area of land and water used for long-term planning of resource management activities. These units, which are typically 5000-400,000 ha in area, are important for designing strategies and patterns for landscape-level biodiversity and for managing a variety of resource values" (British Columbia Forest Practices Board 2007, 58).

13  Most notably *Haida Nation v. British Columbia (Minister of Forests)*, [2004] S.C.J. No. 70; upholding [2002] B.C.J. No. 378 (CA); reversing [2000] B.C.J. No. 2427 (SC).

14  The "New Relationship" agreement with the First Nations is discussed in more detail in Chapter 8.

15  Concerns regarding the future of the agreement remain. On 30 March 2007, US Trade Representative Susan Schwab requested formal consultations to address the validity of several Canadian government forestry programs and the Canadian government's implementation of export taxes and volume constraints as delineated in the agreement. In a public statement released the same day, David Emerson, Canada's minister of international trade, stated that the agreement reached in September 2006 anticipated such ongoing discussions, and that it provided a new framework in which parties could constructively discuss disagreements regarding the implementation of this complex document.

16  *Haida Nation v. British Columbia (Minister of Forests)*, [2002] B.C.J. No. 378 (CA) *supra* note 13. The matter was appealed further to the Supreme Court of Canada: see [2004] S.C.J. No. 70. There, the court confirmed the government's duty to consult but relieved Weyerhaeuser of such a duty. For further explanation of the duty to consult, see also *Taku River Tlingit First Nation v. British Columbia (Project Assessment Director)*, 2004 SCC 74; *Halfway River First Nation v. British Columbia (Ministry of Forests)*, 1999 BCCA 470.

17  *Forest and Range Practices Act*, S.B.C. 2002, c. 69.

## Chapter 4: Hard Bargaining

1  As of June 2005, these were FSC-Pacific Coast, FSC-Rocky Mountain, FSC-Northeast, FSC-Southwest, FSC-Lake States, FSC-Ozark-Ouachita, FSC-Mississippi Alluvial Valley, FSC-Southeast, and FSC-Appalachian. All of the US regional standards have been accredited by FSC-IC, with the exception of FSC-Mississippi Alluvial Valley.

2  Lara Beckett, personal email correspondence, 4 December 2005.

3  The month and year references in the text refer to the minutes of various FSC-BC and ISC steering committee meetings. These were originally published online on FSC-BC's website, but they are no longer available following the site's closure in 2006.

4  The process is described in the minutes as follows:

The representatives of each Chamber in turn described each of the candidates nominated from their respective chambers and made an initial ranking into three broad categories: (1) candidates who did not meet the criteria and qualifications; (2) qualified candidates; (3) candidates the chamber representatives highly recommended. Other members of the Steering Committee asked questions about the candidates and offered further information and observations as each chamber presented its report. In a general discussion following each chamber's report, the Steering Committee as a whole attempted to divide the qualified and highly recommended candidates into three groups: likely Standards Team members, possible members, unlikely members. Following the presentations and discussions of the candidates from all four chambers, the Steering Committee had a general discussion of the top four or five candidates in each of the four Chambers and of the mix of skills and experience they would bring to the Standards team. These were compared to the criteria set forth in the Steering Committee's documents regarding the Standards team. On the basis of this comparison, the Steering Committee then discussed each candidate still in contention and, in consultation with the chamber representatives concerned, gradually reduced the number of likely and possible Standards Team candidates.

5 Note, therefore, that the ST was appointed rather than elected, albeit from a panel nominated by chamber members.

6 Not all forest unions shared the IWA's adversarial stance. The IWA's main union competitor – the much smaller Pulp, Paper and Woodworkers of Canada (PPWC) – was much more supportive of the FSC and played an active role in the FSC-BC social chamber.

7 The two government representatives were Brian Murphy, deputy director and manager of forest development in the Forest Practices Branch, BC Ministry of Forests, and Rod Davis, acting director of the Resource Stewardship Branch, Ministry of Environment. It appears that two considerations drove the standards team's desire to have both government representatives at meetings: the ST valued the forestry and environmental expertise they brought to the table; and the ST rationalized that if only a single representative were permitted, the government would select Brian Murphy from the Ministry of Forests over Rod Davis from the Ministry of the Environment, resulting in unbalanced representation. More generally, although the FSC sought to minimize governmental influence by denying public officials membership in its foundation, it has softened its position over the years. In practice, government officials have played an instrumental role in standards development in several jurisdictions, including, notably, the development of FSC-Sweden's first national standard (see Chapter 5) and the brokering of the United Kingdom's Woodland Assurance Scheme (see Cashore, Auld, and Newsom 2004).

8 Financial constraints were one factor that exerted pressure on the SC to finalize negotiations on the D3 standard in April 2002.

9 FSC-Rocky Mountain completed its D2 in August 2000.

10 The sixth (D6) of eight drafts of the FSC-Pacific Coast standard appeared in January 2002.

11 This is clearly specified in the *National Initiatives manual:* "The Working Group must submit (a) a list of existing regional, national and local standards that were analysed and (b) the feedback on the standard obtained by the Working Group/Standards Writing Group from appropriate other Working Groups. Note, FSC generally expects that standards will be harmonized in 'an upwards rather than a downwards manner'" (FSC-IC 1998a, 12.3.1).

12 The purpose of major failure provisions is to signal to certifiers that some indicators are of such importance that if the operation does not meet them it should not be certified. Notably, the FSC-BC D3 standard contained twenty-nine major failure provisions, substantially curtailing certifier discretion in trading off excellent performance in some areas against poor performance in others.

13 The process is described in the issues analysis report: "The complete database file was made available to the Technical Advisory Team early in October for their use in preparing this report. Through a series of teleconference meetings, the Team agreed on the broad approach and the Table of Contents of the report and then allocated responsibility for preparing drafts of various components of the report to the four Team members. At a face-to-face meeting in Vancouver in late October the team reviewed and revised these drafts and assigned final writing responsibility and timelines" (FSC-BC TAT 2001, 2).

14 Personal interviews 2003.

15  Bourgeois' view of his own role confirms this perspective:

> One way was to say "no" to everything that I thought would not be broadly accepted; or the other one was to be as an advisor to the steering committee, relative to whether or not this would get endorsement by a significant amount of the industry. I chose the latter, because I felt that if I said "no" to everything, then that would be counterproductive to the process, and just total confrontation ... So I took the other approach, being a nice guy. But I would advise them and say, "You're going too far here" or, "This won't make it" ... and the answer would come back, "Well, is it OK with you?" "Is it OK for Lignum?" "Would Lignum buy into it?" I said, "Well, in some cases, yes, Lignum would, I think we could live with that, but that's not *everybody*. And just because Bill says it's OK, doesn't necessarily mean Lignum says it's OK, or that industry says it's OK." In almost every case, my advice was ignored ... And then when it came to the time when I had to make a decision, I said, "Look, I asked you to have these things analyzed, and I'm still in the same spot, and so I will have them analyzed." And I had those studies done, and I used them to inform my decision. (interview with Bill Bourgeois, 2003)

16  As one interviewee noted,

> that was essentially his [Bill Bourgeois'] position. He was signing it on behalf of industry, not on behalf of Lignum, and he in good conscience did not feel he could sign it, because of a couple of significant hitters in the standard ... The stand-level retention was a big hitter. The protected reserves is a big hitter. On the Coast, plantations is a huge issue. In the North, herbicides is a huge issue. We've got issues around demonstrating sustainability of cut like modelling, and I guess that ties in to process/procedure. If you look at process/procedure around this, it's very, very, very expensive to get your inventory and your analysis up to the state that the standard requires ... So there are some big hitters. P3 is a big hitter, a signed protocol, that's a huge hitter for some people, particularly when you look at operations in Williams Lake, where they're influenced by, in some cases, up to fourteen different First Nations. (personal interview 2003)

17  The members of this group included Tembec and Western Forest Products.
18  The NSAC's mandate was to provide advice or guidance on regional standards processes as appropriate; to advise the FSC-Canada working group on areas where the working group might wish to set core requirements for standards developed in Canada; to consult with regional standards groups; to seek technical, legal, or scientific advice on specific issues and advise the FSC-Canada working group on its findings; and to review and address issues raised by FSC-Canada staff. It was the staff's responsibility to monitor ongoing regional standards activities and to seek NSAC's advice as appropriate, to review all standards or revisions to standards, and to provide advice to the FSC-Canada working group (FSC-CAN 2002b).
19  As one interviewee noted:

> They [FSC-Canada board] had a role to play to have to accept and to have to decide on what to do with the [BC] Standard. What kind of recommendation do they make? Either send it back to BC, or are they prepared to accept this outcome and then pass it on through our international accreditation process? They were interested to understand this mediation process that took place, and how they got to the point of a seven-to-one vote on the BC side. They wanted to understand what they should do with this independent report [the impact and effectiveness study commissioned by Bill Bourgeois] that was going to be generated, where does it fit in their assessment of things? ... They had some discussion in April, they reviewed material, they had some more discussion over a teleconference to try to understand the nature of the concern expressed by the one negative vote ... With that, they prepared for a debate of the standard more intensely at the [board meeting held in conjunction with the] annual meeting in June, in Edmonton. We prepared a motion for them to begin that debate that was really framed around how can we propose either some amendment

to the standard to take account of this concern, that might satisfy the Canadian board finding consensus, and they had a concern largely because it was such a large segment of industry who would have to use this thing. At the end of the day, they approved the standard to pass on as it was to International with an abstaining vote and a negative vote. (personal interview 2003)

20 Of the eight-member board, only seven were actually present. Chris McDonnell, representing large industry on the economic chamber, voted against; John Wiggers, representing small industry, abstained; and the remaining five, representing environmental, social, and First Nations chambers, voted in favour.

21 Several interviewees disputed the use of the term *lobby* in this context, viewing the conversations they had with relevant actors in less pejorative terms, such as "consultations," "discussions," "clarifications," or "information seeking."

22 One interviewee commented:

> What was going to be recommended were preconditions: increase the industry buy-in, deal with major failures, deal with the fact that you have untested and unproven requirements in the standard – as requirements not as guidance. And so when that looked like the outcome – and knowing the environment we were going to face – whether the people in BC liked it or not, everybody was saying, "We're looking at another Maritimes project again." Where we may have a standard that – it's a good standard, pareto principle standard, it's way past the 80 percent mark – but has some significant issues remaining and it's not going to get used. (personal interview 2004)

23 Notably, however, as of 2007 FSC-BC remained the only jurisdiction to employ preliminary accreditation, suggesting that the Brazil and PNG cases were not as amenable to its provisions as was argued at the time.

24 Heiko Liedeker asked Jim McCarthy to outline the basic proposal at the first meeting, and this prompted some suspicion that a deal to this effect had already been agreed upon.

25 The reports are listed as "Riparian forest management for the protection of aquatic values: Literature review and synthesis," by M. Carver, PhD; "A comparison of the riparian protection approaches in the Pacific NW and British Columbia," by Ken Zielke, RPF, and Bryce Bancroft, RPBio; "Scientific input and level of caution used to develop recent riparian guidelines in the Pacific Northwest," compiled by Ken Zielke, RPF, and Bryce Bancroft, RPBio; and "Summary of riparian review comments," compiled by Greg Utzig.

**Chapter 5: Beyond British Columbia**

1 A detailed analysis of the evolution of the FSC-Sweden standard has been carried out by Elliott (1999), Elliot and Schlaepfer (2001), and Cashore, Auld, and Newsom (2004). This section draws heavily on their work.

2 Swedish environmental groups linked up with their counterparts in the United Kingdom and Germany to apply pressure on Swedish companies in their two critical export markets (see Cashore, Auld, and Newsom 2004, 197).

3 As Elliott notes:

> Membership was made conditional on a written declaration of support for the Forest Stewardship Council Principles and Criteria (FSC P and C), and a commitment to working constructively to prepare certification standards for Sweden within the FSC framework. At that time neither the forest owners nor the forest industry agreed to these requirements, so they did not join the group, which was made up of NGOs (WWF, SSNC, Friends of the Earth and Greenpeace), the Church and Skogssällskapet (the Forestry Society) ... In January 1996, the forestry companies collectively decided to join the working group under strong pressure from AssiDomän and Kornsnäs which had indicated that they would be prepared to join the group without other companies, if necessary ... This decision placed the forest owners in a difficult situation and by February, they finally agreed to participate in the working group. The group was formally constituted on February 15, 1996. (Elliott 1999, 382-83)

4  These were known, respectively, as the applications, market, environmental and biodiversity, social, and productive and economic subcommittees.

5  The social subcommittee, for example, identified these key fields: workers' rights, workers' training, local communities, rights of indigenous communities, outdoor recreation activities, and best available technologies. For fields addressed in other subcommittees, see Elliott (1999, 385).

6  These standards eventually became the Finland Forest Certification Council standards, which are now accredited to the PEFC.

7  One environmentalist viewed the departure of private forest owners associations from the FSC-Sweden standards negotiation process as "a huge strategic mistake" (see Cashore, Auld, and Newsom 2004, 207). Indeed, Cashore, Auld, and Newsom go on to argue that the failure to accommodate the concerns of Sweden's forest owners in the negotiations on the Swedish FSC standard precipitated the break, which then ineluctably led to the formation of the PEFC. While this may be true, dangers also existed in accommodating private forest owners. Such an accommodation would have required Sweden's environmental and social chamber members to compromise their own interests to a significant extent.

8  According to Elliott,

> this [exotic tree planting] was a difficult issue for SCA [a large Swedish forest products corporation] which in the past had planted significant areas with this species *[Pinus contorta]*. Even though only small areas were being planted in 1997, the company did not want to rule it out as an option for the future. In the final days leading up to June 18, 1997, and even in the final hours of June 18, there was a series of rounds of last-minute lobbying on all sides. The Swedish Forest Industries Association, believing that SCA would not be able to compromise on this issue was even ready to conclude that consensus could not be reached on the standards. The Federation was however put under heavy pressure by AssiDomän who indicated that they were prepared to agree to the standard even if no other company did. Finally, a compromise was found which satisfied all participants except Greenpeace which was opposed in principle to exotic species and fertilization, and therefore found itself unable to support the standard. (1999, 388)

9  This scheme was endorsed by the PEFC in May 2000 (PEFC 2004).

10  The commission of enquiry consisted of three members: Gemma Boetekees, then FSC contact person for the Netherlands and formerly of the Heart of the Wood Campaign, Friends of the Earth, Netherlands; Keith Moore, former chair of the BC Forest Practices Board; and Gregory Weber, a professor of law at the McGeorge School of Law, University of the Pacific in California.

11  The commission of enquiry's report (FSC-AC 2001a), coupled with Cashore and Lawson's in-depth comparative study of the Maritimes and US Northeast standard-development process (2003), are the key sources relied on in the following account of the FSC-Maritimes negotiation process.

12  D1-M was forwarded directly to about five hundred designated individuals and also posted on a website.

13  See Cashore and Lawson (2003, 24-32) for a detailed analysis of the FSC-Northeast standard-development process.

14  The review was carried out by Peter Duinker, a regional forestry expert, who concluded that the MRSC had not reached the "significant agreement" FSC-IC required of it due to J.D. Irving's continued dissent (Cashore and Lawson 2003, 21).

15  There have been few detailed studies of regional standard setting in the United States, with Cashore and Lawson's 2003 study of the US Northeast standard constituting a major exception. As far as we can ascertain, no formal detailed study exists of either the FSC-Rocky Mountain or the FSC-Pacific Coast regional standard-setting processes. The account presented here is based on information contained in the FSC-Rocky Mountain draft standards (FSC-RM 2000, 2001, 2004).

16  For example, a two-day meeting of the FSC-Rocky Mountain regional initiative was held at the University of Montana, Lubrecht Forest, in May 2000 to "revise the first draft of the regional

forest certification standards." There were sixteen public comments on this draft (D1-RM), including observations from working group members, Scientific Certification Systems, and SmartWood. The comments were based not only on D1-RM, but also on results of a "July 1999 field test on the Salish-Kootenai Tribal Forest in western Montana" (FSC-RM 2000).

17 Simpson (1995) defines this region as including "parts of five states in the lower 48 – California, Oregon, and Washington, Idaho, and Montana ... much of the Province of British Columbia ... and finally, the better part of coastal Southeast Alaska" (140).

18 Although we have not uncovered a detailed account of the Boreal process in the academic literature, the process itself generated a large amount of documentation, and we invited our interviewees to comment on it as a model of standard development, generating a variety of useful responses. This account is based on FSC-Canada documentation on the development of the Boreal standard, supplemented by interviews with key informants.

19 In addition to the "core standard with regional variation" model, McEachern studied "geographic/political," "biophysical," "Aboriginal-led," and "centres of expertise" models (McEachern 2000).

20 In discussing the advantages of the core standard model, McEachern (2000, n.p.) notes: "The development of a core standard for issues common across the boreal region could ensure the harmonization of boreal standards up front and avoid a lengthy and difficult harmonization process in the future."

21 Not all members of the ST concur with this hypothesis. For example, according to Jessica Clogg, by the time that the ST was wound down it had become deadlocked over a variety of "bracketed" issues that, due to their complex and political nature, could only be resolved at the FSC-BC steering committee (communication with the authors, 26 September 2006).

### Chapter 6: Tenure, Use Rights, and Benefits from the Forest

1 Volume-based tenures assign rights to the tenure holder to log a specified volume of Crown timber. As such, they are distinct from "area-based" tenures, which assign logging rights over a specified geographical area (see Chapter 3).

2 The remaining tenure types include a number of small, short-term licences designed to meet specific needs and timber licences that revert to the Crown when the old-growth timber they carry has been harvested.

3 Most provinces allocate the majority of their AAC to area-based tenures; six provinces allocate more than 80 percent of their AACs in this way, and three – Ontario, Quebec, and New Brunswick - allocate a full 100 percent. See discussion of the Boreal and Maritimes standards below.

4 *Forest Act*, R.S.B.C. 1996, c. 157, s. 43(3).

5 *Forest and Range Practices Act*, S.B.C. 2002, c. 69.

6 The long-run sustainable yield of a management unit can be estimated by multiplying the operable area of the unit (i.e., the area within which timber management and harvesting is planned) by the average volume of timber this area is capable of growing annually (i.e., the mean annual increment), taking into account operational constraints and planned rotation ages.

7 *Forest Act*, s. 8(8). Prior to an AAC determination, the area in question is the subject of a timber supply review, which includes an estimate of the area's long-term sustainable harvesting level and an analysis of projected timber supplies from the area in the short term (twenty years), medium term (twenty-one to one hundred years), and long term (two hundred years or more). Alternative AAC scenarios are investigated with reference to the timber supply and their environmental, economic, and social implications. In determining long-term harvesting levels and AACs, the actual size of the timber-producing land base of the area is reduced to account for such factors as areas set aside for non-timber uses, environmentally sensitive areas including riparian protective strips, areas that cannot be accessed economically, and areas that will be used for roads and other rights of way.

8 The FSC-BC Technical Advisory Team (TAT), which took over from the standards team after the release of the second draft of the BC standard recognized two pivotal, or "Tier 1," issues relating to Principle 2 – the appropriate way to deal with volume-based tenure and the definition of customary rights.

9    *Innovative Forest Practices Regulation* [BC Reg. 197/97 O.C. 694/97].
10   Details on forest certification progress in each of BC Timber Sales' twelve business areas is available at http://www.for.gov.bc.ca/bcts/.
11   *Water Act*, R.S.B.C. 1996, c. 483, s. 2(1); *Range Act*, R.S.B.C. 1996, c. 39, ss. 5 and 6; *Wildlife Act*, R.S.B.C. 1996, c. 488, s. 2(1).
12   *Allemansrätten* (the right of all men) is customary law in Sweden and other Nordic countries that allows the public, including non-citizens or non-residents, rights of access to all forestland – public or private – to hike through the natural landscape, camp for one night, and gather berries, mushrooms, and wild flowers, provided that there is no property damage resulting in economic loss and that the privacy of the landowner is respected.

### Chapter 7: Community and Workers' Rights
1    For a detailed microhistory of one of these towns, see Rajala 1993.
2    In 1993, the provincial government unveiled the protected area strategy (PAS) as part of its comprehensive land-use planning process. The objective of this strategy was to protect 12 percent of the province's area from industrial activity or development by the year 2000 in order to preserve viable areas representative of the province's main terrestrial, marine, and freshwater ecosystems. Although some argue that certain ecosystems are still inadequately protected, the PAS exceeded its goal, and today approximately 12 million hectares of the province's land base are fully protected from commercial activities.
3    These initiatives had their inception in the CORE (Commission on Resources and Environment) process, which in 1991 established four regional "tables," representing a broad range of stakeholders, to develop regional land-use plans for Vancouver Island, Cariboo-Chilcotin, East Kootenay, and West Kootenay-Boundary. The CORE process was not a resounding success and was superseded in the mid-1990s by local, multi-stakeholder Land and Resource Management Planning (LRMP) committees.
4    Under the 1996 Jobs and Timber Accord, the forest industry was put on notice that it was expected to create an additional 21,000 jobs (a 25 percent increase in its work force) by the year 2000 or face the possibility of losing its cutting rights. By 1998, the accord had suffered an early, but timely, death.
5    In 1991, the Forest Resources Commission (the Peel Commission) recommended that one-third of British Columbia's managed forestland should be held under long-term industrial tenures, one-third should be managed by a Crown corporation as a profit centre, and one-third should be in smaller Crown tenures held by First Nations, communities, and individuals in partnerships or small non-integrated companies.
6    The Technical Advisory Team (TAT) separated the issues relating to the structure and substance of the second draft of the standard into three categories or "tiers." Tier 1 issues were pivotal concerns that, in the opinion of the TAT, had to be resolved before the redrafting could proceed. Tier 2 issues reflected differences of opinion among commentators that had to be considered but, in the opinion of the TAT, were not central to reaching agreement on the standard. Tier 3 issues were matters relating to the structure of the standards document, rather than to the substance of the indicators and verifiers.

### Chapter 8: Indigenous Peoples' Rights
1    *Constitution Act, 1982,* being Schedule B to the *Canada Act 1982* (U.K.), 1982, c. 11, s.35.
2    So far, only two new treaties have emerged out the BC treaty process, both of which have now passed third reading in the BC Legislative Assembly and are awaiting royal assent. These treaties are with the Tsawwassen First Nation (located near the City of Vancouver) and the Maa-Nulth Treaty Society (an organization representing five First Nations located on central Vancouver Island). The only other modern-era treaty, with the Nisga'a Nation, was concluded in 2000. Negotiations leading up to the Nisga'a treaty, however, began before the BC treaty process was initiated and were concluded outside that process.
3    *Haida Nation v. British Columbia (Minister of Forests),* 2004 SCC 73; *Haida Nation v. British Columbia (Minister of Forests),* 2002 BCCA 462; *Taku River Tlingit First Nation v. British Columbia (Project Assessment Director)* 2004 SCC 74.

4   *Calder v. British Columbia (Attorney General)*, [1973] S.C.R. 313 *[Calder]*.
5   *Constitution Act, 1982*.
6   See *supra* note 2.
7   According to the Ministry of Forests, "FRAs can play an important role in a pre-treaty environment by providing increased economic opportunities for First Nations, building community capacity, increasing the level of trust among First Nations, government and other stakeholders, and providing a greater stability on the land-base prior to long term treaty settlements being finalized" (British Columbia Ministry of Forests 2004b, 4).
8   See Ministry of Forests and Range website http://www.for.gov.bc.ca/haa/FN_Agreements.htm (visited 29 February 2008).
9   Although, to date, it appears that some First Nations do not believe the FRO program goes far enough in addressing these concerns (UBCIC 2006).
10  *R. v. Sparrow*, [1990] 1 S.C.R. 1075; *R. v. Van der Peet*, [1996] 2 S.C.R. 507; *Delgamuukw v. British Columbia*, [1997] 3 S.C.R. 1010; *Haida Nation v. British Columbia (Minister of Forests)*, 2004 SCC 73.
11  The costs incurred by First Nations in treaty negotiations are typically funded as a repayable "advance" from the federal government against the final settlement that is achieved.
12  *Huu-ay-aht First Nation v. British Columbia (Minister of Forests)*, 2005 BCSC 697.
13  The 2007 decision of the BC Supreme Court in *Tsilhqot'in v. British Columbia* (noted in Chapter 3) is a case in point. In this case, after a lengthy trial, the court concluded that the plaintiff First Nation had filed evidence adequate to prove Aboriginal title and rights to much of its traditional territory, but it nonetheless declined to grant a binding remedy against the Crown based on narrow technical grounds.
14  See *supra* note 3.
15  The glossary definition states: "*Aboriginal Rights:* A practice, custom or tradition integral to the distinctive culture of the Aboriginal group claiming the right. Often Aboriginal rights, including site specific rights, can be made out even if title cannot: based on *R. v. Van der Peet*, [1996] 2 S.C.R. 507; *R. v. Adams*, [1996] 3 S.C.R. 101."
16  The glossary definition states:

> *Aboriginal Title:* The unique title to the First Nation's lands, territories and resources which arises from occupation before the assertion of British sovereignty, or which arises from and reflects the pattern of land holdings under Aboriginal law. Aboriginal title confers more than the right to engage in site-specific activities. What Aboriginal title confers is the right to the land itself. If a Nation has Aboriginal title, the land may be used for a variety of activities that need not be elements of a practice, custom or tradition integral to the distinctive culture of the Aboriginal group claiming the right: based on *Delgamuukw v. British Columbia*, [1997] 3 S.C.R. 1010.

17  See the FSC-BC preliminary standard glossary definition of "sites of special cultural, ecological, or religious significance," which provides the definition for Criterion 3.3. This definition was unaltered by the revisions that informed the final standard.
18  This flows from the operation of the definitions of "consulting with First Nations" and "joint management agreement," as set out in the glossary.

## Chapter 9: Environmental Values

1   The FSC-BC final standard definition of "principles of conservation biology" begins, "In the context of protected reserve network planning, applicable concepts from conservation biology include: complete ecosystem representation, protection of core habitats to ensure the maintenance of viable populations of all native species in natural patterns of distribution and abundance, sustaining ecological and evolutionary processes and the maintenance of a landscape that is resilient to environmental change" (FSC-BC 2005a, 77).
2   According to Dellert, the *normal*(ized) forest – consisting of fast-growing young trees in an approximately equal distribution of all age classes up to the crop rotation age – is integral to the achievement of sustained yield. As the growth rate slows down in the oldest age classes,

they are harvested, and the area is promptly regenerated with another crop. This cycling is repeated each year so as to maximize growth over time and ensure a crop of mature trees is always available for harvest (1994, 30).

3  Grumbine (1994) identified ten dominant themes of ecosystem-based management: hierarchical context, ecological boundaries, ecological integrity, data collection, monitoring, adaptive management, interagency cooperation, organizational change, humans embedded in nature, and values.

4  The Sloan Commission (Sloan 1945, 127) defined SYFM as a "perpetual yield of wood of commercially usable quality from regional areas in yearly or periodic quantities of equal or increasing volume" (quoted in Gale and Burda 1998).

5  West Coast Environmental Law Association conducted a detailed report that analyzed gaps in British Columbia's system of terrestrial protected areas (WCELA 1997). The report identified significant shortfalls in the province's most northern and northeast protected areas. Although significant improvements occurred over the past decade, gaps still remain. According to the Sierra Club of Canada: "Ten of 13 Forested biogeoclimatic zones have less than 12% protected and the Coastal Douglas Fir, Interior Douglas Fir and Ponderosa Pine zones have less than 5%. These [government] reports do not include the new Protected Areas announced in 2006 as part of the Great Bear Rainforest; however the three biogeoclimatic zones and nine ecosections with the least protection do not occur within this region" (Sierra Club of Canada 2007). Regions that are biodiversity poor tend to be over-represented; those that are biodiversity rich tend to be under-represented (British Columbia Ministry of Water, Land and Air Protection 2002).

6  This cost was attributed to more rigorous inventories, monitoring, and evaluation for a broader range of forest values.

7  At a generic level, FSC-AC (2004a) defines HCVF as forest areas that possess one or more of the following attributes:

   • They contain globally, regionally, or nationally significant concentrations of biodiversity values; this includes large landscape-level forests, within or containing the FMU, in which viable populations of most, if not all, naturally occurring species exist in natural patterns of distribution and abundance.
   • They fall within or contain rare, threatened, or endangered ecosystems.
   • They provide basic services of nature in critical situations.
   • They are fundamental to meeting basic needs of local communities and/or are critical to the traditional cultural identity of local communities.

8  FSC-AC (2004a) defines the "precautionary approach" as a tool for implementing the precautionary principle. In the BC context, where "the forest manager will often be required to act with incomplete knowledge of cause and effect relationships," the precautionary approach places the onus on the manager to protect and ameliorate threats to ecosystem function and integrity (FSC-CAN 2005, 76).

9  FSC-AC (2004a) defines "plantations" as "forest areas lacking most of the principal characteristics and key elements of native ecosystems as defined by FSC-approved national and regional Standards of forest stewardship, which result from the human activities of either planting, sowing or intensive silvicultural treatments." In the BC context, plantations are defined as mapped and designated areas of the FMU that are to be managed over the long term under a plantation management regime (FSC-CAN 2005, 74).

10  British Columbia has 1,138 species in total, comprising 488 species of birds, 468 species of fish, 142 species of mammals, 22 species of amphibians, and 18 species of reptiles (British Columbia Ministry of Water, Land and Air Protection 2003).

11  FSC-AC defines "biological diversity" as "the variability among living organisms from all sources including, *inter alia,* terrestrial, marine and other aquatic ecosystems and the ecological complexes of which they are a part; this includes diversity within species, between species and of ecosystems," referencing the Convention on Biological Diversity (1992) (cited in FSC-CAN 2004a, 3).

12  The standard for small operations (see note 17 below) references the glossary appended to the BC final standard.

13 In large forest areas (i.e., 100,000 hectares or larger).

14 The remaining attributes in the list are forest areas designated threatened or endangered at global, continental, or national levels; mature and old forest, where those age classes are becoming rare due to human activities; forest areas under-represented in protected areas; and plant communities designated endangered, threatened (Red List), or vulnerable (Blue List) by the BC Conservation Data Centre (FSC-BC 2005a).

15 These requirements are articulated in Appendix D, "High conservation value forest assessment framework." The six HCVF categories cover forest areas that (1) contain globally, regionally, or nationally significant concentrations of biodiversity values; (2) contain globally, regionally, or nationally significant large landscape-level forests contained within or containing the FMU, where the viable populations of most if not all naturally occurring species exist in natural patterns of distribution and abundance; (3) contain or are in rare, threatened, or endangered ecosystems; (4) provide basic services of nature in critical situations; (5) are fundamental to meeting basic needs of local communities; and (6) are critical to local communities' traditional cultural identity (FSC-CAN 2005, 100). The standard for small operations has a less-intensive HCVF assessment procedure than the main standards (FSC-BC 2005b, 101).

16 Small operations are subject to different standards, which are outlined in the stand-alone document "Forest Stewardship Council regional certification standards for British Columbia: Small operations standards" (2005b). In the BC context, "small operations" are defined as FMUs that meet the FSC-Canada definition of small and low-intensity managed forests (SLIMFs) or are less than 2,000 hectares in area (FSC-BC 2005b, 4). SLIMF operations must conduct a simplified HCVF assessment comprising nine questions to determine whether any of the six categories of HCVFs are present on the FMU, as outlined in the FSC-BC "Toolkit for small operations appendix" (45-54). In the BC context, FSC-Canada defines SLIMFs as less than 1,000 hectares in area or having an AAC that is less than 5,000 cubic metres and less than 20 percent of the mean annual increment of the productive forest area (4).

17 These requirements are not explicitly entrenched in the FSC-BC "Small operations standards."

18 Based on British Columbia's biogeoclimatic ecosystem classification (BEC) system.

19 Such an inventory collects information on stand structure, frequency and size of live wildlife trees and snags, presence of aquatic habitats, rare ecosystem features and/or other critical habitats, and soils.

20 To categorize ecosystems, BC uses the BEC system. The system is defined as follows in the FSC-BC glossary:

> A hierarchical system that organizes ecosystems at three levels of integration: site, regional and chronological. At the regional scale the system integrates climate, vegetation and zonal site classifications. The zonal or regional climate (reflected by vegetation and soil relationships) defines the basic biogeoclimatic unit, the subzone. Subzones are grouped into biogeoclimatic zones (based on similar climax tree species), and may be further subdivided into variants based on further refinements of climate (e.g., wetter, warmer, snowier), and the presence or absence of particular tree species. At the site level the most commonly used unit is the site series, defined as all land areas within a BEC subzone or variant with similar or equivalent physical properties that will produce similar plant communities at climax. Successional communities are grouped into a series of structural or seral stages for each site series to define the chronological level of integration. (FSC-BC 2005, 59)

21 This does not apply for small operations.

22 Plantations can be located in "previously harvested well-managed natural forest," followed by "unharvested, non old growth forest," and finally "old growth forest," but "only if none of the previous areas are available."

23 In the British Columbia context, RONV is defined as the "range of dynamic changes in natural systems in the last 2,000 years, prior to the influence of European settlers" and includes "consideration of the range of ecosystem conditions such as seral stage distribution,

patch size distribution, stand structure and disturbance regimes (i.e., frequency, intensity, spatial extent and heterogeneity of disturbances)" (FSC-BC 2005a, 77).

24  For example, where large openings are natural occurrences as a consequence of fire, wind, or snow.

25  See Indicators 6.3.4 to 6.3.12 of the final standard.

26  The FSC-BC final standard defines NDTs as follows: "An area that is characterized by a broadly homogeneous natural disturbance regime and range of natural variability; in BC the five NDTs listed below are defined by the FPC Biodiversity Guidebook 1995 (for further information see the Guidebook and also range of natural variability and compatible with natural disturbance regimes)." The five NDTs are NDT 1 – landscapes with rare stand-replacing events; NDT 2 – landscapes with infrequent stand-replacing events; NDT 3 – landscapes with frequent stand-replacing events; NDT 4 – landscapes with very frequent stand-maintaining fires; and NDT 5 – alpine tundra and subalpine parkland landscapes. Note, too, that the quantitative approach proposed in the FSC-BC final standard was a concern for British Columbia's large producers, and they commissioned two reports to assess the effects that requirements in the third draft of the preliminary standard would have on timber supply. One report indicated that the approach would have negative effects on operations not currently practising significant levels of stand retention (Zielke and Bancroft 2002). The other suggested that Indicators 6.3.8 to 6.3.11 would "dictate and constrain the ability of managers to carry out practices best suited to the site" and that "these indicators could have significant impacts on timber supply, depending on existing in-block retention levels" (Spalding 2002, 25). On the other hand, although environmentalists were mostly pleased with the approach taken in the final standard, some would have preferred an outright ban on clear-cut logging.

27  It is this new criterion that prompted Richard Donovan's objection to the FSC-BC final standard as discussed in Chapter 4. *Bis* is Latin for "second." It is used in the FSC-BC final standard to distinguish the new criterion dealing with riparian issues (Criterion 6.5*bis*) from the existing criterion (Criterion 6.5). The version included in "Small operations standards" was greatly simplified in order to make the requirements more acceptable to managers of small operations.

28  As outlined in Table 3 of Appendix B, "Requirements for riparian management" (FSC-BC 2005a, 94-95), and Table 1 of Appendix 2, "Requirements for riparian management on small operations" (FSC-BC 2005b, 58-59), respectively.

29  The frameworks require managers to define the riparian assessment unit; inventory the unit in terms of channel descriptions, aquatic habitats, sediment sources, terrestrial habitat, riparian condition, and natural disturbance; and ensure a complete riparian assessment is carried out by qualified specialists. The final steps in the IRA framework are to develop a riparian design with reserve zone locations, establish management zone strategies, define variables, and verify water budget thresholds, and to implement and monitor the regime's effectiveness in the context of adaptive management (FSC-BC 2005a, Appendix B).

30  For example, for streams classified as S1a (i.e., streams that contain fish or are a community watershed and are less than 100 metres wide), managers must establish a minimum RRZ of 15 metres and a minimum RMZ of 120 metres. For S6b streams (i.e., streams that do not contain fish, are not located in a community watershed, and are either (a) 0.5 to 3 metres wide, not in a domestic watershed, and less than 500 metres upstream of fish-bearing streams or (b) less than 0.5 metres wide), however, the RMZ drops from 120 metres to 20 metres, with 70 percent retention (FSC-BC 2005a, 94-95).

31  The western spruce budworm (*Choristoneura occidentalis*) is "a serious defoliator of interior Douglas-fir (*Pseudotsuga menziesii*)" (British Columbia Ministry of Forests 2004a, 14).

32  Since first detected in 1994, the epidemic has killed an estimated 249 million cubic metres of timber across 28 million acres of BC, with the potential to kill 80 percent of the merchantable pine (*Pinus* spp.) in the Interior before peaking in 2008 (British Columbia Ministry of Forests 2004a).

33  Aerial application of Foray 48B became an important public issue in British Columbia when there were significant protests in response to spraying in urban areas to control the spread

of gypsy moth *(Lymantria dispar)* infestations, especially around the provincial capital of Victoria.

34  In the final standard, Indicator 6.6.2 (Indicator 6.6.1 in the preliminary standard) was revised to increase the phase-out period from two to five years and to allow for the use of pesticides where there are no non-chemical alternatives. These changes were made to bring the final standard in line with FSC-IC policy. They also reflected the concerns of industry, outlined in the Spalding report, which regarded the original proposal of a complete prohibition of chemical use and the two-year phase-out period as going "beyond the benchmark established by FSC-IC" (Spalding 2002, 29).

35  There is a simplified version of Indicator 6.8.1 for small operators. It states that when biological agents are employed in the FMU, "the applicant documents, minimizes, monitors and controls their use in accordance with national laws and internationally accepted scientific protocols."

36  Countries with significant areas of established plantations include Australia, Brazil, China, New Zealand, and Ukraine. Countries that have established notable plantation-forest industries in the last quarter century include China, the United States, the Russian Federation, India, and Japan (Brown 2000).

37  Similar confusion meant it was inherently difficult to capture a simple and easily operational definition under Principle 10 in the BC context. The definition of "plantation" turned out to be one of the few technically weak links in the standards during field testing (Moore 2001).

38  In an early Forest 2020 report, Arseneau and Chiu (2003, 125) suggest that "it is possible, through the strategic use of fast growing, high-yield plantations on an area 1 percent the size of the commercial forest, to produce up to 20 percent of the current level of wood supply" in Canada (125).

39  Although this compromise appears to have satisfied ECSOs, industry's submissions on the preliminary standard suggested that the standards provided under P10 amounted to "a disincentive" and appeared "wholly inconsistent with the intent of the Principle" (Spalding 2002, 33).

40  Examples of ecosystem components are tree species diversity, stand diversity, stand structures and associated habitats resulting from pathogens/physical damage, early successional habitat, snags, mature and old trees, or coarse woody debris.

41  Examples of intensive land management practices/treatments are shortened rotation ages, sanitation treatments, even-aged/single-layer canopy management, broad-scale brushing/weeding, stocking control, mechanical site preparation, fertilization, or pruning.

42  Where such land is not available, then under Indicator 6.10.1, a manager must plant on previously harvested, well-managed forestland; only when both of these land types are unavailable on the FMU can the manager convert unharvested non-old-growth forest to plantation.

43  And the total area must not exceed 30 percent of the THLB of any single BEC variant within the FMU (FSC-BC 2005a, Indicator 10.5.1, 56).

44  In the first FSC-Sweden standard, the language of HCVF is not used. However, the standard does contain a special section on "montane coniferous forests," which it describes as having "a special character that is unlike our ordinary domestic forests. Montane forest on which forestry has had very little impact often has a high biodiversity value, and is very important for reindeer husbandry as well as having its own value in terms of the size of its total area" (FSC-Sweden 1998, Section 5).

45  For montane forest regions, the Swedish standard specifically defines "virgin-type forest" as "areas ... [which] are exempt from forestry measures with the exception of measures intended to promote natural biodiversity. All other types of key habitat are included here" (FSC-Sweden 1998, Section 5.1). The standard sets out a detailed specification of the requirements of virgin-type montane forests, which includes "no, or occasional felling stumps, a continuity of old windthrow, an abundance of wood fungi such as *Fomitopsis rosea, Amylocystis lapponica, Phlebi centrifuga*, an abundance of old-growth windthrow, often 15-20 per hectare. Major differences in age and stratification. Plenty of natural stumps and dead standing trees" (Section 5).

46  Indicator 6.2.1 provides that "at least 5 percent of the productive forest area is exempted from measures other than the management required to preserve and support the natural biological diversity of the habitat. Selection and demarcation areas shall be prioritized according to their importance for biodiversity and representativity in the landscape. (Exemptions may be made for landholdings of less than 20 hectares of productive forest land which have no areas that have, or may in the near future develop, high biodiversity values.) Measures to promote outdoor activities may be taken on condition that the biodiversity values are not harmed."

47  Section 6.5.1, states that "Natural regeneration, for example under shelterwood and seed trees, is used where this method will result in good regeneration of species of tree suitable to the site and to management goals"; and 6.5.5 states that "trees with high biodiversity value should be protected in all measures, and not felled. Cleaning and thinning are carried out in a way that protects to a reasonable extent potentially high biodiversity value trees."

48  The major provisions with respect to riparian management are contained in Section 6.3 on water management. Of these, Section 6.3.2 is most relevant, specifying that "measures relating to water courses and open water areas are normally planned on land while not snow-covered, and carried out in a way that avoids damage," and "roads across water courses shall be constructed and maintained to preserve in the long term the natural level and function of the river beds. As a rule, no new road ditches should flow directly into water courses."

49  Section 6.8.1, therefore, basically adopts existing Swedish law with respect to pesticides. It states: "Substances, including chemical pesticides and herbicides in the Chemicals Inspectorate's class 1 and 2 that are harmful to the environment and health shall not be used for the treatment of forest land. Permetrine treatment is currently exempted (until and including 1999). Plants are to be procured only from nurseries that attempt to minimize the use and effects of chemical pesticides."

50  P10 was formally adopted by the FSC in February 1996, and Criterion 10.9 was added in January 1999 following revisions to P9.

51  According to the revised standard, "plantations are cultivated stands that due to very intensive artificial regeneration and/or silviculture are devoid of most of the principal characteristics and key elements of native forests such as complexity, structure, and diversity. Consequently a substantial part of the natural species diversity is lacking and/or water and nutrient circulation processes are not natural" (FSC-Sweden 2005, 38). This recalls the debate in British Columbia during the D2 field trials over the precise definition of a plantation in the BC context.

52  Under Indicator 6.4.3, the owner/manager must "consult with provincial and WWF representatives to ensure that the selection, locations, and size of reserves are ecologically sound and beneficial to a provincial system of protected areas."

53  For example, taking into consideration the extent to which any given ecosystem (BEC variant) is protected outside the FMU.

54  Natural forest characteristics for ecosites are defined extensively in Appendix 1 of the FSC-M standard, "Natural forests of the Maritime region."

55  Appendix 3 of the FSC-M standard, "Watercourse buffer zone guidelines," requires riparian buffer zones over 30 metres wide "on all sides of all bodies of water and all watercourses with an average width greater than 1 metre" (this drops to 15 metres for riparian areas that are less than 1 metre wide). One-third of the buffer zone (alt: 10 metres) adjacent to the watercourse edge must be a "no logging zone," and forestry activities in the remainder of the buffer zone must "maintain and/or enhance the ability of the buffer zone to provide its natural functions" (48).

56  Indicator 10.1.1 in the FSC-M standard includes an interpretation note, which states: "Plantation in the FSC meaning of the word ... does not necessarily result from planting."

57  The natural forest of the region is outlined in Appendix 1. Indicator 10.4.1 states that exotic tree species may be introduced only after the manager provides clear evidence that this introduction is "limited to no more than 5 percent of an ecosite within a certified area."

58  The prohibition is reiterated in Indicator 10.7.2, which further requires that management "has demonstrated the steps which have been taken, and will be taken, to fulfil this commitment."

59  The FSC-M standard defines a "biocide" as "a toxic material with the potential of causing lethal damage to metabolic systems, and producing effects in all forms of living organisms in a more or less compatible range of exposure, or more generally, a substance potentially lethal to an organism, but not necessarily to all organisms" (29).

60  For example, the FSC-RM Glossary notes a number of characteristics of old-growth forests, including "numbers of large, live trees, canopy conditions, levels of decadence, minimum number and size of snags, and minimum quantities of large down logs and coarse woody debris" (FSC-RM 2001, 29).

61  "Even-aged management" is defined in the FSC-RM standard as "a system of forest management in which stands are produced or maintained with relatively minor differences in age" (27)

62  The FSC-RM standard includes a discretionary note on streamside management zones for "stream segments that support no fish and rarely contribute surface flow to other streams or other bodies of water" and normally have surface flow less than six months of the year (14).

63  Under Indicator 10.5.c, "where forestlands were previously converted to plantations," minimum required percentages are 10 percent for FMUs that are 100 acres or smaller; 15 percent for FMUs of 101 to 1,000 acres; 20 percent for FMUs of 1,001 to 10,000 acres; and 25 percent for FMUs larger than 10,000 acres).

64  Indicator 10.2.a states: "On areas already converted to plantations, even-aged harvests lacking within-stand retention are limited to forty acres or less in size."

65  Other resources may include data and lists generated by "researchers, local native plant societies, and experts, and other NGOs and government agencies" (FSC-PC 2005, 63).

66  Indicator 9.2.a states further that "information from stakeholders' consultations and other public review" must be "integrated into HCVF descriptions and delineations."

67  In these cases, the FSC-PC standard specifies that "harvest blocks in even-aged stands average 40 acres or less," with no individual cutblock larger than 60 acres, noting that although "some modified forms of even-aged management are certifiable under these standards ... many traditional even-aged management practices are not" (FSC-PC 2005, 50).

68  For example, Category A streams (capable of supporting native fish populations and/or providing domestic water supply) require an inner buffer that is at least fifty feet wide from the active high-water mark on both sides of the stream channel and is limited to single-tree selection silviculture, and an outer buffer at least 150 feet from the high-water mark, limited to single-tree or group selection silviculture (Indicator 6.5.o).

69  The explanatory note to the standard elaborates on this point in considerable detail. It states:

> For the proportion of the FMU being maintained in plantation management per 10.5.b, it is not expected that the management of the stands maintains or restores all levels of structure and composition associated with natural forests. Accordingly, some components of the first nine Principles and Criteria either do not apply or require modified interpretation when being applied to plantation forest operations. For the Pacific Coast region, indicators that do not apply to plantation forest stands are 6.3.e.4, 6.3.e.5, 6.3.f.3, and 6.3.f.4. Those that may require modification in interpretation, particularly for plantations located on agricultural soils, are 6.3.e.1, 6.3.e.2, 6.3.e.3, 6.3.f.2, 6.3.f.3, and 7.1.d.2. All other provisions of Principles 1-9 are equally relevant to natural forest and plantation forest operations. (FSC-PC 2005, 42)

70  "For plantations on soils capable of supporting natural forests, the average harvest opening is 40 acres, with a maximum opening of 80 acres" (Indicator 10.2.a). This is in contrast to an upper limit of 60 acres for openings under natural forest management (Indicator 6.3.f.4).

71  Upper limits for the total proportion of an FMU that can be managed under a plantation regime are 70 percent for FMUs from 100 to 1,000 acres; 60 percent for FMUs of 1,001 to 10,000 acres; or 50 percent for FMUs over 10,000 acres (Indicator 10.5.a).

72   FSC-Boreal defines pre-industrial forest as "1. A native forest, which has not been subjected to large scale harvesting or other forms of forest management; 2. A forest area such as existed prior to human settlement in the region occupied by the forest" (FSC-Boreal 2004, 141). Essentially, the pre-industrial forest is understood to be the forest that evolved before large scale harvesting began.

73   A note on the intent of Indicator 6.1.5 states that "the overall intent ... is to ensure that an understanding of the character of the pre-industrial forest is used as a basis, but not the sole basis, for setting management objectives related to the future forest."

74   Noting that "the amount of concern generated by clearcuts tends to increase with their size," the FSC-Boreal standard uses the manager's understanding of the PIC as the "basis for setting patch size targets and moving the forest in that direction" (74).

75   The FSC-Boreal standard specifies that "the applicant completes (or makes use of) a peer-reviewed scientific gap analysis to address the need for protected areas in the eco-region(s) and ecodistrict(s) in which the forest is situated" and uses it "to identify the location and extent of additional protected areas," which may be achieved using the WWF (or an equivalent) gap analysis methodology (Indicator 6.4.1).

76   Such harvesting, however, is "subject to public consultation and only to a limited extent based on a conservation or cultural rationale."

77   In Yukon, riparian guidelines outlined in Appendix 6 of the Boreal standard, "Yukon riparian guidelines," apply.

78   Indicator 6.10.2 requires that the "total area converted to plantations does not exceed 5 percent of the productive forest area."

79   In British Columbia, the main standards and those for small operators require that the total area converted to plantations does not exceed 5 percent of the THLB of the FMU and 10 percent of the FMU, respectively (FSC-BC 2005a, Indicator 10.5.1; FSC-BC 2005b, Indicator 10.5.1).

80   The ecological functions and values encompassed by C6.3 include such things as forest regeneration and succession; genetic, species, and ecosystem diversity; and natural cycles that affect the productivity of the forest ecosystem.

81   The FSC-M standard is similarly explicit with respect to old-growth protection and forest restoration.

82   Clear-cut logging may continue to be used in interior regions subject to large-scale stand-altering disturbances, particularly wildfires.

83   During harvesting, road construction, and all other mechanical disturbances.

84   Appendix 2, "Requirements for riparian management on small operations," of the BC standard for small operations outlines modified requirements for managers, including an ecologically based rationale for riparian reserve/management zones (FSC-BC 2005b, Indicator 6.5*bis*).

85   The FSC's plantation review process recently completed Phase 1 following widespread consultations with FSC members and interested parties. The policy working group submitted its report to the FSC board of directors in November 2006. Phase 2 has now commenced, with technical experts selected to participate on technical expert teams. The FSC board aims to submit final guidelines and tools for forest managers, together with any proposed changes to FSC's principles and criteria, for consideration by the 2008 General Assembly.

86   This is consistent with Hickey's (2004) findings that the average number of requirements documented in a sample of Canadian soft-law standards for sustainable forest management was generally higher than in the United States or Europe.

### Chapter 10: A Political Network of Analysis of FSC Governance

1   The neo-pluralist approach recapitulates elements of the Marxist recognition of the relative power of the bourgeoisie within capitalist modes of production, albeit without its associated focus on class struggle or belief in the merits of a socialized, centralized economy.

2   Sovereignty is best conceptualized as the mutual recognition by states of their and other states' rights to make and enforce all the laws within their territory without external interference.

3 The idea of the FSC generating a global forestry polity is further elaborated in Chapter 12.

4 See, for example, Canada, Natural Resources Canada (2006). There are many criticisms of the three-legged stool analogy. For a succinct summary of objections from a scientific perspective, see Dawe and Ryan (2003).

5 The signatories included a broad range of BC-based, Canadian, and international (USA) ENGOs, businesses (small-scale and consulting), First Nations, and labour organizations, including but not restricted to FSC-IC members.

6 Environmental and social elements mentioned include rare, threatened, and endangered species and forests; conversion to plantations; ecological integrity, biodiversity, and structure in managed forests; ecologically and socially responsible timber harvest levels; chemical pesticides and fertilizers; GMOs; benchmarks for field performance; indigenous peoples' interests; workers and forest-dependent communities; stakeholder interests in program governance and standards setting; and standards-setting procedures that prevent forest companies from modifying standards on a case-by-case basis (Joint NGO statement on forest certification 2003).

7 See http://www.wcel.org/goodwoodwatch/. Identified as a forest conservation (versus legal advocacy) organization on this particular website, curiously it is listed within the FSC as a member of the social chamber.

8 Wildsight (previously EKES) is a recent addition to this particular coalition (since 2005).

9 Friends of Clayoquot Sound is the only organization in this coalition without FSC membership.

10 "Valhalla" and "Raincoast" refer to the Valhalla Wilderness Society and the Raincoast Conservation Society, respectively.

11 Large-scale operations include such firms as Al-Pac, Domtar, International Forest Products, and Tembec. Small-scale operations include Iisaak Forest Resources and woodlot owners/operators like Fred Marshall and Allen Hopwood.

12 These include individuals (i.e., Patrick Armstrong of Moresby Consulting and Rachel Holt of Veridian Ecological Research) and woodlot associations that provide services to members.

13 For example, it appears that Bourgeois did not formally consult with a large-producers network about a negotiating strategy or report back on progress.

14 Large producers were concerned that D2 was "too long, too detailed, and too complex" (FSC-BC TAT 2001, 7), and they argued that the many measurable (versus generic) indicators made it too prescriptive. The TAT also noted that "the industry submission ... recommends thresholds not be set in standards per se, but that indicators describe a general outcome and set a process for determining specific thresholds on a management unit by management unit basis (usually with use of qualified specialists)" (12).

15 The report was one of two commissioned in 2002 by Bill Bourgeois, the large-industry representative on the FSC-BC steering committee, to inform his negotiating position on the emerging FSC-D3 standard. It concluded that D3 would have a negative impact on available AAC. See Chapter 4 for further details.

16 The FSC-AC described consultation on, and development of, this social strategy as their "most important strategic achievement of 2002" (see "A Letter from the Chairperson," FSC-AC 2002c).

17 More recently, the FSC-AC Plantations Review Working Group observed that, "while natural resources management is frequently addressed in the P and Cs, less is said about systems for social management, on promoting and managing the 'social contract,' on how to improve local conditions and on promoting local development" (FSC-AC 2005b). Pursuing a more coherent framework for promoting social values, social chamber representatives from diverse networks recommended a three-pronged approach, progressing from impact assessments to minimize negative effects, through provision of positive benefits (i.e., jobs and access to non-timber forest products), to activities promoting local development and increasing the local social capital. Beyond indigenous issues, a more comprehensive social agenda is now starting to take shape at the FSC-AC level. It considers the disruptive effects that ensue when established large-scale actors withdraw from areas and company land purchases

displace people, along with issues such as profit-sharing, corporate social responsibility, unions, workers' rights, and labour regulations (FSC-AC 2005b).

18  The social chamber contains only 17 percent of FSC-AC membership, split equally between North (56 members) and South (55 members) (see Table 2.1). (Note that at the FSC-AC level, indigenous peoples, including Canada's First Nations, are included in the social chamber.)

19  The social chamber has the least specific membership regulations. It is "limited to not-for-profit NGOs and assigned individuals with a demonstrated commitment to environmentally appropriate, socially beneficial and economically viable forest management." Based upon the FSC Membership List, October 2005 (Document 5.2.2), FSC-Canada and FSC-BC members constituted 15 percent and 9 percent, respectively, of FSC-AC social chamber members.

20  The IWA (Industrial, Wood, and Allied Workers of Canada) was vocally critical of the FSC-BC process, but has since endorsed the final standard.

21  In the FSC-BC standard-development process, Jessica Clogg has served in the social chamber as the lead on P1, a collaborator on P2, and a member of the standards team, the Technical Advisory Team, and the steering committee. At time of writing, she is the FSC-BC steering committee chair.

22  In the FSC-BC standard-development process, Greg Utzig served on both the social and environmental chambers as the lead on P6 and C5.6*bis*, a collaborator on P9, and a member of the unofficial drafting committee for D1, the standards team, and the Technical Advisory Team. He is generally credited for bringing the concept of RONV into the final standard (with technical advice from Rachel Holt of the economic chamber). He has also been involved with the FSC as a member of the FSC-AC drafting team for P9, an auditor, a technical person in drafting the FSC-Canada Boreal standard, and a technical advisor hired by the FSC-BC board of directors to review FSC-Canada/FSC-AC recommended changes to the preliminary standard.

23  Two other FSC regional initiatives have created a fourth chamber (Madagascar and Belgium), but in each case this was a government chamber for civil servants from the forest service (Hauselman 2002).

24  The same point can be made about the FSC-Boreal process, in which Canada's indigenous peoples chamber was equally active.

25  An eighth FSC certificate was awarded to a group of Vancouver Island woodlot owners through their manager, Ecotrust Canada, after the Abusow (2005) report was published. This is not represented in the data presented here but will have effectively reduced these proportions.

26  This First Nations-led joint-venture agreement between the central region Nuu-chah-nulth First Nations (Ma-Mook Natural Resources Ltd., 51 percent) and Weyerhaeuser (49 percent) was the first Aboriginal-managed forestry operation to be FSC certified in British Columbia in 2001 (Iisaak Forest Resources 2000).

27  Tsleil-Waututh First Nation created this corporate entity to manage forestry operations on their Inlailawatash lands (purchased in 2001) (SmartWood 2004).

28  FSC-Canada has collaborated several times with NAFA on outreach, consultation, organizing conferences, and developing policy papers for consideration, tapping into existing networks.

29  The IAC was initially established to provide input to FSC-Canada on issues related to indigenous peoples' rights so as to meet the FSC-AC timeline for the national Boreal standard-development process.

30  A. Peeling, "Review of the application of Principle 3 in the boreal forests subject to treaties and aboriginal rights (of First Nations, Innus and Metis)," legal opinion commissioned by FSC-Canada.

31  The IAC recommended collaborating with such organizations as NAFA, the Sustainable Forest Management Network, the Model Forest Network, and Convention on Biological Diversity-related groups, including the Canadian Indigenous Biodiversity Network.

32  David Nahwegahbow, Russell Collier, and Peggy Smith all played high-profile roles in Canada's indigenous network.

33  In addition to SmartWood, Woodmark Soil Association (UK) is a not-for-profit organization.

34  The focus of this section is on forest management certification rather than chain-of-custody certification. The thirteen companies were KPMG Forest Certification Services (Canada); Fundación vida para el bosque A.C. (VIBO) (Mexico); Scientific Certification Systems (SCS) (US); SmartWood (SW), a program of the Rainforest Alliance (US); Eurecertifor – BVQI Program of BVQI France (France); GFA Terra Systems (GFA) (Germany); Certiquality (Italy); Instituto per la Certificazione ed i Servizi per le Imprese dell'Arrendemento e del Legno (ICILA) (Italy); Institut für Marktökologie (IMO) (Switzerland):, Swiss Association for Quality and Management Systems (SQS) (Switzerland); Skal International (SKAL) (The Netherlands); Woodmark, Soil Association (SA) (UK); and QUALIFOR, SGS South Africa (SGS) (South Africa).

35  They have a particular interest in the criteria for obtaining FSC accreditation and the procedures for suspension and cancellation.

36  McDermott (2003) highlights the role trust and distrust played in the negotiation of the final standard. In her interviews with stakeholder groups, she finds considerable evidence of distrust of "outsider" CBs. Moreover, it is clear from her interviews that most CBs believed D3 set the bar too high to be implemented on a widespread basis.

37  SmartWood's chief consultant during the Tembec TFL14 preliminary certification audit in British Columbia was Keith Moore, who was previously recruited by the FSC-BC steering committee to manage the field testing of D2.

38  FSC-US has received significant foundation support totalling almost US$2 million. This figure does not include a grant of over US$2 million from the Pew Charitable Trusts in 2001 to support FSC-Canada's development of the national Boreal standard.

39  Total direct grants to the CFPC from the five major donors amounted to more than US$2.5 million between 1997 and 2005.

40  This does not include an additional US$50,000 that went directly to the Sierra Club of British Columbia for work on forest certification.

41  Northrop appears to have played a key role in the otherwise very loose administrative structure of the donors network (GrantCraft 2005).

42  Now the Certified Forest Products Council, led by David Ford.

43  The "organic" nature of this polity is partially controlled by the FSC. Applicants indicate their chamber preference on their application forms, but the FSC reserves the right to place them in another chamber if it deems this more appropriate. Notably, those with an economic interest in forestry are placed in the economic chamber based on a range of standard criteria.

## Chapter 11: A Regulatory Analysis of FSC Governance

1  See articles featured in a National Symposium on Second Generation Environmental Policy and the Law, published in *Capital University Law Review* 29, 1 (2001).

2  In its final report, in the face of strenuous advocacy by the ECSO community, the government's External Advisory Committee on Smart Regulation toned down its overt advocacy of performance-based regulation as presumptively superior to more traditional forms of regulation. See note 6 from Chapter 1.

3  This illustration is borrowed from FSC-AC's Accreditation Report (on file with authors).

4  Most American BAT standards give firms the option to be regulated under a quantitative performance-based standard; however, most firms opt to comply by adopting the BAT out of cost- and compliance-related concerns (Latin 1985).

5  A third area of criticism identified in the ABU's report was the vagueness of the language used in the standard, in particular its reliance on adjectival modifiers. Critics contend that such terminology requires that "value judgments" be made in the application of the standard; they say this is undesirable and should be avoided. In its response to this critique, FSC-BC committed to "reducing ... vague and unclear language." However, it noted that terms such as "appropriate, substantial, minimize and significant" appear in the FSC principles and criteria, suggesting that value judgments in the interpretation of these and other terms are an unavoidable feature of standards implementation. The ABU report is no longer available on the FSC-AC or FSC-Canada websites; copies are available on request from the authors.

6   Other factors include the firm's perception of the risk that non-compliance will be detected and the consequences (pecuniary and otherwise) of detection (Coglianese and Lazar 2003).

## Chapter 12: An Institutional Analysis of FSC Governance

1   Recently, for example, FSC-AC set up a Plantation Review Policy Working Group to consult FSC members and the wider forestry community about the FSC's approach to plantations. The working group, which consists of twelve members balanced between chambers and across the North-South divide, is engaging in global consultations prior to finalizing recommendations for FSC-AC's board of directors.

2   All citations to FSC statutes and by-laws in this chapter are to the following two foundational documents: FSC statutes (FSC-AC 2005d) and FSC by-laws (FSC-AC 2006). A thoroughgoing account of FSC's constitution would also involve consideration of the myriad policy statements, manuals, resolutions, and rulings issued by the General Assembly, board, and secretariat. Such an exercise cannot be undertaken here, however, and the chapter has a more modest purpose, which is to use the lens of comparative constitutionalism to examine the formal governance arrangements that are contained in FSC statues and by-laws.

3   The first lines of the preamble to the *Constitution Act, 1867,* are, "Whereas the provinces of Canada, Nova Scotia and New Brunswick have expressed their desire to be federally united into one dominion under the Crown of the United Kingdom of Great Britain and Ireland, with a constitution similar in principle to that of the United Kingdom."

4   FSC regional offices in Africa, Asia, Europe, and Latin America act as service centres for national initiatives and support FSC processes in countries without national initiatives. Canada, China, Russia, and the United States are also considered regions due to their size (FSC-AC 2007a).

5   For example, By-law 71a states that a contact person "shall collaborate with FSC to distribute information regarding the organization and its mission and to promote discussions on certification within the country or region concerned," but they provide little guidance regarding how contact people should be appointed or the nature of their relationship to national FSC members.

6   The tensions between national initiatives and FSC-AC are evident in a recent report by a self-constituted National Initiatives Task Force. Arguing that the FSC is a "tool" used by others to achieve their objectives, the task force makes far-reaching suggestions for change in FSC governance arrangements. At the centre of the task force's concerns, however, is the current FSC-AC revenue model and the absence of any requirements for FSC-AC to distribute centrally collected funds to national initiatives. A central recommendation of the task force is to replace national initiatives with field offices that would be funded centrally (FSC-UK National Initiatives Task Force 2005).

7   FSC-Canada's website no longer provides a list of members of the FSC-BC steering committee. Jessica Clogg is listed as co-chair of the FSC-BC steering committee on an FSC Canada media advisory, dated 30 March 2006, that announces a meeting of FSC members in British Columbia to celebrate FSC-AC's accreditation of the FSC-BC final standard. More recently, an FSC Canada newsletter solicited comments on a discussion paper, specifically "on whether and how certification should apply in regions of B.C. that have been severely impacted by the mountain pine beetle," noting that these "will be shared with members of FSC-Canada as well as the FSC-BC Steering Committee before any decision is taken to revisit the current FSC-BC standards" (FSC-CAN 2007).

8.  In the United States, for example, executive government (the president, secretaries of state, and the wider bureaucracy) is elected separately from Congress (the House of Representatives and the Senate), and the president is not permitted to introduce legislation directly into the Congress; instead, it must be sponsored by a member of Congress. However, in neither presidential nor parliamentary systems is the ideal separation of powers achieved. In parliamentary systems, power is fused because the prime minister and senior ministers hold seats in Parliament rather than being separately elected. In the United States, even though the separation of powers was an explicit concern of the founders, Supreme Court justices are nominated by the White House and appointed by Congress, not elected by the citizenry.

9  Paragraph 72 of the FSC by-laws requires that DRAAC be composed of one member from each of the following six regions: North America (including Mexico); Central and South America and the Caribbean; Europe; Australia and Oceania; Asia; and Africa.

10  Interestingly, the Change Management Team was set up during Maharaj Muthoo's brief period as executive director in early 2001. Despite severe criticism from many members, the team served an important function in identifying many of FSC's shortcomings as an institution.

11  This tension constituted the essence of the hit British television comedy *Yes, Minister*, in which a subversive bureaucracy, personified by Sir Humphrey Appleby, confronted a willing but incompetent political leader in the shape of Jim Hacker, minister of administrative affairs. *Yes, Minister* captured the culture of the "old" approach to managing public service in the 1950s-60s, which was dismantled in the 1980s-90s and replaced by "New Public Sector Management." This movement saw politicians reassert power over bureaucrats by increasing the number of political appointees, removing some of the protections in the contracts of senior civil servants, and hiring from outside the public service entirely. For further discussion of the new managerialism, see Considine and Painter 1997.

12  The Standards Committee was established in response to the Change Management Team's recommendations and was composed of David Nahwegahbow (social chamber), Grant Rosoman (environmental chamber), and Mok Sian Tuan (economic chamber).

13  The decision appears to have been taken by Muthoo alone, without consulting either the FSC-Canada or FSC-AC board. The resulting international furor within various levels of the FSC system prompted Muthoo's resignation only six months after his appointment as FSC-AC ED.

14  Not all commentators view the FSC's shift to Bonn, Germany, as positive, and there are indeed a number of pros and cons. On the positive side are lower rental costs, increased proximity to other international organizations, lower travel costs for European and African members, and greater possibilities of EU and German support. On the negative side is the perception that the FSC has backed away from its commitment to be sensitive to the interests of the South, the costs and disruption of the move itself, and the substantial administrative requirements of Germany's employment regulations.

15  The governmental process is complex, but in most parliamentary systems the following steps are involved: the cabinet agrees to introduce legislation in a particular area; a minister works with his political advisors and senior bureaucrats to draw up a bill and introduce it into parliament (First Reading); there is a Second Reading in which the bill is debated and amended by parliament, either in plenary or committee; there is a Third Reading to confirm final content, consideration by an upper house, possible amendment, and proclamation by a governor general or equivalent; and the new legislation is published in the *Gazette*, establishing the date it comes into force.

16  General Assembly motions are either statutory (and binding) or policy (and advisory). The FSC-AC board generally accords the latter a high level of priority as they have been passed by the membership at large and are therefore viewed as having considerable legitimacy. If policy motions are not implemented, it often has more to do with practical considerations related to cost than to relative degrees of enthusiasm and non-enthusiasm.

17  Synnott (2005, 23) writes: "According to the [interim board] Minutes, he [James Cameron] warned that a decision-making mechanism that depended on frequent General Assemblies and a relatively open membership would be unwieldy, expensive, and a severe limitation to the flexibility and decision-making capacity of the secretariat, and it would put at risk the balance of stakeholder interests."

18  Matthew Wenban-Smith served as head of the FSC Policy and Standards Unit until January 2006, when he left to take up the position of director of forest certification with the British consulting company Dovetail Partners Inc. (http://www.dovetailinc.org/who.html).

19  Although the purpose of FSC-IC is to implement its certification and accreditation system for "environmentally sustainable, economically viable and socially beneficial" forestry, the statutes and by-laws are silent on exactly how this shall be done.

20  Information from FSC-Canada letter to Liedeker (FSC-IC ED), November 2002.

21  Weber (2005a) notes that the only case to go through the entire procedure set out in the IDRP was the PT Diamond Raya dispute, which took forty-two months from the first

request for an "informal resolution" (25 April 2002) to the board's statement of decision on 8 November 2005.

22 Meidinger is a professor of law at the State University of New York at Buffalo.

23 Section 81 provides that the board "shall consider any amendment to these By-laws proposed by a member in writing and seconded by two other members. *If the Board agrees* to the proposal the amendment shall be submitted to vote by the next General Assembly" (our emphasis).

24 Several FSC-BC representatives travelled to the 2002 General Assembly in Oaxaca, Mexico, to propose resolutions embedding the precautionary principle and upward harmonization in the FSC-AC principles and criteria in an effort to strengthen the case for D3. Another group, spearheaded by Chris Elliott of the World Wildlife Fund (WWF), were intent on rectifying a perceived fault in FSC-Canada's process.

25 See Statutory Motion 6, which defined consensus as the "consent" of each of the three chambers.

26 The text of the "International non governmental organisations accountability charter" (2005) is available online at www.ingoaccountabilitycharter.org/download/ingo-accountability-chater-eng.pdf.

### Chapter 13: Theorizing Regulation and Governance within and beyond the FSC

1 Indeed, Meidinger (2000, 226) goes on to contend that the FSC and its competitors are engaged in a competition "to become long-term institutions for shaping public policy."

2 Moreover, recent research by Overdevest and Rickenbach (2005) suggests that the "market dimension" of certification, at least in terms of its role in generating price premiums and market access, is less important to certificate holders than its function in signalling to a wide variety of audiences the acceptability of goods produced. Drawing on Beck's work on the risk society, they argue that certification "alleviates concern about risks ... generated by 'non-reflexive' production systems." Recent research by Cashore et al. (2006) focusing on Bolivia, Uganda, and eastern Europe also highlights a range of non-market rationales for employing certification, including as a strategy to monitor and enforce actor commitments in relation to a carbon sequestration project (Uganda) and biodiversity conservation (Bolivia) and as a strategy to legitimate state forestry agencies in the face of widespread pressure to privatize (Poland).

3 Freeman draws attention to the "linguistic conundrum" this approach presents for analysts who wish to use the terms without being trapped by what they signify. Drawing on Potter Stewart's practical epistemology, Freeman then argues that, as is the case with pornography, we "know" a public (private) agency when we see it. "'Public' refers to organisations that we associate with state power and of which we typically expect 'public-regarding' behaviour, such as government agencies. 'Private' refers to organisations that we associate with the pursuit of profit, such as firms, or ideological goals, such as environmental organisations" (Freeman 2000, 550).

4 The policy was changed in 2003 to permit forest managers from state-owned forest companies to participate in the economic chamber, but by then the damage had been done, with many states actively establishing competitor schemes or providing tacit support to industry-created alternatives.

5 According to the SFB website, "there are 5 environmental nonprofit CEO's, 5 forest industry CEO's, and 5 members of the broader forestry community. Through the balance of these stakeholders, the SFB can maintain their goal of continual improvement of the SFI Standard, and ensure that the Standard is protecting the economic, environmental and social needs of our forests and communities" (SFB 2006a).

6 See Indicator 1.6.1, for example, which states that "the manager has a publicly available, written commitment to adhere to the FSC-BC Regional Standards over the long term, which is signed off by the board of directors and/or equivalent senior authority, and included in the management plan"(FSC-BC 2005a).

7 Note, however, that, as discussed under the "Range of values being regulated" section later in this chapter, many other values are not addressed under such standards.

8  See "Understanding Canada's forest certification schemes: A complete guide to filing appeals and complaints" (SLDF 2005).
9  There is no separate institutional architecture for CSA's forest-certification system, which is fully integrated into the broader structure of CSA International. The latter hosts a web page that describes the CSA's Sustainable Forest Management Program and contains a list of documents and contacts (http://www.csa-international.org/product_areas/forest_products_marking/default.asp?language=english).
10  For a detailed study of different models of democracy see Held 1996.
11  Mainstream democratic theorists often fall back on the existence of an *ethnos,* based either on shared blood lines *(ius sanguinis)* or historical ties to a particular place *(ius solis),* as the key constituent of the demos. City-states of the past resembled this account of the demos, but today the account is far more problematic as the political unit (the polis) often embraces more than one ethnos (as in Canada), or the ethnos extends well beyond the boundaries of the polis (as in Israel). There are perhaps no states where the polis and the demos perfectly coincide.
12  The classic work in the realist democratic tradition is Robert Dahl's *Who governs?* (1961). Dahl contrasts idealistic conceptions of democracy with his realist conception of "polyarchy."
13  Brasset and Higgot (2003, 34-35) make a similar observation with respect to their argument that a "world polity" is coming into being. They note that "even if states and non-state actors are in disagreement about the norms and principles that are emerging, by the very fact that they contest the nature of these principles and practices in global assemblies and other instances of global public space, it has the consequences (unintended as it may be) of furthering the development of a global polity."
14  Corporatism guarantees that interests can be directly and powerfully represented in the policy-making process. In itself, however, it does not guarantee that the policy outcome will be satisfactory to all interests. The decision-making rules are crucial. If these rules are based on a simple majority, minorities will encounter considerable difficulty having their interests met. They will have "voice" but little power within the corporatist decision making structure. Significantly, the simple-majority arrangement was in place when FSC-Canada recommended British Columbia's draft standard to FSC-IC. This was at odds with FSC's vision that all major interests – economic, social, and environmental – should be accommodated in the development of a standard. Following the 2002 General Assembly, this deficiency in FSC's organizational arrangements was eliminated when members passed Motion 6, which required a supermajority for national initiative decision making, including at least 50 percent support from each chamber.
15  See the article by Pons Rafols and Brander (2005) for an account of the existence of such a "stewardship council" model.
16  Critics of corporatism come from across the democratic political spectrum. Direct democrats criticize corporatism because it replaces the unmediated representation of individual interests (at town hall meetings, for example) with group-based interest mediation. Pluralists worry about corporatism's capacity to restrict "new interests" from forming, forcing groups to be represented by existing interest mediation arrangements. Representative democrats tend to see corporatism as distorting the "will of the people" by promoting extra-parliamentary interest mediation.

## Chapter 14: Reflections on the Nature and Significance of the FSC-BC Case

1  For example, as the FSC-BC steering committee was developing its standard it had only a few precedents to build on (Sweden, the Martimes, and the Rocky Mountains) in terms of working out how long it might take to develop a standard. Both the Swedish and the Rocky Mountain experience suggested standards could be developed relatively quickly and unproblematically; the Maritimes debacle demonstrated the opposite. In planning its own standards development timeline, the SC used the considerable financial resources available to it to support both the ST and the TAT during the early, and arguably "easier," phases of negotiations. Consequently, by early 2002, when negotiations entered the later and "harder" phase, the SC found itself under considerable financial pressure as donor

funding ran out. The shortage of funds placed considerable pressure on the SC to complete negotiations, even though it was clear that the negotiated standard did not have the backing of the large-industry members. Better budgetary management would have reserved a significant portion of donor funding for the later bargaining phase – to ensure, for example, that needed studies and field trials could be undertaken on the draft standards impact and feasibility.

2  Although we consider the modularity concept especially useful in the regulatory context, we recognize the challenges associated with implementing the concept in complex institutional settings. Clearly, social institutions such as the FSC cannot be built or rebuilt overnight. Yet the modularity concept serves to remind us of the need for a creative, interactive approach to governance.

3  Only the Boreal standard offers a definition of "local" in the context of workers and communities, and it is, arguably, significantly narrower than that employed in the BC standard (see Chapter 7).

4  See, for example, several of the contributions to the online FSC Objectives conference held in the run-up to the November 2002 FSC General Assembly meeting (FSC Objectives 2002).

# References

**Interviews**

Liviu Amariei, independent forest certification auditor. Former FSC regional director for Europe; former head of the Accreditation Business Unit, FSC-AC.

Patrick Armstrong, Moresby Consulting Ltd. FSC-BC standards team (economic chamber nominee); FSC-BC technical advisory team (economic chamber nominee)

Lara Beckett (née Lamport), eco-forestry advocate, Prince George, British Columbia. Co-ordinator, FSC-BC initiative (1996-1999).

Arnold Bercov, forestry officer, Pulp, Paper and Woodworkers of Canada. FSC-Canada working group (social chamber).

Bill Bourgeois, principal of New Directions Resource Management Ltd. Former vice-president, environment and government affairs, Lignum Ltd, Vancouver, British Columbia. FSC-BC steering committee (economic chamber).

Jessica Clogg, staff counsel, West Coast Environmental Law. FSC-BC standards team (social chamber nominee); FSC-BC technical advisory team (social chamber nominee).

Russell Collier, member of the Gitxsan Nation. FSC-Canada working group (First Nations chamber).

Richard Donovan, chief of forestry and director of the SmartWood of the Rainforest Alliance, Richmond, Vermont.

Denise English, Wildsight (Golden branch). Former chair, FSC-Canada working group (environmental chamber)

Troy Hromadnik, chief forester, Tembec Ltd., British Columbia. Joint chair, FSC-BC chapter; FSC-BC steering committee (economic chamber nominee [alternate to Bill Bourgeois]).

Robert Hrubes, senior vice president, Scientific Certification Systems (SCS). Former member of FSC-IC board.

Rod Krimmer, woodlotter, Big Lake Ranch, British Columbia. FSC-BC steering committee (social chamber).

Marcelo Levy, Responsible Forestry Solutions, Toronto. Former director, standards program, FSC-Canada.

Jim McCarthy. former executive director, FSC-Canada, Toronto.

Chris McDonnell, manager of environmental and Aboriginal relations, Tembec Ltd, Montreal, Quebec. Former chair, FSC-Canada working group (economic chamber)

Brian Murphy, executive director, Ministry of Sustainable Resource Management. Former deputy director and manager forest development, Forest Practices Branch, Ministry of Forests, Victoria, British Columbia; ex-officio, non-voting member, FSC-BC standards team.

David Nahwegahbow, IPC partner Nahwegahbow, Corbiere, Barristers and Solicitors, Nipissing Indian Reserve, North Bay, Ontario. Former chair, FSC-AC board (social chamber north)

Peter Robertnz, coordinator, Forest and Trade Network Sweden. Former secretary, FSC-Sweden.
Grant Rosoman, forest campaigner, Greenpeace New Zealand. Former chair, FSC-AC board (environmental chamber north).
Peggy Smith, professor, Faculty of Forestry, Lakehead University, Thunder Bay, Ontario. Advisor to FSC-Canada's Indigenous Advisory Council; FSC-Canada working group (First Nations chamber)
Tamara Stark, forest campaign manager, Greenpeace China. Formerly forest campaign co-ordinator, Greenpeace Canada, Vancouver, British Columbia; FSC-BC steering committee (environmental chamber).
Ananda Lee Tan, director, Canadian Reforestation and Environmental Workers Society. Former chair, FSC-BC chapter (social chamber).
Asa Tham, forest manager, Church of Sweden. Former vice-chair, FSC-AC board (economic chamber north)
Greg Utzig, Kootenai Nature Investigations Ltd, Nelson, British Columbia. Member, D1 committee; FSC-BC standards team (environmental chamber nominee); FSC-BC technical advisory team (environmental chamber nominee).
Matthew Wenban-Smith, director of certification program, Dovetail Partners Inc. Former head, Policy and Standards Unit, FSC-AC, Bonn.

## Works Cited

Abbott, K.W., and D. Snidal. 2001. International "standards" and international governance. *Journal of European Public Policy* 8, 3: 345-70.
–. 2000. Hard and soft law in international governance. *International Organization* 54, 3: 421-56.
Abusow International. 2007. Certification status report: British Columbia-SFM. June. http://www.certificationcanada.org/_documents/status_reports/SFM_BC_Data_Forest_Certification_Status_Report_details.pdf.
Abusow, K. 2005. Canadian forest management certification status report: December 15, 2005 (BC data only). Report prepared for the Canadian Sustainable Forestry Certification Coalition by Abusow International. http://www.certificationcanada.org/_documents/english/dec20-05_bc_forest_certification_details.pdf.
Ackerman, B.A., and R. Stewart. 1985. Reforming environmental law. *Stanford Law Review* 37, 5: 1333-66.
AF&PA (American Forest and Paper Association). 2006. Sustainable forestry initiative: Background. http://www.afandpa.org/Content/NavigationMenu/Environment_and_Recycling/SFI/SFI.htm.
–. 2002. 2002-2004 Sustainable Forestry Initiative (SFI) program, standard and verification procedure. www.aboutsfi.org/miscPDFs/H-sfistandard02.pdf.
–. 2000. 5th annual Sustainable Forest Initiative progress report. http://www.afandpa.org/forestry/sfi/Final_Standard.pdf.
Ajani, J. 2007. *The forest wars*. Carlton, VIC: Melbourne University Press.
*Alternatives Journal*. 2004. *Alternatives* interviews Martin von Mirbach on the Forest Steward-ship Council's new guidelines for good logging practices in the boreal forests. *Alternatives Journal* 30, 3: n.p.
Anderson T., and D. Leal. 1991. *Free market environmentalism*. Boulder, CO: Westview Press.
Ansell, C. 2000. The networked polity: Regional development in western Europe. *Governance* 13, 3: 303-33.
Arsenau, C., and M. Chiu. 2003. Canada – A land of plantations. Paper presented at UNFF Intersessional Experts Meeting on the Role of Planted Forests in Sustainable Forest Management, 25-27 March, Wellington, New Zealand.
Atkinson, M., and W. Coleman. 1989. Strong states and weak states: Sectoral policy networks in advanced capitalist economies. *British Journal of Political Science* 19, 1: 47-67.
Ayres, I., and J. Braithwaite. 1992. *Responsive regulation: Transcending the deregulation debate*. New York and Oxford: Oxford University Press.

Baharuddin, H.J., and M. Simula. 1994. *Certification schemes for all timber and timber products.* Yokohama, Japan: International Tropical Timber Organization.

Bartley, T. 2007. How foundations shape social movements: The construction of an organizational field and the rise of forest certification. *Social Problems* 54, 3: 229-55.

BC Parks. 2007. BC Parks statistics. http://www.env.gov.bc.ca/bcparks/facts/stats.html.

BC Stats. 2006. British Columbia manufacturing shipments. http://www.bcstats.gov.bc.ca/data/dd/handout/mannaics.pdf.

–. 2005. The new relationship. http://www.gov.bc.ca/arr/down/new_relationship.pdf.

Benhabib, S., ed. 1996. *Democracy and difference: Contesting the boundaries of the political.* Princeton, NJ: Princeton University Press.

Bernstein, S., and B. Cashore. 2004. Non-state global governance: Is forest certification a legitimate alternative to a global forest convention? In *Hard choices, soft law: Voluntary standards in global trade, environment and social governance,* ed. J.J. Kirton and M.J. Trebilcock, 33-63. Aldershot, UK: Ashgate.

–. 1999. World trends and Canadian forest policy: Trade, international institutions, consumers and transnational environmentalism. *Forestry Chronicle* 75: 34-38.

Berry, F., R. Brower, S. Ok Choi, W. Xinfang Goa, H. Jang, M. Kwon, and J. Ward. 2004. Three traditions of network research: What the public management research agenda can learn from other research communities. *Public Administration Review* 64: 539-52.

Blatter, J. 2001. Debordering the world of states: Towards a multi-level system in Europe and a multi-polity system in North America? Insights from border regions. *European Journal of International Relations* 7, 2: 175-209.

Boetekees, G., K. Moore, and G. Weber. 2000. Forest stewardship standards for the Maritime region: Commission of enquiry. http://www.fsccanada.org/ word_doc/COE_final_report.doc.

Borrows, J. 1994. Constitutional law from a First Nation perspective: Self-government and the Royal Proclamation. *University of British Columbia Law Review* 28: 1-47

Boyd, R. 1994. Smallpox in the Pacific Northwest. *BC Studies* 101. 5-40.

Braithwaite, J., and P. Drahos. 2000. *Global business regulation.* Cambridge, UK: Cambridge University Press.

Brassett, J., and R. Higgott. 2003. Building the normative dimension(s) of a global polity. *Review of International Studies* 29, Suppl. S1: 29-55.

British Columbia Forest Practices Board. 2007. Glossary of forest management terms, July 2007. Victoria, BC: FPB. http://www.for.gov.bc.ca/hfd/library/documents/glossary/Glossary_July2007.pdf.

–. 2006. A review of the early Forest Stewardship Plans under FRPA. http://www.fpb.gov.bc.ca/special/reports/SR28/SR28.pdf.

–. 2004. Evaluating mountain pine beetle management in BC: Special report. FPB/SR/20 FPB. http://www.fpb.gov.bc.ca/special/reports/SR20/SR20.pdf.

British Columbia Ministry of Aboriginal Relations and Reconciliation. 2008. The new relationship. Aboriginal people and the government of British Columbia: Building a healthy and prosperous future together. http://www.gov.bc.ca/arr/newrelationship/down/new_relationship_brochure.pdf.

British Columbia Ministry of Forests and Range. 2008a. Apportionment system, provincial summary report. Report ID APTR032, 20 June. http://www.for.gov.bc.ca/hth/apportionment/Documents/Aptr032.pdf.

–. 2008b. Status of community forest agreements, April 2008. http://www.for.gov.bc.ca/hth/community/Documents/status-table-april-2008.pdf.

–. 2006a. *The state of British Columbia's forests 2006.* Victoria, BC: Ministry of Forests and Range. http://www.for.gov.bc.ca/hfp/sof/2006/pdf/sof.pdf.

–. 2006b. 100th First Nation signs forest agreement with province. News release, 19 January.

British Columbia Ministry of Forests. 2004a. Cut level increased in response to beetle epidemic. News release, 14 September. http://www2.news.gov.bc.ca/nrm_news_releases/2004FOR0040-000707.htm.

–. 2004b. Forest and Range Agreements FAQs. Aboriginal Affairs Branch. http://www.for.gov.bc.ca/haa/Docs/ Public_Q&A_Oct27_2004.htm.

–. 2003a. *Summary of forest health conditions in British Columbia*. Victoria, BC: Ministry of Forests. http://www.for.gov.bc.ca/ftp/HFP/external/!publish/Aerial_Overview/2003/report/F%20H%202003%20Final.pdf.

–. 2003b. British Columbia forests and their management. http://www.coastforest.org/pdf/bc_forest_management.pdf.

–. 2003c. Number of licences as of March 31, 2003 by forest region and volume committed as of March 31, 2003 by forest region. Unpublished tables.

–. 2003d. BC forest management certification status report. Victoria, BC: Ministry of Forests (December). http://www.for.gov.bc.ca/het/certification/CertificationStatusReport.pdf.

–. 2002a. Forestry revitalization plan. http://www.for.gov.bc.ca/mof/plan/frp.

–. 2002b. Timber supply analysis in British Columbia. Forest Analysis Branch.

British Columbia Ministry of Sustainable Resource Management. 2004. Working forest policy will bring jobs and investment. http://www2.news.gov.bc.ca/nrm_news_releases/2004SRM0027-000624.htm.

British Columbia Ministry of Water, Land and Air Protection. 2003. Status and trends of protected areas. MWLAP State of Environment Reporting.

–. 2002. Is BC's rich ecosystem diversity protected? MWLAP State of Environment Reporting.

British Columbia Treaty Commission. 2001a. Annual report 2001: The year in review.

–. 2001b. Looking back, looking forward: A review of the BC treaty process. http://www.bctreaty.net/files_2/pdf_documents/review_bc_treaty_process.pdf.

Brown, C. 2000. *The global outlook for future wood supply from forest plantations*. Rome: Food and Agriculture Organization.

Brundtland, G., ed. 1987. *Our common future: The World Commission on Environment and Development*. Oxford: Oxford University Press.

Bun, Y., and I. Bewang. 2006. Forest certification in Papua New Guinea. In *Confronting sustainability: Forest certification in developing and transitioning countries*, ed. B. Cashore, F. Gale, E. Meidinger, and D. Newsom. New Haven, CT: Yale Forestry and Environmental Studies Publication Series.

Burda, C., and F. Gale. 1998. Trading in the future: An examination of BC's commodity export strategy in forest products. *Society and Natural Resources* 11, 6: 555-68.

Cameron, R. 2005. Standards Australia limited: Standards development governance review. http://www.standards.org.au/strategy.asp?contentid=3.

Campbell, C. 2005. Current performance of the BC forest and paper industry. Paper presented at the Vancouver Board of Trade Policy Forum, BC Forestry: Present Status and Future Prospects, 12 May. http://www.boardoftrade.com/events/presentations/Campbell12may05.pdf.

Canada. External Advisory Committee on Smart Regulation. 2004. Smart regulation for Canada. http://www.pco-bcp.gc.ca/smartreg-regint/en/08/rpt_fnl.pdf.

Canada. Industry Canada. 2006. Canadian trade by industry – NAICS: British Columbia. http://strategis.gc.ca/sc_mrkti/tdst/tdo/tdo.php.

Canada. Ministry of Indian and Northern Affairs. 2005. Band classification manual May 2005. http://www.ainc-inac.gc.ca/pr/pub/fnnrg/2005/bandc_e.pdf.

–. 2002. Basic departmental data. www.ainc-inac.gc.ca/pr/sts/bdd02/bdd02_e.html.

Canada. Natural Resources Canada. 2006. Overview: Sustainable development – Walking the talk. http://www.nrcan.gc.ca/mms/poli/tlk_e.htm.

Canada Privy Council Office. 2005. Government directive on regulating. http://www.regulation.gc.ca/default.asp?Language=E&Page=thegovernmentdirectiveon (visited 27 June 2006).

CanFor. 2003. "Customer Update." Certifying Our Forests 5, 2 (November). http://www.canforpulp.com/_resources/sustainability/CertifyingV5N2.pdf.

Cashore, B. 2003. Perspective on forest certification as a policy process: Reflections on Elliott and Schlaepfer's use of the advocacy coalition framework. In *Social and Political Dimensions of Forest Certification*, ed. Errol Meidinger, C. Elliott, and G. Oesten, 219-33. Remagen-Oberwinter, Ger.: Verlag.

–. 2002. Legitimacy and the privatisation of environmental governance: How non-state market-driven (NSMD) governance systems gain rule-making authority. *Governance* 15, 4: 503-29.

Cashore, B., G. Auld, and D. Newsom. 2004. *Governing through markets: Forest certification and the emergence of non-state authority.* New Haven, CT: Yale University Press.

Cashore, B., F. Gale, E. Meidinger, and D. Newsom. 2006. *Confronting sustainability: Forest certification in developing and transitioning countries.* New Haven, CT: Yale Forestry and Environmental Studies Publication Series.

Cashore, B., G. Hoberg, M. Howlett, J. Rayner, and J. Wilson, eds. 2001a. *In search of sustainability: British Columbia forest policy in the 1990s.* Vancouver: UBC Press.

–. 2001b. Conclusion: Change and stability in BC forest policy. In *In search of sustainability: British Columbia forest policy in the 1990s,* ed. B. Cashore, G. Hoberg, M. Howlett, J. Rayner, and J. Wilson, 232-57. Vancouver: UBC Press.

Cashore, B., G.C. van Kooten, I. Vertinsky, G. Auld, and J. Affolderbach. 2005. Private or self regulation? A comparative study of forest certification choices in Canada, the United States and Germany. *Forest Policy and Economics* 7, 1: 53-69.

Cashore, B., and J. Lawson. 2003. Comparing forest certification in the U.S. Northeast and the Canadian Maritimes. Orono, ME: Canadian-American Center, University of Maine. http://environment.yale.edu/cashore/pdfs/2003/03_private_po_maritimes.pdf.

CCFM (Canadian Council of Forest Ministers). 2005. National Forestry Database Program, compendium, 2005-10. http://www.nfdp.ccfm.org/compendium/data/2005_10/tables/tab82je.php.

–. 2004. National Forestry Database Program, compendium, 2004-10. http://www.nfdp.ccfm.org/compendium/data/2004_10/tables/.

Charnovitz, S. 2002. The law of PPMs in the WTO: Debunking the myth of illegality. *Yale Journal of International Law* 27: 59-110.

Chertow, M.R., and D.C. Esty, eds. 1997. *Thinking ecologically: The next generation of environmental policy.* New Haven, CT: Yale University Press.

Clogg, J. 2003. Provincial forestry revitalization plan – Forest Act amendments: Impacts and implications for BC First Nations. West Coast Environmental Law Research Foundation.

–. 1999. Comments on the public consultation draft of the Forest Stewardship Council regional certification standard for British Columbia. http://www.wcel.org/wcelpub/1999/wrapper.cfm?docURL=http://www.wcel.org/wcelpub/1999/13013.html.

Coglianese, C., and D. Lazar. 2003. Management-based regulation: Prescribing private management to achieve public goals. *Law and Society Review* 37: 691-730.

Coglianese, C., J. Nash, and T. Olmstead. 2003. Performance-based regulation: Prospects and limitations in health, safety and environmental protection. *Administrative Law Review* 55: 705-28, http://repositories.cdlib.org/csls/lss/10/.

Cohen, J.L., and A. Arato. 1994. *Civil society and political theory.* Cambridge, MA: MIT Press.

Colchester, M. 2000. Indigenous peoples and the new "global vision" on forests. Forest Peoples Program for the World Bank/WWF. http://greatrestoration.rockefeller.edu/21Jan2000/Colchester.htm.

Collier, R., R. Diabo, J. Gladu, P. Smith, and V. Peachey. 2001. Draft strategic directions paper on indigenous peoples and Forest Stewardship Council certification. Report prepared by an Indigenous Writing Committee for Review and Comment.

Collier, R., B. Parfitt, and D. Woolard. 2002. A voice on the land: An indigenous peoples' guide to forest certification in Canada. Report prepared for Ecotrust Canada and the National Aboriginal Forestry Association.

Considine, M., and M. Painter. 1997. *Managerialism: The great debate.* Carlton South, Australia: Melbourne University Press.

Constance, D., and A. Bonanno. 2000. Regulating the global fisheries: The World Wildlife Fund, Unilever, and the Marine Stewardship Council. *Agriculture and Human Values* 17, 2: 125-39.

Counsell, S., and K. Lorass, eds. 2002. *Trading in credibility: The myth and reality of the Forest Stewardship Council.* London: The Rainforest Foundation. http://www.rainforest

foundationuk.org/s-Trading%20in%20Credibility-%20Myth%20and%20Reality%20of%20the%20FSC.

CSA (Canadian Standards Association). 2003. Sustainable forest management: Requirements and guidance. Program document 2809-02. http://www.csa-international.org/product_areas/forest_products_marking/program_documents/CAN_CSA_Z809-02O_English.pdf.

CSFCC (Canadian Sustainable Forestry Certification Coalition). 2007. Our members. http://www.certificationcanada.org/english/who_we_are/our_members.php.

Cummins, A. 2004. The Marine Stewardship Council: A multi-stakeholder approach to sustainable fishing. *Corporate Social Responsibility and Environmental Management* 11, 2: 85-94.

Cutler, C., V. Haufler, and T. Porter, eds. 1999. *Private authority and international affairs.* Albany, NY: State University of New York Press.

Dahl, R. 1961. *Who governs? Democracy and power in the American city.* New Haven, CT: Yale University Press.

Dawe, N., and K. Ryan. 2003. The faulty three-legged stool model of sustainable development. *Conservation Biology* 17, 5: 1458-60.

Dellert, L. 1998. Sustained yield: Why has it failed to achieve sustainability? In *The wealth of forests: Markets, regulation, and sustainable forestry,* ed. C. Tollefson, 255-77. Vancouver: UBC Press.

–. 1994. Sustained yield forestry in BC: The making and breaking of a policy (1900-1993). Master's thesis, York University, Ontario.

Dicken, P. 2003. *Global shift: Reshaping the global economic map in the 21st century.* 4th ed. New York: Guildford Press.

Dicken, P., P.F. Kelly, K. Olds, H. W.-C. Yeung. 2001. Chains and networks, territories and scales: Towards a relational framework for analyzing the global economy. *Global Networks* 1, 2: 89-112.

Donovan, R. 2005. FSC British Columbia standard: Serious concerns. Memorandum to Matthew Wenban-Smith, Alistair Monument, and A. Droste; stakeholder comment regarding the accreditation of the BC standard, received 25 August 2005. http://www.fsccanada.org/SiteCM/U/D/0286EA301747CE3C.pdf.

Dorsen, N., M. Rosenfeld, A. Sajo, and S. Baer. 2003. *Comparative constitutionalism: Cases and materials.* Eagen, MN: Thomson West.

Drushka, K. 1999. *In the bight: The BC forest industry today.* Madeira Park, BC: Harbour Publishing.

Drushka, K., B. Nixon, and R. Travers, eds. 1993. *Touch wood: BC forests at the crossroads.* Madeira Park, BC: Harbour Publishing.

Dudley, N., J.-P. Jeanrenaud, and F. Sullivan. 1995. *Bad harvest: The timber trade and the degradation of the world's forests.* London: Earthscan.

Eba'a Atyi, R., and M. Simula. 2002. Forest certification: Pending challenges for tropical timber. ITTO Technical Series No. 19. Yokohama, Japan: ITTO.

Ecotrust Canada. 2005. Forest and range agreements: Opportunities and challenges for increasing FSC certification for First Nations in BC. Report prepared for Donovan Woollard, Ecotrust Canada, by Deb MacKillop, Karen Wu, and John Cathro.

Elliott, C. 1999. Forest certification: Analysis from a policy network perspective. PhD diss., Ecole Polytechnique Federale de Lausanne, Switzerland. http://library.epfl.ch/en/theses/?nr=1965.

Elliott, C., and R. Schlaepfer. 2001. The advocacy coalition framework: Application to the policy process for the development of forest certification in Sweden. *Journal of European Public Policy* 8, 4: 642-61.

Eng, M., A. Fall, J. Hughes, T.L. Shore, W.G. Riel, P. Hall, and A. Walton. 2005. Provincial-level projection of the current mountain pine beetle outbreak: An overview of the model (BCMPB-2) and results of year 2 of the project. Report prepared for the Canadian Forest Service and BC Forest Service. http://www.for.gov.bc.ca/hre/ topics/mpb.htm.

Eritja, Mar Campins, ed. 2004. *Sustainability labelling and certification labelling.* Madrid: Marcial Pons.

Estonian Fund for Nature. 2005. Report on harmonization exercise of FSC standards in the Baltic Region. http://www.elfond.ee/fail.php?id_fail=696.

Ferguson, Y., R. Mansbach, R. Danemark, H. Spruyt, B. Buzan, R. Little, J. Gross Stein, and M. Mann. 2000. What is the polity? A roundtable. *International Studies Review* 2, 1: 3-31.

FERN (Forests and the European Union Resource Network). 2004. Footprints in the forest: Current practice and future challenges in forest certification. http://www.fern.org/pubs/reports/footprints.pdf.

Fiorino, D.J. 2006. *The new environmental regulation.* Cambridge, MA: MIT Press.

Fisher, R., and W. Ury. 1981. *Getting to yes: Negotiating agreement without giving in.* New York: Penguin Books.

ForestEthics. 2003. British Columbia's endangered forests: What government and industry aren't telling you. http://forestethics.org/downloads/end_for_report.pdf.

Franson, M.A.H., R.H. Franson, and A.R. Lucas. 1982. *Environmental standards: A comparison of Canadian standards, standards-setting processes and enforcement.* Edmonton, AB: Environmental Council of Alberta.

Freeman, J. 2000. The private role in public governance. *New York University Law Review* 75, 101: 543-675.

Freeman, J., and D. Farber. 2005. Modular environmental regulation. *Duke Law Journal* 54, 4: 795-912.

Friends of Clayoquot Sound. 2006. Markets campaigns. http://www.focs.ca/logging/marketscampaigns.asp.

FSC (Forest Stewardship Council). 2007. FSC accredited certification bodies. http://www.fsc.org/keepout/en/content_areas/32/1/files/5_3_1_2007_08_29_FSC_Accredited_CBs.pdf.

–. 2006. News and notes, December. http://www.fsc.org/keepout/en/content_areas/63/36/files/FSC_PUB_20_04_11_2006_12_03_1_.pdf.

FSC-AC (Forest Stewardship Council, Asociación Civil). 2007a. Contact FSC. http://www.fsc.org/en/about/contact_fsc.

–. 2007b. Letter from Robert Waack, Chairman, FSC Board of Directors, 15 November.

–. 2006. Forest Stewardship Council A.C. Bylaws. http://www.fsc.org/keepout/en/content_areas/77/84/files/1_1_FSC_By_Laws_2006.pdf.

–. 2005a. FSC pesticides policy: Guidance on implementation. FSC guidance document FSC-GUI-30-001.

–. 2005b. FSC plantations review: Report from the first policy working group meeting, 8 April. http://www.fsc.org/plantations/docs/FSC%20PWG%20meeting%20report%202005-03%20-%20Eng.PDF.

–. 2005c. Report on the closure of conditions issued to the preliminary accreditation of the FSC regional certification standards for British Columbia, Canada. ABU-REP-34-CAN-2005-09-23-BC. http://www.fsccanada.org/SiteCM/U/D/BC5BEED22A955A1A.pdf.

–. 2005d. Statutes. Document 1.3. Revised June. http://www.fsc.org/keepout/en/content_areas/77/83/files/1_3_FSC_Statutes_2005.pdf.

–. 2004a. FSC glossary of terms. FSC-STD-01-002. http://www.fsccanada.org/SiteCM/U/D/F9F230AFAD457FCF.pdf.

–. 2004b. Conversion of plantation to non-forest land. FSC advice note. FSC-ADV-30-602. http://www.fsc.org/keepout/en/content_areas/77/136/files/FSC_ADV_30_602_EN_Conversion_of_plantation_to_non_forest_land_2004_03_29.pdf.

–. 2004c. Structure and content of forest stewardship standards. FSC-STD-20-002, Version 1.0, March. http://fsccanada.org/SiteCM/U/D/20A2743DEEB4A92A.pdf.

–. 2003a. Accreditation report for the regional certification standards for British Columbia, Canada. ABU-REP-32-06-2003-FSS-ACC-F-CA-BC.

–. 2003b. Forest Stewardship Council: Because forests matter. http://www.fsc.org/en/about/contact_fsc.

–. 2003c. FSC policy for preliminary accreditation of national/regional forest stewardship standards (2003). FSC-POL-10-002. http://www.fsc.org/keepout/en/content_areas/77/

133/files/FSC_POL_10_002_EN_Preliminary_Accreditation_of_National_Regional_FS_ Standards_2003.pdf.

–. 2003d. Policy for preliminary accreditation of national/regional forest stewardship standards.

–. 2002a. FSC International Standard: FSC principles and criteria for forest stewardship. http://www.fsc.org/keepout/en/content_areas/77/134/files/FSC_STD_01_001_V4_0_EN_ FSC_Principles_and_Criteria.pdf.

–. 2002b. FSC General Assembly 2002 – final motions. http://www.fsc.org/en/about/ documents/Docs_cent/32 (visited February 2006).

–. 2002c. FSC annual report 2002. http://www.fsc.org/keepout/en/content_areas/88/1/ files/FSC_1640_e.pdf (visited 28 June 2006).

–. 2002d. Chemical pesticides in certified forests: Interpretation of the FSC principles and criteria. http://www.fsc.org/keepout/en/content_areas/102/1/files/FSC_POL_30_601_ FSC_Chemical_Pesticides_Policy__2002_.pdf.

–. 2002e. Interpretation of FSC Criterion 10.9. http://www.fsccanada.org/SiteCM/U/ D/DF4ADE2CA81389CD.pdf.

–. 2001a. Forest stewardship standards for the Maritime forest region: Commission of enquiry: Final report. http://www.fsccanada.org/word_doc/COE_final_report.doc (visited February 2004).

–. 2001b. Change Management Team report. Submitted by Bruce Caberle, Pierre Hausel-mann, Michael Rogers, Michael Northrop, Wilma Gormley, and Olof Johannson, special advisor to CMT. FSC document reference BM-22.5a., 8 May.

–. 1998a. FSC national initiatives manual: First secretariat'draft, by I.J. Evison.

–. 1998b. Forest Stewardship Council interim dispute resolution protocol, document 1.4.3, Bonn, Ger. FSC. www.fsc.org/keepout/en/content_areas/39/1/files/1_4_3_Interim_ dispute_resolution_protocol.pdf.

FSC-BC (Forest Stewardship Council of British Columbia). 2005a. Forest Stewardship Council regional standards for British Columbia: Main Standards. http://www.fsccanada. org/SiteCM/U/D/48B4F585905BF469.pdf.

–. 2005b. Forest Stewardship Council regional certification standards for British Columbia: Small operations standards. http://www.fsccanada.org/SiteCM/U/D/9DD640BC791A8F78. pdf.

–. 2005c. FSC-BC guidance: A companion document to the FSC regional standards for British Columbia. http://www.fsccanada.org/SiteCM/U/D/1BD8555831CD7ECF.pdf.

–. 2003. Forest Stewardship Council (FSC) regional certification standards for British Columbia: FSC-BC preliminary standards, July 11.

–. 2002a. Social chamber working group workshop: Summary report. Prepared for the FSC-BC Regional Initiative Steering Committee.

–. 2002b. Regional certification standards for British Columbia: Draft 3.

–. 2001. Forest Stewardship Council (FSC) regional certification standards for British Colum-bia, Draft 2. Prepared by the FSC-BC Standards Team for the FSC-BC Regional Initiative Steering Committee.

FSC-BC ISC (Forest Stewardship Council of British Columbia, Interim Steering Committee). Various years. Minutes of the FSC-BC Interim Steering Committee and Steering Commit-tee. Formerly online at www.fsc-bc.org.

FSC-BC ST (Forest Stewardship Council of British Columbia, Standards Team). 2000. FSC-BC Standards Team workplan. Appended to FSC-BC ISC minutes for 27-28 February.

FSC-BC TAT (Forest Stewardship Council of British Columbia, Technical Advisory Team). 2001. Second draft public comments and field testing: Issues analysis report.

FSC-Boreal. 2004. National boreal standard. http://www.fsccanada.org/pdf_document/ BorealStandard_Aug04.pdf.

FSC-CAN (Forest Stewardship Council of Canada). 2007. Input welcomed on salvage logging. *Branching Out* (November), http://www.fsccanada.org/november07.htm.

–. 2006. Celebrating a landmark consensus on forest management. FSC-Canada Media Advisory, 30 March.

–. 2005. Update on revisions to the preliminary regional standard for British Columbia, March.

–. 2003. FSC Canada's boreal forest standard unanimously approved.

–. 2002a. National Boreal standards process. http://www.canopees.org/fsc_qc/documents/BorealRoadMap.pdf.

–. 2002b. Developing an FSC Boreal standard for Canada: Options for decision-making and reconciling inputs.

–. 2001. Concept paper: Addressing the application of Principle 3, indigenous rights, and indigenous involvement in the Forest Stewardship Council.

–. Annual Reports. Various Years. http://www.fsccanada.org/Governance.htm.

FSC-CAN IAC (Forest Stewardship Council of Canada, Indigenous Advisory Council). 2003. Report of the Indigenous Advisory Council, March 3.

FSC-IC ABU (Accreditation Bureau Unit). 2003. Accreditation report for the regional certification standards for British Columbia, Canada. ABU-REP-32-06-2003-FSS-ACC-F-CA-BC. Archived with authors.

FSC-M (Forest Stewardship Council of Canada, Maritime Regional Initiative). 2003. Certification standards for best forestry practices in the Maritime forest region. Accredited by FSC-AC, March 2003. http://www.fsccanada.org/SiteCM/U/D/178DA46EC93D9624.pdf.

FSC Objectives. 2002. FSC Objectives conference: An independent threaded E-conference on the future objectives of the Forest Stewardship Council, FSC. http://www.fscobjectives.org/index.htm.

FSC-PC (Forest Stewardship Council Pacific Coast). 2005. Pacific Coast (USA): Regional forest stewardship standard: Draft 9.0. Accredited by FSC-IC, July 2003. http://www.fscus.org/images/documents/2006_standards/pcwg_9.0_NTC.pdf.

–. 2002. Pacific Coast (USA): Regional forest stewardship standard: Draft 8.01.

FSC-RM (Forest Stewardship Council Rocky Mountains). 2004. Revised final forest stewardship standard. Accredited by FSC-IC, September 2001. http://www.fscus.org/images/documents/standards/STND_RM_final_V2.PDF.

–. 2001. Forest stewardship standard for the Rocky Mountain region (USA), August draft.

–. 2000. Second draft of Rocky Mountain regional standards, August 2000.

FSC-Sweden (Forest Stewardship Council Sweden). 2005. Swedish FSC standard for forest certification: Draft for endorsement. (Second revised FSC-Sweden standard, English version). Not accredited by FSC-IC as of December 2007. http://www.fsc-sweden.org/Portals/0/Documents/Swedish%20FSC%20Standard_050907.pdf.

–. 1998. Swedish FSC standard. Accredited by FSC-IC, November 1998. http://www.fsc-sweden.org/Portals/0/Fsc-eng.pdf_1.pdf.

FSC-US. 2001a. FSC-U.S. national indicators for forest stewardship. http://www.fscus.org/images/documents/FSC_National_Indicators.pdf.

–. 2001b. First plantation forestry certification awarded in the U.S. FSC News and Views (November).

FSC-UK National Initiatives Task Force. 2005. National initiatives task force recommendations report.

Gale, F. 2004. The consultation dilemma in private regulatory regimes: Negotiating FSC regional standards in the United States and Canada. *Journal of Environmental Policy and Planning* 6, 1: 57-84.

–. 2002. *Caveat certificatum:* The case of forest certification. In *Confronting consumption,* ed. T. Princen, M. Maniates, and K. Conca, 275-99. Boston, MA: MIT Press.

–. 1998a. Constructing civil society actors: An anatomy of the environmental coalition contesting the tropical timber trade regime. *Global Society* 12: 343-61.

–. 1998b. *The tropical timber trade regime.* Basingstoke, UK: St. Martin's Press.

Gale, F., and C. Burda. 1998. The pitfalls and potential of eco-certification as a market incentive for sustainable forest management. In *The wealth of forests: Markets, regulation, and sustainable forestry,* ed. C. Tollefson, 278-96. Vancouver: UBC Press.

Gale, F., and M. Haward. 2004. Public accountability in private regulation: Contrasting models of the Forest Stewardship Council (FSC) and Marine Stewardship Council (MSC).

Refereed paper presented to the 2004 Australasian Political Studies Association Conference, Adelaide, Australia.

Gale, R., and F. Gale. 2006. Accounting for social impacts and costs in the forest industry, British Columbia. *Environmental Impact Assessment Review* 26: 139-55.

Gereffi, G. 1994. The organization of buyer-driven global commodity chains: How US retailers shape overseas production. In *Commodity chains and global capitalism,* ed. G. Gereffi and M. Korzeniewicz, 95-122. Westport, CT: Praeger.

Ghazali, B.H., and M. Simula. 1998. Timber certification: Progress and issues. Report prepared for ITTO. http://www.gtz.de/de/dokumente/en-d3-timber-certification-progress-and-issues-itto.pdf.

–. 1996. Timber certification in transition. Report prepared for ITTO. http://www.itto.or.jp/live/PageDisplayHandler?pageId=203.

–. 1994. Certification schemes for timber and timber products. Report prepared for ITTO. Yokohama, Japan: ITTO.

Gibson, R.B., ed. 1999. *Voluntary initiatives: The new politics of corporate greening.* Peterborough, ON: Broadview Press.

GrantCraft. 2005. Funders collaborative: Sustainable forestry. http://www.grantcraft.org.

Green, P., J. Joy, D. Sirucek, W. Hann, A. Zack, and B. Naumann. 1992. Old-growth forest types of the northern region. Unpublished report for USDA Forest Service, Northern Region.

Grumbine, R.E. 1999. Reflections on "What is ecosystem management?" *Conservation Biology* 11, 1: 41-47.

–. 1994. What is ecosystem management? *Conservation Biology* 8, 1: 27-38.

Gunningham, N., and P. Grabosky. 1998. *Smart regulation: Designing environmental policy.* Oxford: Oxford University Press.

Gunningham, N., and D. Sinclair. 2002. *Leaders and laggards: Next-generation environmental regulation.* Sheffield: Greenleaf.

Guthman, J. 1998. Regulating meaning, appropriating nature: The codification of California organic agriculture. *Antipode* 30: 135-54.

GWW (Good Wood Watch). 2005. Good Wood Watch: Keeping forest certification honest. http://www.wcel.org/goodwoodwatch/.

Haas, P. 1992. Introduction: Epistemic communities and international policy coordination. *International Organization* 46: 1-35.

Hahn, R.W., and R.N. Stavins. 1991. Incentive-based environmental regulation: A new era from an old idea? *Ecology Law Quarterly* 18: 1-42.

Haley, D. 2002. Community forests in British Columbia: The past is prologue. *Trees, Forests and People* 46: 54-61.

–. 1996. Paying the piper: The cost of the British Columbia *Forest Practices Code.* Paper prepared for the conference Working with the Forest Practices Code, Vancouver.

Haley, D., and J. Leitch. 1992. The future of our forests – Report of the British Columbia Forest Resources Commission: A critique. *Canadian Public Policy* 28: 47-56.

Haley, D., and M.K. Luckert. 1998. Tenures as economic instruments for achieving objectives of public forest policy in British Columbia. In *The wealth of forests: Markets, regulation, and sustainable forestry,* ed. C. Tollefson, 123-51. Vancouver: UBC Press.

–. 1990. *Forest tenures in Canada: A framework for policy analysis.* Ottawa: Ministry of Supply and Services Canada.

Hamilton, R.C. 1993. Characteristics of old-growth forests in the intermountain region. Report for USDA Forest Service, Intermountain Region.

Hammond, H. 1991. *Seeing the forest among the trees: The case for holistic forest use.* Vancouver: Polestar.

Hammond, S. 1993. B.C. groups attend wood products certification founding assembly. *BC Environmental Reporter* 4: 38.

Haufler, V. 1999. Negotiating international standards for environmental management systems: The ISO 14000 standards. A case study for the UN Vision Project on Global Public Policy Networks, http://www.globalpublicpolicy.net/fileadmin/gppi/ Haufler_ISO_14000.pdf.

Hauselman, T. 2002. Voting procedure for the election of a multi-stakeholders working group. Pully, Switzerland: PIEC Consulting for the World Bank/WWF Alliance. http://www.piec. org/PathFinder/Pathfinder_portal/Instruments_Engl/B4-elections/web/B4-screen.pdf.

Hayter, R. 2000. *Flexible crossroads: The restructuring of British Columbia's forest economy.* Vancouver: UBC Press.

Held, D. 1996. *Models of democracy.* 2nd ed. Palo Alto, CA: Stanford University Press.

Hessing, M., and M. Howlett. 1999. *Canadian natural resource and environmental policy: Political economy and public policy.* Vancouver: UBC Press.

Hickey, G.M. 2004. Monitoring and information reporting for sustainable forest management in North America and Europe: Requirements, practices and perceptions. PhD diss., University of British Columbia.

Hirsch, D.D. 2001. Symposium introduction: Second generation policy and the new economy. *Capital University Law Review* 29: 1-20.

Hoberg, G. 2002. Finding the right balance: Designing policies for sustainable forestry in the new era. Paper presented at the Jubilee Lecture Series, Faculty of Forestry, University of British Columbia.

–. 2001. Don't forget government can do anything: Policies towards jobs in the BC forest sector. In *In search of sustainability: British Columbia forest policy in the 1990s*, ed. B. Cashore, G. Hoberg, M. Howlett, J. Rayner, and J. Wilson, 207-31. Vancouver: UBC Press.

Hock, T. 2001. The role of eco-labels in international trade: Can timber certification be implemented as a means of slowing deforestation? *Colorado Journal of International Environmental Law and Policy* 12: 347-65.

Horne, G., and C. Powell. 1995. British Columbia local area economic dependencies and impact ratios. Report for Government of British Columbia, Ministry of Finance and Corporate Relations.

Howlett, M. 2001. Policy venues, policy spillovers, and policy change: The courts, Aboriginal rights, and British Columbia forest policy. In *In search of sustainability: British Columbia forest policy in the 1990s*, ed. B. Cashore, G. Hoberg, M. Howlett, J. Rayner, and J. Wilson, 120-40. Vancouver: UBC Press.

Howlett, M., A. Netherton, and M. Ramesh. 1999. *The political economy of Canada: An introduction.* Oxford: Oxford University Press.

Humphreys, D. 2006. *Logjam: Deforestation and the crisis of global governance.* London: Earthscan.

–. 2004. Redefining the issues: NGO influence on international forest negotiations. *Global Environmental Politics* 4: 51-74.

Hunter, J. 1998. MacBlo decides to end clearcutting in old-growth coast forests. *Vancouver Sun,* 10 June, A1-A2.

Ilsaak Forest Resources. 2000. http://www.iisaak.com/about.html.

IISD/UNEP (International Institute for Sustainable Development/United Nations Environment Programme). 2000. *Environmental trade: A handbook.* Stevenage, UK, and Winnipeg, MT: UNEP and IISD.

Information on Certified Forest Sites Endorsed by the Forest Stewardship Council (FSC). http://www.certified-forests.org/index.htm.

IRMA (The Initiative for Responsible Mining Assurance). 2007. About IRMA. http://www. responsiblemining.net/about.html.

ISEAL (International Social and Environmental Accreditation and Labelling Alliance). 2006a. Code of good practice for setting social and environmental standards. P005. Public version (January). http://www.isealalliance.org/index.cfm?fuseaction=document. showDocumentByID&nodeID=1&DocumentID=212.

–. 2006b. Guidance on the application of the ISEAL code of good practice for setting social and environmental standards. P020. Working Draft 3 (March). http://www.isealalliance. org/index.cfm?fuseaction=document.showDocumentByID&nodeID=1&DocumentID= 220.

ISO (International Organization for Standardization). 2006. ISO members. http://www.iso. org/iso/en/aboutiso/isomembers/index.html.

–. 1996. ISO 14001: Environmental management systems – Specification with guidance for use. Mississauga, ON: Canadian Standards Association.

ITTO (International Tropical Timber Organization). 1992. Guidelines for the sustainable management of natural tropical forests. Yokohama, Japan: ITTO. http://www.itto.or.jp/live/Live_Server/147/ps01e.doc.

IUCN (International Union for Conservation of Nature and Natural Resources). 1997. *Indigenous peoples and sustainability: Cases and actions.* Gland, Switzerland: IUCN.

Joint NGO statement on forest certification. 2003. Why FSC is the only forest certification system worthy of support. http://www.wcel.org/goodwoodwatch/PDF/jointstatement.pdf.

Jordan, A., R. Wurzel, and A. Zito. 2005. The rise of "new" policy instruments in comparative perspective: Has governance eclipsed government? *Political Studies* 53: 477-96.

–. 2003. *New instruments of environmental governance: National experiences and prospects.* London: Frank Cass Publishers.

Katzenstein, P. 1987. *Policy and politics in West Germany: The growth of a semi-sovereign state.* Philadelphia, PA: Temple University Press.

Keck, M., and K. Sikkink. 1998. *Activists beyond borders: Advocacy networks in international politics.* Ithaca, NY: Cornell University Press.

Kennett, S.A. 1993. Hard law, soft law, and diplomacy: The emerging paradigm for intergovernmental cooperation in environmental assessment. *Alberta Law Review* 31: 644-61.

Keohane, R. 1984. *After hegemony: Cooperation and discord in the world political economy.* Princeton, NJ: Princeton University Press.

Kersbergen, K. van, and F. van Waarden. 2004. Governance as a bridge between disciplines: Cross-disciplinary inspiration regarding shifts in governance and problems of governability, accountability and legitimacy. *European Journal of Political Research* 43: 143-71.

Kew, M. 1993-94. Anthropology and First Nations in British Columbia. *BC Studies* 100: 78-105.

Kirton, J.J., and M.J. Trebilcock. 2004. Introduction: Hard choices and soft law in sustainable global governance. In *Hard choices, soft law: Voluntary standards in global trade, environment and social governance,* ed. J.J. Kirton and M.J. Trebilcock, 3-33. Aldershot, UK: Ashgate.

Kjaer, M.M. 2004. *Governance.* Cambridge, UK: Polity Press.

Kooiman, J. 2003. *Governing as Governance.* London: Sage.

–., ed. 1993. *Modern governance: New government-society interactions.* London: Sage.

Kuert, W. 1997. The founding of ISO. In *Friendship among equals,* ed. J. Latimer. Geneva: ISO. http://www.iso.org/iso/founding.pdf.

Landes, R. 1987. *The Canadian polity: A comparative introduction.* Scarborough, ON: Prentice-Hall Canada.

Latin, H. 1985. Ideal versus real regulatory efficiency: Implementation of uniform standards and "finetuning" regulatory reforms. *Stanford Law Review* 37: 1267-32.

–. 1982. Environmental deregulation and consumer decision making under uncertainty. *Harvard Environmental Law Review* 6: 187-240.

Latour, B. 2005. *Reassembling the social: An introduction to actor network theory.* Oxford: Oxford University Press.

Levy, M., P. Roberntz, and N. Hagelberg. 2003. Report on harmonization: Boreal standards. Draft report for FSC-Canada, June.

Lindblom, C. 1982. The market as prison. *Journal of Politics* 44: 324-36.

Lipschutz, R. 2000. Crossing borders: Global civil society and the reconfiguration of transnational political space. *GeoJournal* 52: 17-23.

–. 1992. Reconstructing world politics: The emergence of global civil society. *Millennium: Journal of International Relations* 21, 3: 389-420.

Luthiers Mercantile Catalogue. 1993. W.A.R.P. http://members.shaw.ca/strings/warp.htm, accessed June 2008.

Mackendrick, N. 2005. The role of the state in voluntary environmental reform: A case study of public land. *Policy Sciences* 38: 21-44.

MacKinnon, A. 1998. Biodiversity and old-growth forests. In *Conservation biology principles for forested landscapes,* ed. J. Voller and S. Harrison, 146-84. Vancouver: UBC Press.

Mason, M. 1999. *Environmental democracy.* London: Earthscan.

Marchak, M.P., S. Aycock, and D. Hebert. 1999. *Falldown: Forest policy in British Columbia.* Vancouver: Ecotrust Canada.

Marchak, P. 1983. *Green gold: The forest industry in British Columbia.* Vancouver: UBC Press.

Mathias, J., and G.R. Yabsley. 1991. Conspiracy of legislation: The suppression of Indian rights in Canada. *BC Studies* 89: 34-45.

McDermott, C.L. 2003. Personal trust and trust in abstract systems: A study of Forest Stewardship Council-accredited certification in British Columbia. PhD diss., University of British Columbia.

McEachern, G. 2000. Examination of potential models for developing Forest Stewardship Council standards for Canada's boreal forests. Report for FSC-Canada.

MCEP (Mining Certification Evaluation Project). 2006. *Mining Certification Evaluation Project: Final report.* January. http://www.minerals.csiro.au/sd/Certification/MCEP_ Final_Report_Jan2006.pdf.

McGarrity, K., and G. Hoberg. 2005. The beetle challenge: An overview of the mountain pine beetle epidemic and its implications. An issue brief. http://www.policy.forestry.ubc. ca/PDF/The%20Beetle%20Challenge%20-(3).pdf.

McKee, C. 2000. *Treaty talks in British Columbia: Negotiating a mutually beneficial future.* Vancouver: UBC Press.

Mehl, M.S. 1992. Old-growth descriptions for the major forest cover types in the Rocky Mountain region. In *Old-growth forests in the Southwest and Rocky Mountain regions: Proceedings of a workshop,* ed. M.R. Kaufmann, W.H. Moir, and R.L. Bassett, 106-20. USDA Forest Service General Technical Report RM-213.

Meidinger, E. 2006a. The administrative law of global private-public regulation: The case of forestry. *European Journal of International Law* 17: 47-87. http://www.law.buffalo.edu/ eemeid/scholarship/FCAdlaw.pdf.

–. 2006b. Multi-interest self-governance through global product certification programs. *Buffalo Legal Studies Research Paper Series* 2006 0016. http://www.law.buffalo.edu/eemeid/ scholarship/Onati.pdf.

–. 2006c. Beyond Westphalia: Competitive legalization in emerging transnational regulatory systems. *Buffalo Legal Studies Research Paper Series* 2006-0019. http://www.law.buffalo. edu/eemeid/scholarship/Zurich.pdf.

–. 2000. "Private" environmental regulation, human rights and community. *Buffalo Environmental Law Journal* 7: 123-238. http://www.law.buffalo.edu/homepage/eemeid/ scholarship/hrec.pdf.

Meidinger, E., C. Elliott, and G. Oesten. 2003. *Social and political dimensions of forest certification.* Remagen-Oberwinter, Ger.: Verlag.

Mercer, W.M. 1944. Growth of ghost towns. Report for the Government of British Columbia, Department of Trade and Commerce, Bureau of Economics and Statistics.

Meridian Institute. 2001. Description of the Forest Stewardship Council Program. *Comparative Analysis of the Forest Stewardship Council and Sustainable Forestry Initiative Certification Programs,* Vol. II. http://www2.merid.org/comparison/FSC_SFI_Comp_Analysis-Volume_II.pdf.

M'Gonigle, M., and B. Parfitt. 1994. *Forestopia: A practical guide to the new forest economy.* Madeira Park, BC: Harbour Publishing.

Miller, J.R. 1991. *Skyscrapers hide the heavens: A history of Indian-White relations in Canada.* Toronto: University of Toronto Press.

Moore, K. 2001. Field-testing Draft 2 of the FSC regional certification standards for BC: Report 1 – The field testing process. Nelson, BC: FSC-BC.

Monture-Angus, P. 1999. *Journeying forward: Dreaming First Nations' independence.* Halifax: Fernwood Publications.

Morais, H.V. 2002. The quest for international standards: Global governance vs. sovereignty. *Kansas Law Review* 50: 779-821.

MSC (Marine Stewardship Council). 2007. Annual report 2006/07. London: MSC. http:// www.msc.org/assets/docs/MSC_Annual_report_2006-07_EN.pdf.

–. 2006, 2007. Certified fisheries.

Muckle, R.J. 1998. *The First Nations of British Columbia: An anthropological survey*. Vancouver: UBC Press.

NAFA (National Aboriginal Forestry Association). 2001. An overview of certification and application of FSC's P3, indigenous rights.

–. 1996. Discussion paper: Assessment of the need for Aboriginal compliance with sustainable forest management and forest product.

Noss, R.F. 1997. A big picture approach to forest certification: A report for the World Wildlife Fund's Forests for Life campaign in North America. Washington, DC: World Wide Fund for Nature.

O'Brien, R., A.M. Goetz, Jan Aart Scholte, and Marc Williams. 2000. *Contesting global governance: Multilateral economic institutions and global social movements*. New York: Cambridge University Press.

Overdevest, C., and M. Rickenbach. 2005. Forest certification and institutional governance: An empirical study. Paper presented at 2005 annual meeting of the American Sociological Association, 13-16 August, Philadelphia, PA.

Oxford Forestry Institute. 1991. Incentives in producer and consumer countries to promote sustainable development of tropical forests. Report for ITTO.

Parminter, J. 1998. Natural disturbance ecology. In *Conservation biology principles for forested landscapes*, ed. J. Voller and S. Harrison, 1-41. Vancouver: UBC Press.

Pattberg, P. 2005. What role for private rule-making in global environmental governance? Analysing the Forest Stewardship Council (FSC). *International Environmental Agreements* 5, 2: 175-89.

Pearce, M., G. Behie, and N. Chappell. 2002. The effects of aerial spraying with Bacillus thuringiensis Kurstaki on area residents. *Environmental Health Review* 2, 1: 19-22.

Pearse, P. 2001. *Ready for change: Crisis and opportunity in coast forest industry*. Report to the Minister of Forests on the British Columbia coastal forest industry. Vancouver (November). http://www.for.gov.bc.ca/hfd/library/documents/phpreport/.

–. 1976. Timber rights and forest policy in British Columbia: Report of the Royal Commission on Forest Resources. 2 vols.Victoria, BC: Queen's Printer.

Pedersen, L. 2003. Allowable annual cuts in British Columbia: The agony and the ecstasy. Paper presented at the Jubilee Lecture Series, Faculty of Forestry, University of British Columbia, 20 March, Vancouver, BC. http://www.for.gov.bc.ca/hts/pubs/jubilee_ubc.pdf.

–. 2001. Letter to Mr. Marty Horswill, Standards Development Committee, FSC-BC regional initiative, from Larry Pedersen, Chief Forester, BC Ministry of Forests, 7 September 2001. File 10050-07.

Peeling, A.C. 2002. Review of the application of Principle 3 in the boreal forests subject to treaties and Aboriginal rights (of First Nations and Métis). Legal opinion commissioned by the FSC-Canada.

PEFC (Pan European Forest Certification Council). 2006a. About PEFC. http://www.pefc.org/internet/html/about_pefc/4_1137_498.htm.

–. 2006b. PEFC in figures. http://www.pefc.org/internet/html/about_pefc/4_1137_515.htm.

–. 2006c. Members and schemes. http://www.pefc.org/internet/html/members_schemes/4_1120_59.htm.

–. 2006d. Members and schemes: Sustainable Forestry Initiative. http://www.pefc.org/internet/html/members_schemes/4_1120_59/5_1246_326/5_1123_1190.htm.

–. 2005. Rules for standard setting. http://www.pefc.org/internet/resources/5_1177_288_file.1672.pdf .

–. 2004. Swedish forest certification scheme. http://www.pefc.org/internet/html/members_schemes/4_1120_59/5_1246_324/5_1123_1126.htm.

–. 1999. PEFC council statutes. http://www.pefc.org/internet/html/documentation/4_1311_399.htm.

Pellizzoni, L. 2004. Responsibility and environmental governance. *Environmental Politics* 13: 541-66.

Percival, R.V. 1996. *Environmental regulation: Law, science and policy.* 2nd ed. New York: Aspen Publishers.

Phillips. J., and C. Wetherell. 1995. The great Reform Act of 1832 and the political modernization of England. *American Historical Review* 100, 2: 411-36.

Pierre, J., and B.G. Peters. 2000. Is there a governance theory? Paper presented at the IPSA conference in Quebec City, 1-5 August.

Pildes, R.H., and C.R. Sunstein. 1995. Reinventing the regulatory state. *University of Chicago Law Review* 62: 1-130.

Pinto de Abreu, J., and M. Simula. 2004. Report on the procedures for the implementation of phased approaches to certification in tropical timber producing countries. Report prepared for ITTO.

Pojar, J., and D. Meidinger. 1991. *Ecosystems of British Columbia.* Victoria, BC: Ministry of Forests. http://www.for.gov.bc.ca/hfd/pubs/docs/Srs/Srs06/chap3.pdf.

Pollock, K.B. 1996. The IWA speaks out on job loss. Letter to the editor. *BC Environmental Report* 7: 30.

Pollution Probe. 2002. Environmental non-governmental organization (ENGO) participation in national standard setting. www.pollutionprobe.org/publications/datepublished. htm.

Pons Rafols, X., and L. Brander. 2005. The stewardship council model: A comparison of the FSC and the MSC. *ILSA Journal of International and Comparative Law* 11: 637-62.

Ponte, S. 2006. *Marine Stewardship Council certification and the South African hake industry.* Tralac Working Paper No. 9/2006. http://www.tralac.org/pdf/20060829_PonteMSCcertification.pdf.

Potts, T., and M. Haward. 2001. Sustainability indicator systems and Australian fisheries management. *Maritime Studies* 117: 1-11.

Powell, W.W. 1990. Neither market nor hierarchy: Network forms of organization. *Research in Organizational Behaviour* 12: 295-336.

Rainforest Alliance. 2003. Sustainable Tourism Stewardship Council: Raising the standards and benefits of sustainable tourism and ecotourism certification. Version 8.6. Washington, DC: Rainforest Alliance.

Rajala, R. 1993. *The Legacy and the challenge: A century of the forest industry at Cowichan Lake.* Lake Cowichan: Lake Cowichan Heritage Advisory Committee.

Rametsteiner, E. 2000. Sustainable forest management certification: Frame conditions, system designs and impact assessment. Vienna, Austria: Ministerial Conference on the Protection of Forests in Europe.

Rauch, J. 2001. Business and social networks in international trade. *Journal of Economic Literature* 39: 1177-1203.

Raustiala, K. 2002. The architecture of international cooperation: Transgovernmental networks and the future of international law. *Virginia Journal of International Law* 43: 1-92.

Read, M. 1991. An assessment of claims of "sustainability" applied to tropical wood products and timber retailed in the U.K., July 1990-January 1991. Godalming, Surrey: World Wide Fund for Nature-UK.

Rhodes, R. 2000. Governance and public administration. In *Debating governance: Authority, steering, and democracy,* ed. J. Pierre, 54-90. Oxford: Oxford University Press.

–. 1997. *Understanding governance: Policy networks, governance, reflexivity and accountability.* Maidenhead, UK: Open University Press.

–. 1996. The new governance: Governing without government. *Political Studies* 44: 652-57.

Rhone, G., D. Clarke, and K. Webb. 2005. Two voluntary approaches to sustainable forestry practices. In *Voluntary codes: Private governance, the public interest and innovation,* ed. K. Webb, 249-72. Ottawa, ON: Carleton Research Unit for Innovation, Science and Environment. http://www.carleton.ca/spa/VolCode/Ch9.pdf.

Rosenau, J. 1992. *Governance without government: Order and change in world politics.* Cambridge: Cambridge University Press.

Rotman, L.I. 1996. *Parallel paths: Fiduciary doctrine and the Crown-Native relationship in Canada.* Toronto: University of Toronto Press.

Sabatier, P., and H. Jenkins-Smith. 1999. The advocacy coalition framework: An assessment. In *Theories of the policy process,* ed. P. Sabatier, 117-68. Boulder, CO: Westview Press.

Sasser, E.N. 2003. Gaining leverage: NGO influence on certification institutions in the forest products sector. In *Forest policy for private forestry: Global and regional challenges,* ed. L. Teeter, B. Cashore, and B. Zhang, 229-44. London, England: CABI Publishing.

SCC (Standards Council of Canada). 2005. *Canadian standards strategy: Update 2005-2008.* Ottawa: SCC. http://www.scc.ca.

Schmitter, P. 1974. Still the century of corporatism? *Review of Politics* 36: 85-131.

Schneider, V. 2002. Private actors in political governance. In *Participatory governance: Political and societal implications,* ed. Jürgen R. Grote and Bernard Gbikpi, 245-64. Opladen, Germany: Leske and Budrich.

Schout, A., and A.J. Jordan. 2005. Coordinated European governance: Self-organizing or centrally steered? *Public Administration* 83: 201-20.

SFB (Sustainable Forestry Board). 2006a. About the Sustainable Forestry Board. http://www.aboutsfb.org/aboutsfb.cfm.

–. 2006b. SFB membership. http://www.aboutsfb.org/index.htm.

–. 2006c. SFB working committees. http://www.aboutsfb.org/committees.cfm.

–. 2004. 2005-2009 sustainable forestry initiative. http://www.sfiprogram.org/miscPDFs/SFBStandard2005-2009.pdf .

Shaw, M. 1992. Global civil society and global responsibility: The theoretical, historical and political limits of "international relations." *Millennium* 23: 421-34.

Sierra Club of Canada. 2007. National forest policy strategy report card for provinces and territories. http://www.sierraclub.ca/national/programs/biodiversity/forests/nfs/database/index.php.

Simpson, D. 1995. The bioregional basis of forest certification: Why Cascadia? *International Journal of Ecoforestry* 11: 140-43.

Sizer, N. 2000. Perverse habits: The G8 and subsidies that harm forests and economies. Washington, DC: World Resources Institute. http://www.globalforestwatch.org/english/pdfs/perverse_habits.pdf.

Slaughter, A.M. 2004. *A new world order.* Princeton, NJ: Princeton University Press.

SLDF (Sierra Legal Defence Fund). 2005. Understanding Canada's forest certification schemes: A complete guide to filing appeals and complaints. http://ecojustice.ca/publications/reports/understanding-canadas-forest-certification-schemes-a-complete-guide-to-filing-appeals-and-complaints/attachment.

Sloan, G. 1945. Report of the commissioner relating to the forest resources of British Columbia. Victoria: C.F. Banfield, King's Printer, 1945.

SmartWood. 2004. Forest management public summary for Inlailawatash Holdings Ltd: Certification code SW-FMCOC-1311. http://www.rainforest-alliance.org/programs/forestry/smartwood/documents/inlailawatash.pdf .

Spalding, S. 2002. An assessment of the utility, usability and cost implications associated with the Forest Stewardship Council regional certification standards for British Columbia (Draft 3, April 22, 2002) when applied to large-scale forestry operations. A report prepared for Dr. Bill Bourgeois, FSC-BC ISC.

Stanbury, W.T., and I. Vertinsky. 1998. Governing instruments for forest policy in British Columbia: A positive and normative analysis. In *The wealth of forests: Markets, regulation, and sustainable forestry,* ed. C. Tollefson, 42-77. Vancouver: UBC Press.

Statistics Canada. 2001. Population reporting an Aboriginal identity, by age group, by provinces and territories (2001 Census) (Newfoundland and Labrador, Prince Edward Island, Nova Scotia, New Brunswick). http://www40.statcan.ca/cgibin/getcans/sorth.cgi?lan=eng&dtype=fina&filename=demo40a.htm&sortact=2&sortf=2.

Steinzor, R.L. 2001. Myths of the reinvented state. *Capital University Law Review* 29: 223-43.

Stevenson, M.L. 2000. Legal memorandum regarding Principle 3 of the Forest Stewardship Council (FSC's) principles and criteria. FSC-Canada.

Stewart, R.B. 2001. A new generation of environmental regulation. *Capital University Law Review* 29: 21-182.

Stiglitz, J. 1997. *Economics*. 2nd ed. New York: W.W. Norton.

Stoker, G. 1998. Governance as theory: Five propositions. *International Social Science Journal* 155: 17-28.

Stokke, O. 2004. Labelling, legalisation and sustainable management of forestry and fisheries. Paper presented to the Fifth Pan-European Internatonal Relations Conference, 9-11 September, The Hague, Netherlands. http://www.sgir.org/conference2004/papers/Stokke %20-%20Environmental%20Labelling.pdf.

Swan, P. 2002. Governing at a distance: An introduction to law, regulation, and governance. In *Law, regulation, and governance*, ed. M. Mac Neil, N. Sargent, and P. Swan, 1-27. Oxford: Oxford University Press.

Synnott, T.J. 2005. Some notes on the early years of FSC. FSC-IC. http://www.fsc.org/keepout/ en/content_areas/45/2/files/Notes_on_the_early_years_of_FSC_by_Tim_Synnott.pdf.

–. 2000. GMOs: Genetically modified organisms: Draft interpretation for FSC. FSC-POL-30-602. http://www.fsc.org/keepout/en/content_areas/z77/29/files/FSC_POL_30_602_GMO_ Policy_Paper_BM_19_22_2000_05.pdf.

Theys, J. 2002. Environmental governance: From innovation to powerlessness. In *Participatory governance: Political and societal implications*, ed. Jürgen R. Grote and Bernard Gbikpi, 213-44. Opladen, Ger.: Leske and Budrich.

Thompson, B. 2000. The continuing innovation of citizen enforcement. *University of Illinois Law Review* 1: 185-236.

–. 1996. Foreword: The search for regulatory alternatives *Stanford Environmental Law Journal* 15: vii-xxii.

Thompson, G., and C. Pforr. 2005. Policy networks and good governance: A discussion. Working Paper Series, School of Management, Curtin University of Technology, Perth, Australia. http://espace.lis.curtin.edu.au/archive /00000947/01/2005-1_Thompson_&_Pforr.pdf.

Tollefson, C. 2004. Indigenous rights and forest certification in British Columbia. In *Hard choices, soft law: Voluntary standards in global trade, environment and social governance*, ed. J.J. Kirton and M.J. Trebilcock, 93-120. Aldershot, UK: Ashgate.

–, ed. 1998. *The wealth of forests: Markets, regulation, and sustainable forestry*. Vancouver: UBC Press.

Treib, O., H. Bähr, and G. Falkner. 2005. Modes of governance: Towards a conceptual clarification. In *European governance papers (EUROGOV)*. No. N-05-02. http://www.con-nex-network.org/eurogov/pdf/egp-newgov-N-05-02.pdf.

UBCIC (Union of British Columbia Indian Chiefs). 2006. UBCIC letter to the premier: Interim agreement on forest and range operations. http://www.ubcic.bc.ca/News_ Releases/UBCICNews02060601.htm.

UNCTAD/IISD (United Nations Conference on Trade and Development and International Institute for Sustainable Development). 2006. Outline of research priorities for research in the coffee sector. Geneva and Winnipeg: UNCTAD and IISD.

Unerman, J., and B. O'Dwyer. 2004. Enron, WorldCom, Andersen et al.: A challenge to modernity. *Critical Perspectives in Accounting* 15, 6-7: 971-93.

United Nations. 1993. Draft United Nations declaration on the rights of indigenous peoples. Sub-commission on Human Rights. http://www.unhchr.ch/huridocda/huridoca. nsf/(Symbol)/E.CN.4.SUB.2.RES.1994.45.En.

Voller, J. 1998. Riparian areas and wetlands. In *Conservation biology principles in forested landscapes*, ed. J. Voller and S. Harrison, 98-129. Vancouver: UBC Press.

Von Bernstorff, J. 2003. Democratic global Internet regulation? Governance networks, international law and the shadow of hegemony. *European Law Journal* 9, 4: 511-26.

Walby, S. 2003. The myth of the nation-state. *Sociology* 37, 3: 529-46.

Walter, E. 2005. Decoding codes of practice: Approaches to regulating the ecological impacts of logging in British Columbia. *Journal of Environmental Law and Practice* 15: 143-85.

–. 2003. From civil disobedience to obedient consumerism? Influences of market-based activism and eco-certification on forest governance. *Osgoode Hall Law Journal* 41, 2-3: 531-63. http://www.ohlj.ca/archive/articles/41_23_walter.pdf.

WCELA (West Coast Environmental Law Association). 1997. Background paper presented at BC endangered species protection workshop, June 24-25, Vancouver. http://www. wcel.org/wcelpub/1997/11918.html.

Webb, K., ed. 2004. *Voluntary codes: Private governance, the public interest and innovation.* Ottawa: Carleton Research Unit for Innovation, Science and Environment.

Weber, G. 2005a. Application of the FSC interim dispute resolution protocol to the PT Diamond Raya dispute. Pacific McGeorge Institute for Sustainable Development, McGeorge School of Law, University of the Pacific, Sacremento, California.

–. 2005b. Report of the dispute resolution workshop, Manaus, Brazil. FSC General Assembly.

Wendt, A. 1999. *Social theory of international politics.* Cambridge: Cambridge University Press.

White, A., X. Sun, K. Cany, J. Xu, C. Barr, E. Katsigris, G. Bull, C. Cossalter, and S. Nilsson. 2006. *China and the global market for forest products: Transforming trade to benefit forests and livelihoods.* http://www.forest-trends.org/documents/publications/China%20and %20the%20Global%20Forest%20Market-Forest%20Trends.pdf.

Wilson, J. 2001. Experimentation on a leash: Forest land use planning in the 1990s. In *In search of sustainability: British Columbia forest policy in the 1990s,* ed. B. Cashore, G. Hoberg, M. Howlett, J. Rayner, and J. Wilson, 31-60. Vancouver: UBC Press.

–. 1998. *Talk and log: Wilderness politics in British Columbia, 1965-96.* Vancouver: UBC Press.

Wood, S. 2002-3. Environmental management systems and public authority in Canada: Rethinking environmental governance. *Buffalo Environmental Law Journal* 10: 129-52.

World Bank. 2001. A revised forest strategy for the World Bank Group. http://www.iucn. org/themes/fcp/activities/publications/worldbank_forestpolicynov02.pdf.

Young, Oran. 1989. International cooperation: Building regimes for natural resources and the environment. Ithaca, NY: Cornell University Press.

Zielke, K., and B. Bancroft. 2002. AAC implications of the FSC Draft 3 standards. A report for Bill Bourgeois of the FSC-BC steering committee. Victoria and West Vancouver: Symmettree Consulting.

–. 2001. Achieving FSC certification in British Columbia: A forest sector submission in response to Forest Stewardship Council (FSC) regional certification standards for British Columbia, Draft 2. A report for the Forest Sector group of forest companies. Victoria and West Vancouver: Symmettree Consulting.

# Index

Printed and bound in Canada by Friesens

Set in Stone by Artegraphica Design Co. Ltd.

Copy editor: Audrey McClellan

Proofreader: Lesley Erickson

Indexer: David Luljak

## ENVIRONMENTAL BENEFITS STATEMENT

**UBC Press** saved the following resources by printing the pages of this book on chlorine free paper made with 100% post-consumer waste.

| TREES | WATER | ENERGY | SOLID WASTE | GREENHOUSE GASES |
|---|---|---|---|---|
| 13 | 4,638 | 9 | 596 | 1,117 |
| FULLY GROWN | GALLONS | MILLION BTUs | POUNDS | POUNDS |

Calculations based on research by Environmental Defense and the Paper Task Force. Manufactured at Friesens Corporation